The Physics, Clinical Measurement and Equipment of Anaesthetic Practice *for the FRCA*

SECOND EDITION

by

Patrick Magee
Consultant in Anaesthesia
Royal United Hospital, Bath, UK

Mark Tooley
Head of department of Medical Physics and Bioengineering
Royal United Hospital, Bath, UK

OXFORD
UNIVERSITY PRESS

Great Clarendon Street, Oxford OX2 6DP

Oxford University Press is a department of the University of Oxford.
It furthers the University's objective of excellence in research, scholarship,
and education by publishing worldwide in
Oxford New York
Athens Auckland Bangkok Bogotá Buenos Aires Cape Town
Chennai Dar es Salaam Delhi Florence Hong Kong Istanbul Karachi
Kolkata Kuala Lumpur Madrid Melbourne Mexico City Mumbai Nairobi
Paris São Paulo Shanghai Singapore Taipei Tokyo Toronto Warsaw
with associated companies in Berlin Ibadan

Oxford is a registered trade mark of Oxford University Press
in the UK and in certain other countries

Published in the United States
by Oxford University Press Inc., New York

© Oxford University Press, 2011

The moral rights of the authors have been asserted
Database right Oxford University Press (maker)

First edition published 2005
Second edition published 2011

All rights reserved. No part of this publication may be reproduced, stored in a retrieval
system, or transmitted, in any form or by any means, without the prior permission in
writing of Oxford University Press, or as expressly permitted by law, or under terms
agreed with the appropriate reprographics rights organization. Enquiries concerning
reproduction outside the scope of the above should be sent to the Rights Department,
Oxford University Press, at the address above

You must not circulate this book in any other binding or cover
and you must impose this same condition on any acquirer

British Library Cataloguing in Publication Data
Data available

Library of Congress Cataloging in Publication Data

Library of Congress Control Number: 2011933735

Typeset in Minion
by Cenveo Publisher Services, Bangalore, India
Printed in Great Britain
on acid-free paper by
CPI Group (UK) Ltd, Croydon, CR0 4YY

ISBN 978-0-19-959515-0

10 9 8 7 6 5 4 3 2 1

Oxford University Press makes no representation, express or implied, that the drug
dosages in this book are correct. Readers must therefore always check the product
information and clinical procedures with the most up-to-date published product
information and data sheets provided by the manufacturers and the most recent
codes of conduct and safety regulations. The authors and the publishers do not
accept responsibility or legal liability for any errors in the text or for the misuse or
misapplication of material in this work. Except where otherwise stated, drug dosages
and recommendations are for the non-pregnant adult who is not breast-feeding.

The Physics, Clinical Measurement and Equipment of Anaesthetic Practice *for the FRCA*

This book is dedicated to our families:
Marie-Suzanne, Lucia, Rosie, William and Patrick Magee
and
Stella, Alissa and Sebastian Tooley

Foreword

In my experience, many healthcare professionals struggle to understand physics and physical principles. Most will have only a rudimentary knowledge of the working principles of the equipment that they use every day, and this applies to physicians, surgeons or nurses, as well as anaesthetists. Patrick Magee and Mark Tooley have very successfully written a textbook that helps the reader to understand the wide range of potentially complex topics relating to the physics, clinical measurement and equipment of anaesthetic practice. The content of this book covers the whole of the domain of physics and clinical measurement at both basic and intermediate levels within the Royal College of Anaesthetists (RCoA) 2010 Curriculum for a Certificate of Completion of Training (CCT) in Anaesthetics. This makes it the ideal learning tool for these topics for both the Primary and Final examination for Fellowship of the Royal College of Anaesthetists (FRCA). The section of the book on statistics covers the RCoA basic level curriculum for this topic and thus covers the required statistical knowledge for the Primary FRCA. Many of the topics in this book will also be essential reading for those working in intensive care, particularly those who plan to sit the UK Diploma in Intensive Care Medicine or the European Diploma in Intensive Care Medicine. Although the book has been compiled to ensure full coverage of UK and European anaesthesia and intensive care medicine curricula, the topic is generic and the contents are equally appropriate for healthcare professionals worldwide.

The contents of the book follow a logical sequence that takes the reader from mathematics and statistics through electricity, fluids and gases to imaging techniques and the function of a variety of items of equipment including anaesthetic machines and ventilators. The whole book is presented with great clarity: the pages are uncluttered, the tables are easy to read and each chapter is comprehensively illustrated with simple, clear line diagrams that convey succinctly the principles of each topic. The line diagrams of vaporisers and ventilators are particularly informative. Lists of references and further reading at the end of each chapter guide the reader to further information if required.

I have had the pleasure of working with the authors for many years. It is hard to imagine anyone else with more knowledge, experience and enthusiasm for physics and clinical measurement in relation to anaesthesia. Dr Magee obtained a degree in Engineering Science and Biomechanics before going on to train in medicine and specialise in anaesthesia. He is an examiner for the Primary FRCA and is therefore well informed about the level and scope of the knowledge required for that examination. Not surprisingly, Dr Magee is in big demand as a teacher and lecturer in physics and clinical measurement. Professor Tooley is a medical physicist (originally he trained as a electronic engineer), who has spent a significant part of his career working alongside clinical and academic anaesthetists. He is perhaps uniquely placed to understand the application of physics and clinical measurement to anaesthesia.

This book is not just about providing information to pass exams. It will be a valuable resource for those well-established clinicians who wish to refresh their knowledge and understanding of physics, clinical measurement and anaesthetic equipment. This may reflect the need to learn about new monitors or other items of equipment for use in their own clinical practise or it may be driven by the need to enhance knowledge and confidence before attempting to teach trainees. In any event, I expect this book will be in strong demand from trainees, trainers and other experienced clinicians.

Jerry Nolan FRCA FRCP FCEM FFICM
Consultant in Anaesthesia and Intensive Care Medicine
Royal United Hospital Bath, UK

Preface

The aim of this book is to help junior anaesthetists who are about to sit the diploma examinations for the Fellowship in Anaesthesia and Intensive Care of the Royal College of Anaesthetists, the European Academy of Anaesthesiology and other equivalent international professional anaesthetic qualifications. It is also intended to be useful revision in the physical sciences for more senior anaesthetic colleagues.

The book will also be of interest to operating department assistants, physician assistants (anaesthesia), hospital-based biomedical engineers and clinical scientists, manufacturers' representatives and those involved in the manufacture, marketing and use of anaesthetic equipment in the UK, the EU and further afield.

The contents are based on the syllabus of the FRCA exam in mathematics, physics, clinical measurement and monitoring, equipment and safety.

Anaesthetists are a relatively technophilic group of clinicians, but many have either no higher education in mathematics and physics, or at least a decade has elapsed since they have received any such education. In comparison with the other basic sciences of physiology and pharmacology, physical sciences are less popular and less accessible.

The book starts with mathematics, statistics and background physics, not only to enhance the understanding for what follows in the book, but also because these basic sciences are fundamental to many other aspects of medical science. Areas, which in the authors' teaching experience, candidates have found particularly troublesome, such as electricity and electrical safety, have been discussed in some detail in order to clarify. We have endeavoured to make each chapter independent, although frequent reference is made to other chapters where appropriate.

Dr Patrick Magee BSc, MSc, MB BS, FRCA, MIEE, AMIMechE, is a Consultant Anaesthetist at the Royal United Hospital Bath, who originally graduated in Engineering Science and Biomechanics. He is an honorary Senior Lecturer at the University of Bath in the department of Mechanical Engineering.

Professor Mark Tooley BSc, MSc, PhD, CEng, CSci, FIET, FIPEM, FInstP, ARCP, is a Consultant Clinical Scientist (Medical Physics) at the Royal United Hospital Bath. He is head of the department of Medical Physics and Bioengineering and Director of Research and Development at the Hospital.

He is a visiting Professor at the University of the West of England and an honorary Senior Lecturer at the University of Bath.

The Royal College of Anaesthetists have agreed to the use of FRCA in this publication's title but this does not imply any endorsement of this title or its contents.

Contents

1 Mathematics *1*
Graphs *1*
Trigonometry *5*
Calculus *8*
Powers, logarithms, 'e' *14*
Further reading *21*

2 Statistics *22*
Introduction *22*
Categorical data *22*
Numerical data *23*
Quantifying variability *26*
Probability *29*
The Normal distribution *30*
Hypothesis testing *34*
Parametric methods *35*
Non-parametric methods *36*
Diagnostic tests *38*
Meta-analysis *41*
Correlation and regression *42*
Comparing groups of categorical data *44*
Evidence based medicine *46*
References *46*
Further reading *46*

3 Background physics *47*
Introduction *47*
Dimensions and units *47*
Input/output, linearity, drift, hysteresis *50*
Frequency response *51*
Calibration of transducers *53*
Atomic structure *55*
Basic mechanics *56*
References *59*
Further reading *59*

4 Electricity, magnetism and circuits *60*
Introduction *60*
Electrical properties *60*
Magnetism *62*
Basic electrical circuit rules *63*
Bridge circuits *65*
Alternating and direct voltage/current *66*
Capacitors and inductors *68*
Transformers *70*
Other electrical analogies *71*
Further reading *72*

5 Electronics and biological signal processing *73*
Introduction *73*
Filters *73*
Amplifiers *76*
Signal processing *81*
Digital processing *82*
Electrodes and transducers *84*
References *87*
Further reading *88*

6 Electrical safety *89*
The electricity supply *89*
Electrical safety features *90*
Earth faults *94*
Microshock *95*
Circuit breakers *97*
Earth free supplies *98*
Other safety issues of electricity *98*
References *100*
Further reading *100*

7 Behaviour of fluids (liquids and gases, flow and pressure) *101*
Fluid flow behaviour *101*
Theory of flow measurement devices *104*
Pressure *108*
Partial pressure *110*
Diffusion *112*
Osmosis *113*
References *115*

8 Thermodynamics: heat, temperature and humidification *117*
   Introduction *117*
   The gas laws *117*
   Specific heat capacity *122*
   Latent heats of vaporisation and of crystallization *123*
   Temperature *123*
   Thermometry *124*
   Heat transfer *125*
   Keeping patients warm *125*
   References *129*

9 Solubility, vaporisation and vaporisers *131*
   Introduction *131*
   Solubility of molecules in liquids *131*
   Solubility of gases in gases *134*
   Vaporisation *134*
   The basis of vapour production by a vaporiser *137*
   Vaporisers *139*
   References *147*
   Further reading *147*

10 Ultrasound and Doppler *148*
   Introduction *148*
   Properties of ultrasound *148*
   Visualisation of needle positioning *152*
   Doppler *153*
   Measurement of absolute blood flow *157*
   High power ultrasound *158*
   Conclusions and future directions *158*
   References *159*
   Further reading *159*

11 Principles and standards of anaesthetic monitoring *160*
   Introduction *160*
   Recommendations for standards of monitoring *163*
   References *168*

12 Blood pressure measurement *170*
   Introduction *170*
   Non-invasive blood pressure measurement *170*
   Direct measurement of blood pressure *177*
   Components of the monitoring system *178*

Resonance and damping *182*
Other blood pressure measurements *189*
References *190*
Further Reading *191*

13 Cardiac output measurement *192*
The pulmonary artery catheter *192*
Echocardiography *195*
Pulse contour analysis *197*
Transthoracic electrical impedance *198*
References *199*

14 Gas pressure, volume and flow measurement *201*
Introduction *201*
Gas pressure measurement *201*
Gas volume and flow measurement *203*
References *214*
Further reading *215*

15 Pulse oximetry *216*
Pulse oximetry *216*
References *220*

16 Respiratory gas analysers *222*
Introduction *222*
Refractometry *223*
Infrared spectroscopy *226*
Mass spectrometry *228*
Raman spectroscopy *231*
Piezoelectric gas analysis *232*
Ultraviolet gas analysis *233*
The nitrogen meter *233*
Paramagnetic oxygen analyser *233*
Polarography and fuel cells *234*
Preoperative exercise testing *236*
References *237*
Further reading *238*

17 Blood gas analysis *239*
The pO_2 electrode *239*
The pH electrode *239*
The pCO_2 electrode *241*
Derived variables from a blood gas machine *241*

Errors in blood gas measurement *242*
Temperature and blood gas analysis *242*
Transcutaneous blood gas analysers *242*
Intravascular blood gas analysers *243*
References *243*
Further reading *243*

18 Electrophysiology and stimulation *244*
Introduction *244*
The electrocardiogram *244*
The electroencephalogram *248*
Evoked responses *252*
Monitoring neuromuscular blockade and the electromyogram *256*
References *259*
Further reading *260*

19 Monitoring depth of anaesthesia *261*
Introduction *261*
Depth of anaesthesia monitors using the EEG *261*
Depth of anaesthesia monitors using the AER *267*
Depth of anaesthesia monitors using the ECG *267*
Emerging devices *268*
References *268*

20 Pacemakers and defibrillators *270*
Introduction *270*
Pacemakers *270*
Defibrillators *274*
References *279*
Further reading *279*

21 Surgical diathermy *280*
Introduction *280*
Current density *280*
Surgical effects and associated waveforms *281*
Electrodes *283*
Safety of diathermy *284*
Interference with other equipment *285*
References *286*

22 Gas supply and the anaesthetic machine *287*
Supply of anaesthetic gases *287*

Pressure regulators, gas flow control and anaesthetic machine safety features *293*
The anaesthetic machine and equipment checklist *296*
References *299*

23 Airway management devices *301*
Introduction *301*
The artificial airway *302*
The facemask *303*
The laryngeal mask airway *304*
The I-gel *307*
The cuffed oropharyngeal airway *308*
The endotracheal tube *308*
References *313*

24 Aids to intubation *316*
The laryngoscope *316*
Fibre optics *318*
Bougies and catheters *320*
References *321*

25 Breathing systems *322*
Introduction *322*
The Mapleson classification of semi-closed rebreathing systems *323*
Humphrey ADE system *329*
Venturi systems *330*
The circle system *330*
References *335*

26 Artificial ventilators *338*
Respiratory mechanics during ventilation *338*
Ventilator mechanics *339*
Some ventilator types *345*
Intensive care ventilators *352*
Manual resuscitators *358*
Portable ventilators *358*
Other methods of gas exchange *359*
References *362*

27 Intravenous pumps and syringe drivers *364*
Intravenous pumps and syringe drivers *364*
References *368*
Further reading *368*

28 Environmental safety *369*
   Fire and explosion *369*
   Atmospheric pollution and anaesthetic gases *370*
   Scavenging systems *371*
   References *374*
   Further reading *374*

29 Imaging and radiation *375*
   Introduction *375*
   Radioactive decay *375*
   Production of X-rays *377*
   Imaging *378*
   Biological effects of radiation and radiation protection *380*
   Magnetic resonance imaging *382*
   Lasers *388*
   References *391*
   Further reading *391*

30 Cleaning and sterilisation of equipment *392*
   Cleaning and sterilisation of equipment *392*
   Further reading *393*

31 Medical training using simulators *394*
   Introduction *394*
   Simulation centre environments *395*
   Model driven patient simulators *395*
   Surgical skills labs *399*
   Hybrid training *399*
   Virtual reality training *399*
   References *400*
   Further reading *400*

   Index *401*

Chapter 1

Mathematics

This chapter contains: graphs (straight line, parabola, other graphical relationships); trigonometry; calculus (differentiation, integration); powers, logarithms and 'e'.
The chapter links with: Chapters 2, 3, 5 and 7.

Graphs

The straight line

Graphs are used to represent pictorially or clarify a relationship between two variables, say x and y, or t and *function* (t). If x and y are related by the *linear* equation $y = mx + c$, then the graphical relationship is a straight line as in Figure 1.1, where m is the slope, a constant value for a straight line and c is the value of y when $x = 0$. Figure 1.2 shows the relationship when m and c take negative values. Note that if $c = 0$, the line passes through the origin, 0, of the x–y axes.

Examples of other linear relationships that can be represented in this way include the following (the bracketed symbols are the variables equivalent to x and y above):

- $v = u + at$; (variables are t and v), where u is the starting velocity of an object that is subjected to acceleration a for time t, after which its velocity is v.

- $F = m\dfrac{d^2 x}{dt^2}$; (variables are $\dfrac{d^2 x}{dt^2}$ and F), where a force F acts on a mass m, causing an acceleration $\dfrac{d^2 x}{dt^2}$.

- $Q = k\Delta P$; (variables are ΔP and Q), where, in laminar fluid flow, flowrate Q is proportional to pressure drop ΔP.

- $V_o = V_i - iR$; (variables i and V_o), where the output voltage of an electrical circuit V_o results from an input voltage V_i and a current i flowing through a resistance R.

- $\omega t = \theta$; (variables t and θ), where an object travelling at angular velocity ω, travels through an arc of angle θ after time t.

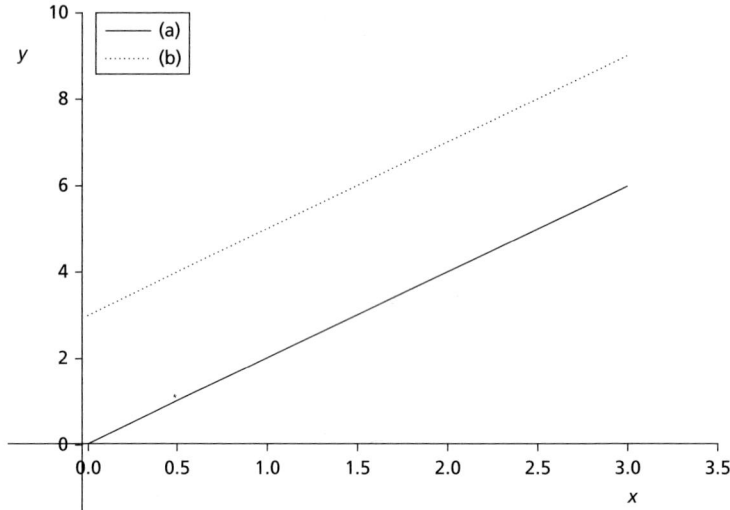

Fig. 1.1 The straight line $y = mx + c$: (a) $y = 2x$; (b) $y = 2x + 3$.

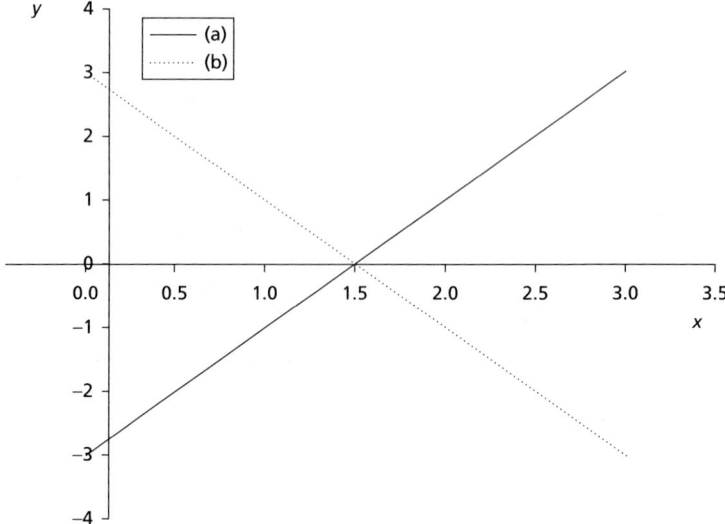

Fig. 1.2 More straight lines; (a) $y = 2x - 3$; (b) $y = 3 - 2x$.

The parabola

The relationship between two variables need not be linear as those drawn previously, but the one may be related to the square of the other, such as $y = mx^2$, producing a parabolically shaped curve, passing through the origin 0.

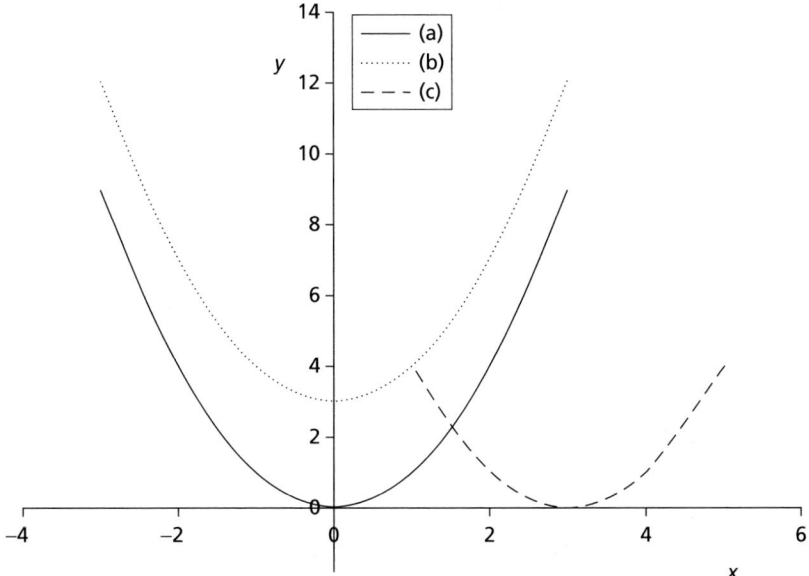

Fig. 1.3 Parabolas (a) $y = mx^2$; (b) $y = mx^2 + c$ (c) $y = m(x - a)^2$.

If the curve is displaced upwards by an amount equal to c on the y-axis, the equation becomes $y = mx^2 + c$, and if displaced sideways along the positive x-axis by an amount equal to a the equation becomes $y = m(x - a)^2$. These are shown graphically in Figure 1.3.

Examples of such relationships might include the following:

$\Delta P = kQ^2$ where ΔP and Q are the pressure drop and flow respectively in a region of turbulent flow.

$V = k'X^2$ where V is the velocity at a distance X radially from the long axis of a pipe in which there is laminar flow.

Other graphical relationships

The relationship between two variables may be even more complex, but nevertheless expressible in mathematical form. For example, if y is expressed as a polynomial

$$y = m_1 x^3 + m_2 x^2 + m_3 x + c,$$

then it will look as in Figure 1.4.

Computers will now fit such curves to experimental data points and produce the appropriate equation describing the relationship. The purpose of

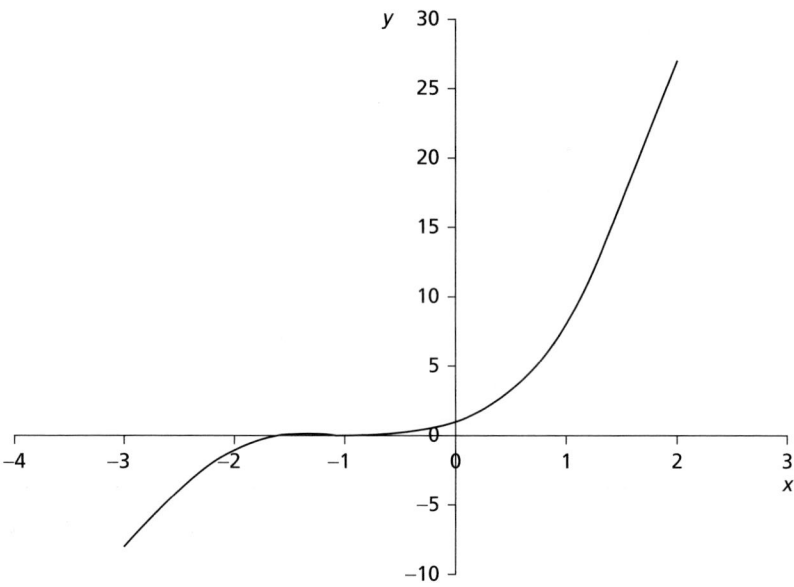

Fig. 1.4 The polynomial $y = m_1 x^3 + m_2 x^2 + m_3 x + c$.

such graphical information is to predict the way the variables will behave under different circumstances. If the point at which this information is required lies on the curve within the data point set, it is called 'interpolation', and if it lies on the curve outside the data points it is called 'extrapolation'; extrapolation is necessarily potentially less accurate than interpolation, unless the data are known to fit the curve exactly.

The rectangular hyperbola

A circle is a special case of an ellipse, and a hyperbola differs from an ellipse only in a single sign in its algebraic equation. A special case of a hyperbola is a *rectangular hyperbola*, whose equation is of the form $xy = k$, and it is worth dwelling on briefly, since there are numerous examples in anaesthesia. These include the following:

- $PV = nRT$, the universal gas law, where P is pressure, V is volume, T is the absolute temperature of the gas, and the other symbols are constants.
- $VF = K(P_{E,CO_2})^{-1}$, where V is the patient's minute ventilation, F is fresh gas flow in a breathing system, and $K(P_{E,CO_2})$ is a constant whose value depends on the value of P_{E,CO_2}.

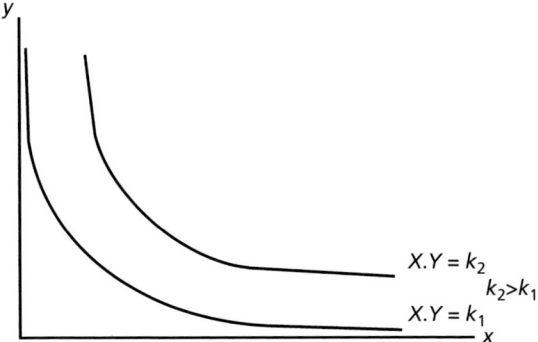

Fig. 1.5 Rectangular hyperbolae of the form $xy = k$, where $k_2 > k_1$.

Another example is the shape of the strength/duration curve of excitation of a muscle or nerve. The rectangular hyperbola resembles the curve in Figure 1.5.

Trigonometry

Figure 1.6 shows a triangle with one right angle (90°) between two sides of length x and y, with the third side of length r, from which some simple and widely applicable relationships can be stated.

If the angle between, say x and r, is designated θ, then the ratios of different side lengths to each other, namely y/r, x/r, and y/x are defined as follows:

$$\sin\theta = \frac{y}{r}, \quad \cos\theta = \frac{x}{r}, \quad \tan\theta = \frac{y}{x}.$$

Some useful numerical examples include $\sin 30° = \cos 60° = \frac{1}{2}$, $\sin 90° = \cos 0° = 1$, $\tan 45° = 1$. Inverse ratios of *sin, cos* and *tan* are, respectively, *cosec, sec* and *cotan*.

In the triangle OAP, if OP is also the radius of length r of a circle, an arc of which is shown, and on which a point P lies, then the coordinates of point P can be described as both (x, y) and (r, θ), where θ is the angle formed with the x-axis by the circle radius. Looking at Figure 1.6, where point P lies on a curve at a constant distance r from the origin 0, the following relationship for any right angled triangle is deduced:

$$x^2 + y^2 = r^2.$$

From Figure 1.6, this is identical to $r^2 \cos^2\theta + r^2 \sin^2\theta = r^2$ or $\cos^2\theta + \sin^2\theta = 1$, another important trigonometric relationship.

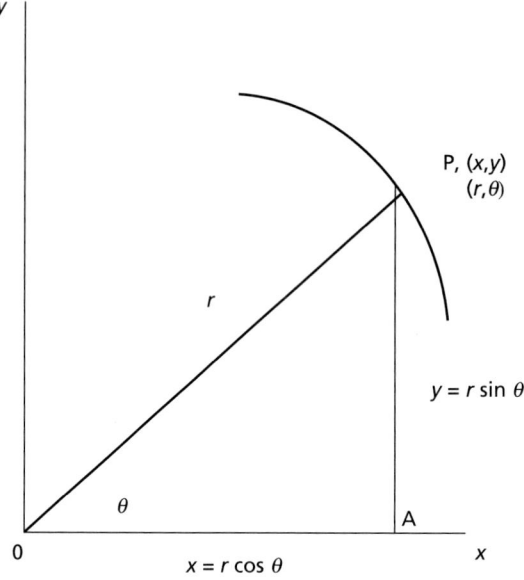

Fig. 1.6 A point P expressed as both (x, y) and (r, θ).

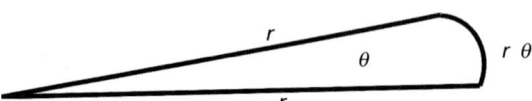

Fig. 1.7 A small arc of a curve of radius r, subtended by a small angle θ.

Figure 1.7 shows a small segment or arc of a curve, not necessarily a part of a circle, whose radius of curvature is r, and the two ends of which subtend a small angle, θ.

Where θ is expressed in radians, the length s of the small arc is given by $s = r\theta$. Where $\theta = s/r$ is very small, it approximates to $\sin \theta$. If all such small arcs are added together to form a circle of radius r, $\theta = 360°$, and the circumference is given by $2\pi r$. Thus $360° = 2\pi$ radians.

As shown in Figure 1.8, if a point P describes a circle by rotating round its circumference at constant angular velocity ω, and it sweeps out an angle θ in time t, then $\theta = \omega t$; furthermore the path of the vertical amplitude y can be traced on a screen with a time base (a cathode ray oscilloscope) and its motion

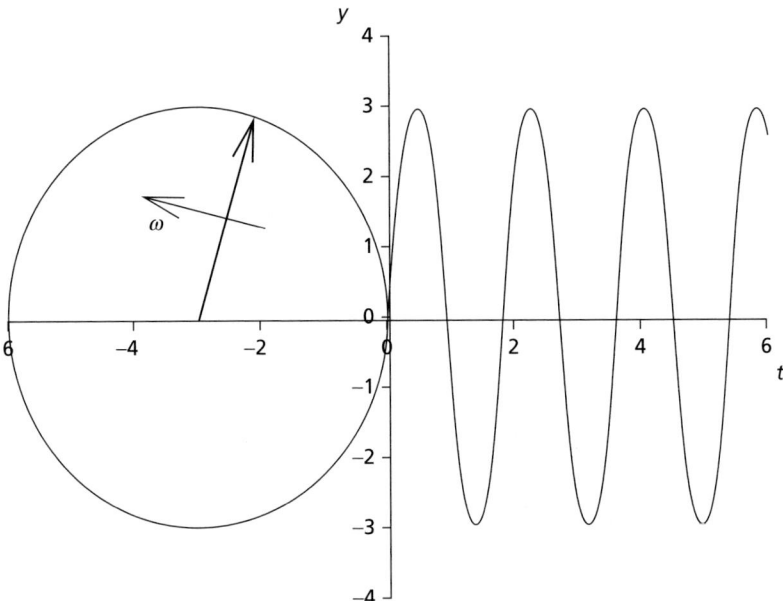

Fig. 1.8 Simple harmonic motion. A point rotating around a circle on a moving horizontal time axis traces out a sine wave.

is given by $y = A \sin \omega t$, where A is the maximum amplitude, occurring when $\theta = \omega t = \pi/2$ radians or 90°.

The path traced is a *sine wave*, described by the same equation.

Physical examples of such sine waves include electromagnetic radiation, alternating current, clock pendulum movement, a vibrating mass and spring system, and the path swept out by a lung function generator. The motion generated in this way is called *simple harmonic motion*.

Different wave forms, with different amplitudes and frequencies, can be superimposed. These are harmonics that make up a more complex, non-sinusoidal waveform, such as the arterial pressure or ECG waveform. Conversely, these complex waveforms can be analysed by being broken down into their constituent sinusoidal components of different frequencies and amplitudes, some of which may be out of phase with each other. This is known as *Fourier analysis*, the significance of which is that in order to reproduce a waveform faithfully, a piece of monitoring equipment has to be able to reproduce all components, from lowest to highest amplitudes, and from lowest to highest frequencies, with minimum distortion and phase shift. This will be further discussed in Chapter 3.

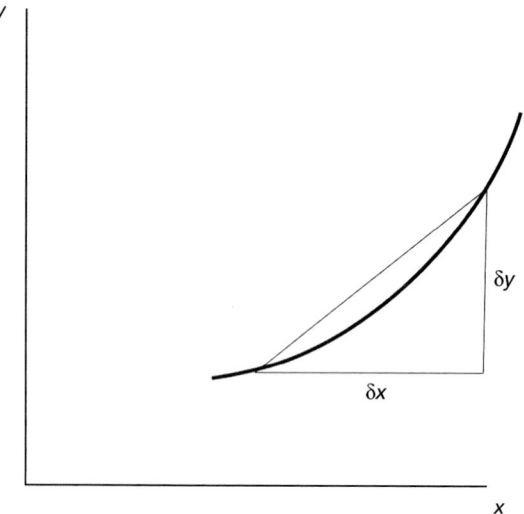

Fig. 1.9 Instantaneous slope δy of a curve $f(x)\delta y/\delta x$.

Calculus

Differentiation

This term refers to the rate of change of a function with respect to a variable. The slope of a straight line, or the rate of change of y with respect to x, has already been described as m, a constant. But, if the line is a curve where y is a function of x, i.e. $y = f(x)$, with a varying slope, then the slope at a given point is the instantaneous rate of change of y with respect to x, or algebraically:

$$\text{Slope } f(x) = \delta y/\delta x,$$

where δy and δx are very small changes in x and y respectively.

This is shown in Figure 1.9. As the intervals of change δx become vanishingly small the term $\delta y/\delta x$ becomes a calculable term called the *first differential* of $f(x)$, or dy/dx. Looking at specific functions, if

$$y = mx + c \quad \text{then} \quad dy/dx = m$$

$$y = kx^2 + c \quad \text{then} \quad dy/dx = 2kx.$$

In general, if $y = a(x+b)^n + c$ then $dy/dx = na(x+b)^{[n-1]}$.

The derivation of these expressions can be found in any calculus text.

A physical example where this is relevant is in an expression to estimate distance travelled with respect to time; its rate of change or differential is

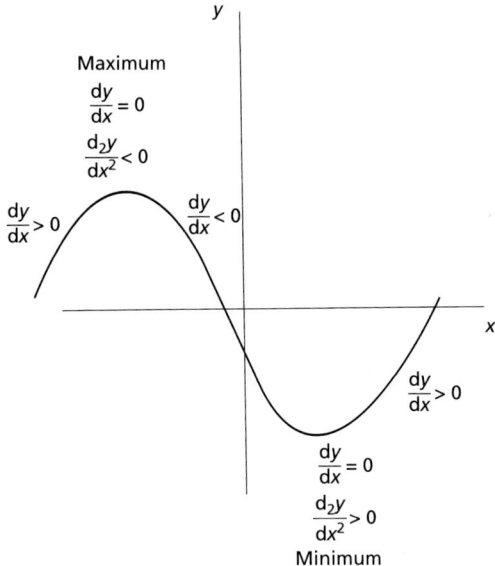

Fig. 1.10 Maxima and minima on curves.

velocity. The function can be differentiated a second time to give acceleration. For example, if the distance travelled in time t is y, where $y = mt + c$, then the velocity, $dy/dt = m$, a constant, and the acceleration $d^2y/dt^2 = 0$, since differentiating the constant m gives zero. Also, information can be gleaned about maxima and minima on curves where the slope dy/dx is zero, and where the rate of change of the slope d^2y/dx^2 is negative or positive respectively. This is shown graphically in Figure 1.10.

Consider a practical example of this. The velocity v of fluid flowing in a tube of diameter $2a$, where the flow is laminar, shown in Figure 1.11, is described by

$$v = V_{max} \frac{(a^2 - x^2)}{a^2},$$

where V_{max} is the maximum value of v, occurring at $x = 0$ in the midline of the tube, and where $v = 0$ at the tube walls, where $x = \pm a$.

$$dv/dx = \frac{V_{max}}{a^2}(-2x) = 0$$

when $x = 0$, i.e. the tube midline, and

$$d^2v/dx^2 = \frac{V_{max}}{a^2}(-2) < 0,$$

i.e., a maximum value (rather than a minimum).

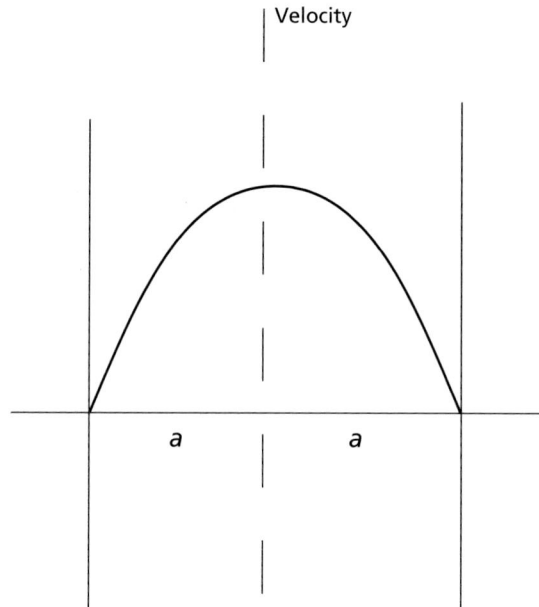

Fig. 1.11 Velocity profile across a tube in which fluid is flowing.

Integration

This is the opposite process to differentiation. If we integrate d^2y/dx^2 with respect to x, we get dy/dx, and if we integrate again we get y. If we integrate yet again, we are calculating the area under the x–y curve.

Figure 1.12 shows a short section of a curve, with two points identified (x, y), and $\{(x + \delta x), (y + \delta y)\}$. The area of the elemental rectangle shown is thus $(y + \tfrac{1}{2}\delta y).\delta x \cong y.\delta x$, ignoring the smallest terms.

The sum of all such elemental rectangles, ignoring small terms, is written as $\Sigma y \delta x$.

As $\delta x \Rightarrow 0$, this becomes the integral of y with respect to x, written mathematically as $\int y\,dx$, the area under the curve between certain limits of x, for example, if

$$y = xn, \text{ then } \int y\,dx = \frac{x^{(n+1)}}{(n+1)} + \text{constant}.$$

The constant is necessary, since differentiating a constant gives zero, so back integration needs the constant to be re-established. Its value is determined by integrating between defined limits of x.

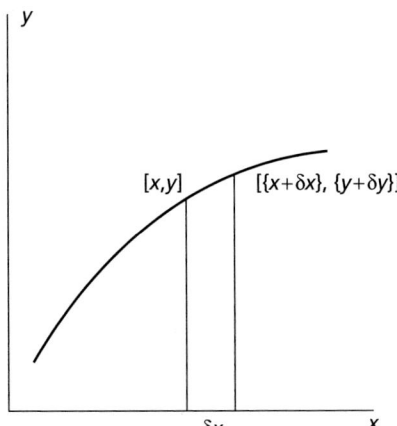

Fig. 1.12 Area under a small element of curve.

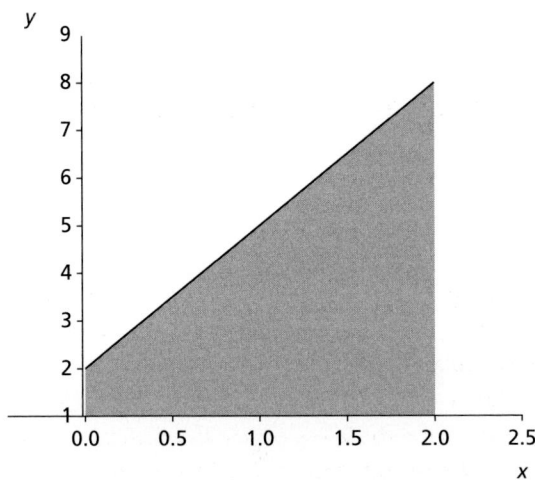

Fig. 1.13 Area under the straight line $y = 3x + 2$, no limits specified.

For example, suppose $y = 3x + 2$, shown in Figure 1.13, then

$$\int y.dx = \int (3x+2)\,dx = 3\frac{x^2}{2} + 2x + constant$$

As the figure shows, some additional information is needed to calculate the constant, since it is otherwise unknown how many elemental rectangles are being summed together in order to calculate the area under the curve. This is

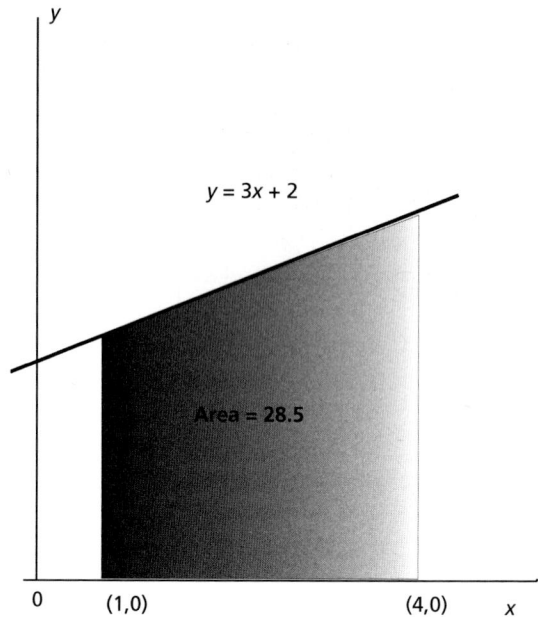

Fig. 1.14 Area under the straight line $y = 3x + 2$, limits specified.

usually dealt with by stipulating boundaries, say between $x = 1$ and $x = 4$. Then the area under the curve is:

$$\int_{x=1}^{x=4} (3x+2)dx = \left[\frac{3x^2}{2} + 2x\right]_{x=4} - \left[\frac{3x^2}{2} + 2x\right]_{x=1} = 28.5.$$

Figure 1.14 shows this.

A clinical example of the use of integration is in the use of the thermodilution catheter to measure cardiac output. The computer associated with the device plots the curve of temperature change (shown in Figure 1.15) as measured at the thermistor tip of the catheter against time, and the expression for cardiac output is of the form

$$cardiac\ output = \frac{constant_1 \times thermal\ capacity\ of\ the\ injectate}{constant_2 \times \int (temperature\ change).d(time)}.$$

Notice therefore that the larger the integral (area under the curve), the smaller the cardiac output because area under the curve is the denominator. Figure 1.15 shows the plotted curve.

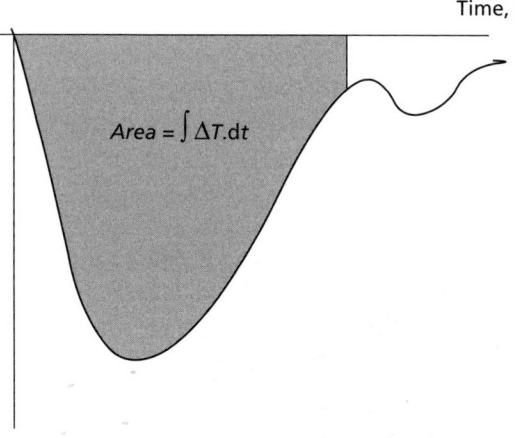

Fig. 1.15 Area under a thermodilution curve, used to calculate cardiac output.

Differential equations

The idea of describing rates of change of processes using differential notation has already been introduced. A simple differential equation is shown below; it says 'the rate of decrease of a substance y with respect to a variable t is proportional to the amount of y which is left':

$$\frac{dy}{dt} = -ky \quad \text{or} \quad \frac{dy}{dt} + ky = 0,$$

whose solution is $y = A_1 e^{-k.t}$.

This is a mathematical way of describing an exponential process, of which there are many examples in medicine such as plasma drug concentration as redistribution occurs; the negative exponent in the first equation indicates a decreasing process rather than an increasing one.

The concept of simple harmonic motion, which has also been introduced, can be expressed in words as 'acceleration is proportional to displacement and opposes it' and it can be expressed in terms of a differential equation as:

$$\frac{d^2y}{dt^2} = -ky, \text{ whose solution is } y = A \sin \omega t.$$

This describes, for example the motion of a mass on a spring. More generally, the behaviour of engineering and physiological systems can be described by similar differential equations such as:

$$A\frac{d^2y}{dt^2} + B\frac{dy}{dt} + Cy = D,$$

where the first, second and third terms on the left hand side are frequently acceleration, velocity and displacement terms respectively, and the term on the right hand side is often a forcing function, which might be zero.

This can describe, for example, the motion set up in a vibrating catheter system to reproduce an arterial waveform with damping. These sorts of differential equations often have solutions that contain multiple exponential expressions.

Powers, logarithms, 'e'

Very large or very small numbers can be conveniently expressed as powers. Since humans use 10 as the computing base, numbers are expressed as powers of 10, i.e.

$$10^0 = 1, 10^1 = 10, 10^2 = 10 \times 10 = 100,$$
$$\ldots 10^6 = 10 \times 10 \times 10 \times 10 \times 10 \times 10 = 1\,000\,000.$$

The verbal prefix 'kilo-' means 10^3, and 'mega-' means 10^6.
Likewise,

$$\frac{1}{10} = 10^{-1}, \frac{1}{1000} = 10^{-3} \text{ (prefix 'milli-')},$$

$$\frac{1}{1\,000\,000} = 10^{-6} \text{ (prefix 'micro-')}, 10^{-9} \text{ (prefix 'nano-')}.$$

These apparently very small quantities appear in electronics and biochemistry, often because of the way these units are derived. For example the unit of electrical capacitance is the farad; it is a very large unit, defined as the charge in coulombs stored in a conductor per unit voltage across it ($C = Q/V$); therefore picofarads (10^{-12}) are often preferred. Likewise the unit of resistance is the ohm (Ω), and it is a small unit; we therefore often refer to megaohms (10^6).

If $y = 1\,000\,000 = 10^6$, then 6 is the power to which 10 is raised to get $1\,000\,000$. In this case therefore, 6 is the *logarithm to base 10* of 10^6. If we replace 6 with the algebraic term b, then in general terms: if

$$y = 10^b,$$

then

$$\log_{10} y = b \cdot \log_{10} 10 = b \cdot 1 = b.$$

In words this would read 'log to base 10 of y equals b (times log to base 10 of 10)'. Log to a base of itself is 1. Logarithms therefore allow easier number representation and manipulation, particularly of very large or very small numbers.

For example, if

$$y = 100\,000\,000 = 10^8 = 10^2 \times 10^6,$$
$$\log_{10}(10^2 \times 10^6) = \log_{10} 10^2 + \log_{10} 10^6$$
$$= 2\log_{10} 10 + 6\log_{10} 10$$
$$= 2 \times 1 + 6 \times 1$$
$$= 8$$
$$= \log_{10} 10^8.$$

Hence powers of 10 multiplied together result in a number whose powers are added, i.e.

$$10^a \cdot 10^b = 10^{(a+b)}, \text{ and } \log_{10} 10^{(a+b)} = (a+b).$$

Likewise

$$\log_{10}(a \cdot b) = \log_{10} a + \log_{10} b$$

and

$$\log_{10}(a/b) = \log_{10} a - \log_{10} b.$$

For example,

$$\log_{10}(10^6/10^2) = \log_{10} 10^6 - \log_{10} 10^2$$
$$= 6 - 2$$
$$= 4$$
$$= \log_{10} 10^4.$$

An example of the use of logarithms in a clinical measurement setting is the *decibel*, a logarithmic ratio of power or amplitude (voltage or current) in relation to the electronic gain (amplification) of biomedical electronic equipment (see Chapter 4).

By definition

$$\text{power gain in decibels (dB)} = 10\log_{10} \cdot \frac{output\ power}{input\ power}$$

and

$$\text{voltage (or current) gain} = 20 \log_{10} \cdot \frac{\text{output voltage (or current)}}{\text{input voltage (or current)}}.$$

Graphical use of logarithms can be made by representing $\log x$ plotted graphically against $\log y$ (a log–log plot) or x against $\log y$ (semi-log). Pharmacological variables are often presented in this way.

Negative logarithms also have a meaning:

$$-\log_{10} a = \log_{10}(1/a) = \log_{10}(a^{-1}).$$

In acid base balance, the concept of pH uses negative logarithms: by definition

$$\text{pH} = -\log_{10} [H^+].$$

Normal hydrogen ion concentration,

$$[H^+] = 40 \text{ nanomol l}^{-1}$$
$$= 40 \times 10^{-9} \text{ mol l}^{-1}.$$

Therefore,

$$-\log_{10}[H^+] = -\{\log_{10} 40 + \log_{10} 10^{-9}\}$$
$$= -\{\log_{10} 10 + \log_{10} 4 + \log_{10} 10^{-9}\}$$
$$= -\{1.0 + 0.6020 + (-9)\} \approx 7.40.$$

Instead of base 10, it is possible to choose other number bases, which can be used to express both large numbers and logarithms. For example if a base of 2 is chosen (binary),

$$2 \times 2 \times 2 = 2^3 = 8$$

and

$$\log_2 8 = 3.$$

If a general base a ($a > 1$) is chosen, such that $y = a^x$, which is a general exponential function, then it can be shown that the gradient, dy/dx of the curve is of the form:

$$\frac{dy}{dx} = m \cdot y = m \cdot a^x,$$

where m is a constant with different values depending on the base, a.

If $a \leq 2$, then $m < 1$; and if $a \geq 3$, then $m > 1$. Imagine a base then, which has the property $m = 1$. This would mean that the rate of change of this function at every point is equal to the value of the function at that point.

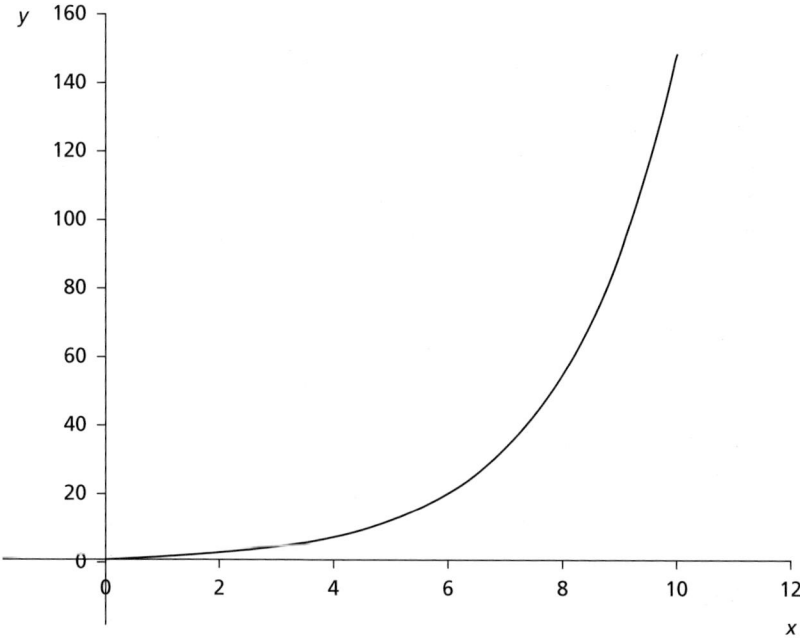

Fig. 1.16 Curve $y = e^{0.5x}$.

This base is the natural exponential, e, where e = 2.7183. This seems a bizarre choice at first, but mathematically e also has the property of being represented by the very tidy power series:

$$e^x = 1 + x + \frac{x^2}{2 \times 1} + \frac{x^3}{3 \times 2 \times 1} + \cdots \frac{x^n}{n(n-1)\ldots 2 \times 1}.$$

Furthermore the graphical representation of e^x very adequately describes many changes in physiological and engineering systems. The property that $m = 1$ for e means the rate of change of this function is equal to the amount left and can be expressed mathematically as

$$\frac{d}{dx}(e^x) = e^x.$$

Such a curve is shown in Figure 1.16.

Logarithms to base 'e' can be used, when they are called natural logarithms, $\log_e x$ or $\ln x$, i.e.

$$\text{if } x = e^y, \text{ then } y = \log_e x \text{ or } \ln x.$$

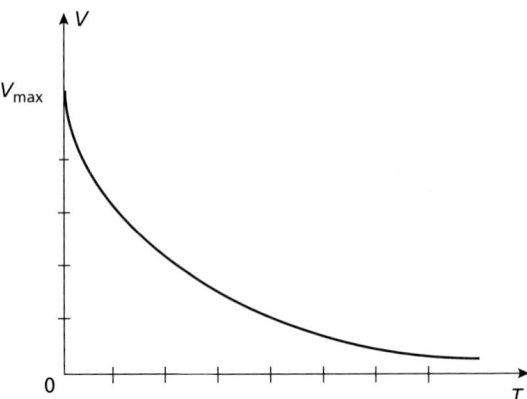

Fig. 1.17 Decaying exponential curve with the equation $V = V_{max} \cdot e^{-0.5t}$.

The fact that the rate of change of a substance is proportional to the amount of substance present is sufficient to define an exponential function, and describes the behaviour of a range of natural phenomena. For example, a rectangular bath full of water starts to empty as the plug is removed; the (volumetric) rate of emptying of the bath is proportional to the height of water 'h' left in the bath at time 't', and this is proportional to the pressure, $\rho g h$, which is driving the emptying process, where ρ and g are constants (see later).

Expressed mathematically,

$$-\frac{d}{dt}(volume\ left) = constant_1 \times \rho.g.h,$$

the minus sign indicating volume decrease with height; i.e.

$$K_1 \frac{dh}{dt} = -K_2 . h.$$

The curve describing this behaviour is shown in Figure 1.17, and can be described by an equation like $V = V_{max} \cdot e^{-kt}$ where V is volume, V_{max} is maximum volume, k is a constant and t is time. It could equally well apply to lung emptying.

Lung filling also occurs in an exponential fashion, and Figure 1.18 shows the form of the curve whose equation is $V = V_{max}(1 - e^{-kt})$.

The reduction in plasma concentration of a drug or dye, or the fall in blood temperature after a thermodilution bolus, also follow an exponential pattern prior to any recirculation. The half life, $t_{1/2}$, is the time taken for the concentration of the drug to fall from C_{max} to $½.C_{max}$. The time constant τ,

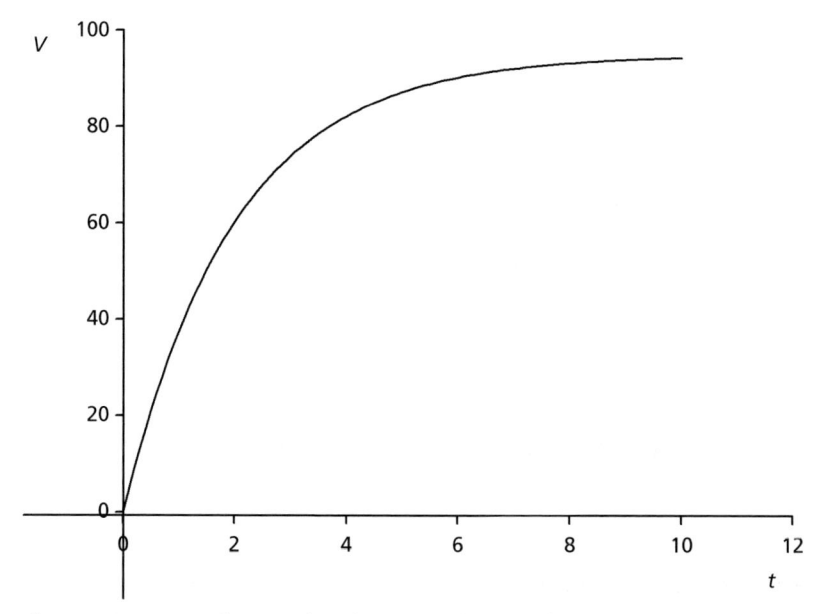

Fig. 1.18 Exponential curve describing lung filling with the equation $V = V_{max}(1 - e^{-0.5t})$.

is the time elapsing for the function to fall to zero if the rate of decay had remained at its initial value, which corresponds to falling to a value of C_{max}/e on the exponential curve. Figure 1.19 shows this.

To calculate $t_{\frac{1}{2}}$:

if $C = C_{max} \cdot e^{-k \cdot t}$, then when $C = \frac{1}{2} \cdot C_{max}$,

$$\frac{1}{2} = e^{-kt_{1/2}},$$

therefore

$$t_{\frac{1}{2}} = -\frac{1}{k} \cdot \log_e \cdot \frac{1}{2}$$

$$= 0.693/k.$$

The slope of the exponential decay curve is $dC/dt = -kC_{max}e^{-kt} = -kC_{max}$ when $t = 0$. The tangent to this curve has the equation $C = C_{max}(1 - kt)$. τ, the time constant, is defined by the point on this line at which $C = 0$, so $\tau = 1/k$, at which time $C = C_{max}/e$ on the curve.

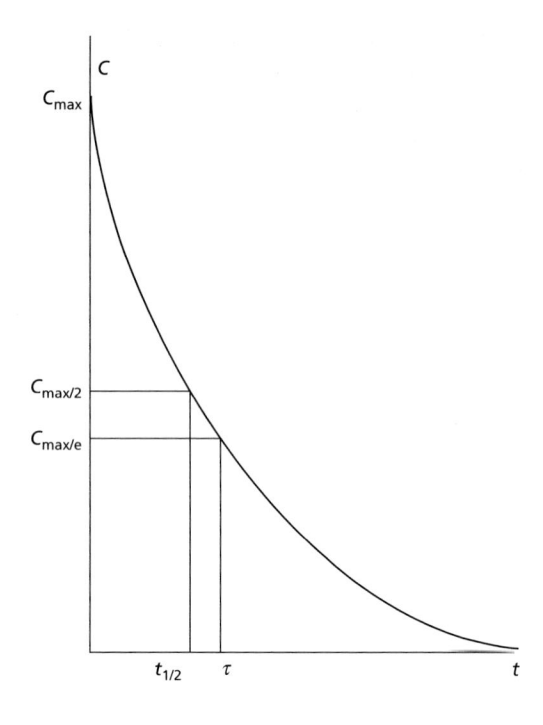

Fig. 1.19 Exponential decay in drug plasma concentration C, and its relationship to time constant and τ (tau) and half life $t_{1/2}$.

Fig. 1.20 Exponential decay of drug concentration in a multi-compartment pharmacokinetic model, with equation $C = C_1 e_1^{-k_1 t} + C_2 e_2^{-k_2 t} + C_3 e_3^{-k_3 t} + C_4 e_4^{-k_4 t}$.

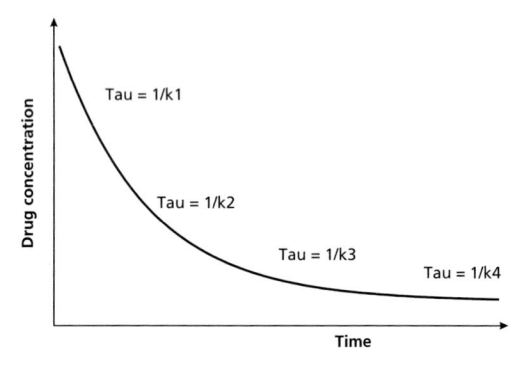

As a rule of thumb, after one time constant there is a fall to 0.37 C_{max} of the initial value; after 2τ, C has become 0.135 C_{max}; after 3τ, C is 0.05 C_{max}. The degradation process is normally considered complete after four time constants.

Note that if the equation $C = C_{max} \cdot e^{-kt}$ is written in logarithmic form:

$$\log_e C = -kt \cdot \log_e C_{max},$$

then the resulting semi-log plot of $\log_e C$ plotted against t is a straight line with a (negative) slope $-k$, which makes its interpretation easier.

Many pharmacokinetic and other processes are complex, involving multiple exponential curves, whose equations might be something like:

$$y = A_1 e^{-k_1 \cdot t} + A_2 e^{-k_2 \cdot t} + \cdots,$$

shown graphically in Figure 1.20, which might represent a drug concentration in a multi-compartment model with different time constants.

Further reading

Cruickshank, S. (1998) *Mathematics and Statistics in Anaesthesia*, Oxford Medical Publications, Oxford.

Hill, D. W. (1976) *Physics Applied to Anaesthesia*, Butterworths, London.

Chapter 2

Statistics

This chapter contains: categorical and numerical data, quantifying variability, probability, the Normal distribution, hypothesis testing (parametric and non-parametric), diagnostic tests, meta-analysis, correlation and regression, comparing categorical data, and evidence based medicine.
The chapter links with: Chapters 1, 3, 5 and 19.

Introduction

This chapter will provide background to enable the reader to understand basic statistics and be able then to follow more complex statistical ideas. Although statistics is more than the mere analysis of data, it is a subject largely about data, so this will be discussed first. Data can be categorical or numerical, and in these two classifications there are various different types of data.

Categorical data

Data that can be put into two categories

This is the allocation of the individual to one of two categories. Often these relate to the presence or absence of some attribute. These data also have many other names such as binary, dichotomous and attribute data. Examples of such categorisations for patients include:

- Male/Female
- Smoker/Non-smoker
- Anaesthetist/Surgeon
- Married/Single.

Each of these can be only be one or the other – they could be coded '1' or '0' to be *binary* (or on, off). For example male $= 0$, female $= 1$, or vice versa.

Data that have more than two categories

Many classifications require more than two categories, such as: blood group, type of doctor, country of birth. Also the two categories, such as described previously, might be expanded into several categories. For example the married/single could be expanded to: married/single/divorced/separated/widowed. This sort of data is called *nominal* data where there are several categories, but with no logical order.

When there is a natural order (such as in seniority), the data are then called *ordinal* data. For example, anaesthetists could be divided into: 'Foundation year 1', 'Foundation year 2', 'speciality doctor', consultants', 'senior consultants' and 'clinical directors'. Ordinal data can be reduced to two categories, with possibly a considerable loss of information (e.g. 'senior doctors', 'junior doctors').

Numerical data

Discrete numerical data are where the observation takes exact numerical values. Counts or events are discrete values. For example: number of children, number of ectopic beats in a time period and so on. Continuous (or analogue) data are usually obtained by some form of measurement. Examples are body temperature, blood pressure, height and weight. These values have an infinite number of possibilities, depending on the measurement interval, and variation. Although there are infinite possibilities, measurement systems usually round the continuous data up, or down, to discrete values. Blood pressure is often rounded up to the nearest 5 mmHg, for example.

Linear and non-linear scales

Continuous data, such as heart rate can be displayed as a point on a linear scale. A linear scale increases at a constant rate and each interval between adjacent numbers is the same (as shown in Figure 2.1(a)). In some measurements, where a wide range of values is encountered, it can be useful to display numbers in a non-linear way. For example, in flowmeters capable of large flows, it would impossible to measure low flows with any degree of confidence. The normal form of non-linear scale is the logarithmic scale in which each increment on the scale corresponds to a proportional increase in the amount measured (as shown in Figure 2.1(b)). This can allow accuracy at small values but also large values can be displayed. Logarithms are discussed in more detail in Chapter 1.

Graphical displays of data

It is normally only possible to automatically display data graphically if the variable concerned has an electrical signal associated with it. This signal is normally digitised and displayed on a computer screen as a straight forward signal against time, or in the more complex frequency domain or, where

(a)

60	70	80	90

Heart rate in beats per minute

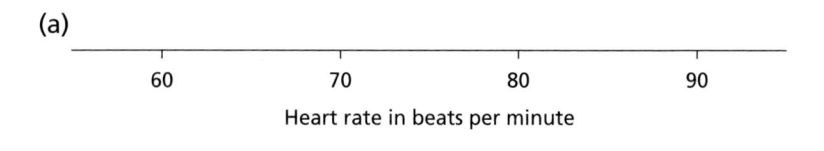

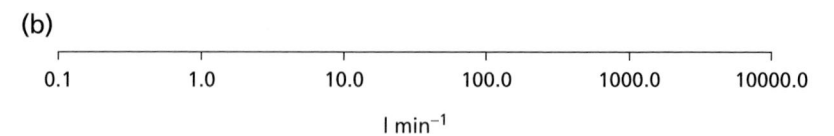

(b)

0.1	1.0	10.0	100.0	1000.0	10000.0

l min^{-1}

Fig. 2.1 Examples of linear (a) and non-linear (b) scales.

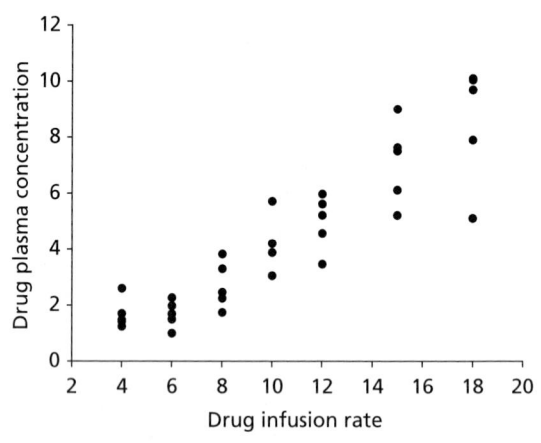

Fig. 2.2 Example of a scatter diagram. The y-axis is plasma concentration and the x-axis is infusion rate.

relevant, in other display forms such as histograms or pie diagrams. The digitisation and computers are described in Chapter 5. Data can also be presented as a scatter diagram. This is a one of a number of ways and is a simple graph where the values of one variable are plotted against those of another. An example of this is shown in Figure 2.2. In should be noted that in this example drug concentration is a continuous variable whereas drug dose is a discrete one.

Simple graphical ways of summarising a complete set of observations are by means of a histogram and a pie chart. In the histogram, the number (or frequency) of observations is plotted for different values or groups of values. Figure 2.3 shows a histogram of the number of children being given immunoglobulin M plotted against the plasma IgM concentrations.

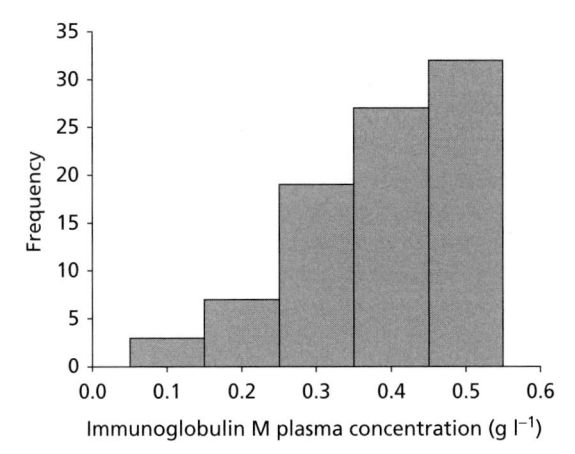

Fig. 2.3 Example of a frequency histogram of IgM plasma concentrations [Isaacs et al., 1983].

A pie chart is very useful for displaying the proportions that go to make up a whole, in the form of a circle split up into sectors. For example using some of the data for the histogram, the proportion of children having five different serum concentrations can be shown as a pie chart, as in Figure 2.4.

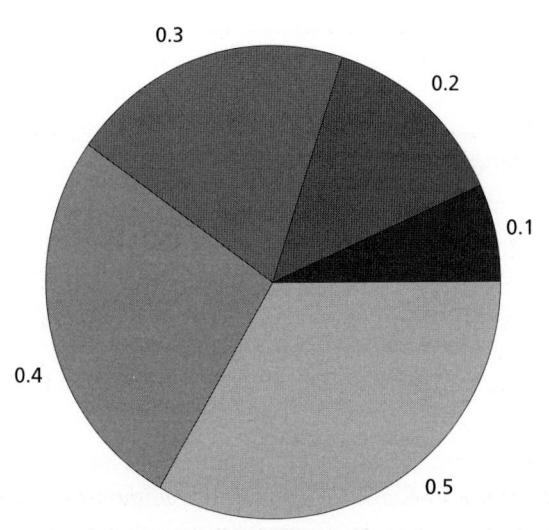

Fig. 2.4 Pie chart showing five plasma concentrations from Figure 2.3 and the proportion of children having those levels. The numbers in each sector are the plasma concentrations in g l^{-1}.

Quantifying variability

In anaesthesia, when measurements are taken from patients regarding their height, weight, blood pressure, etc, there is wide range of values recorded and a wide spread of values between patients. Parameters are needed to describe these data sets. The simplest method used for describing a range of observations of a continuous variable (e.g. height) is to calculate the average.

Averages

The most used average to do this is the *arithmetic mean*, normally referred to as simply the *mean*. This is the sum of all the observations divided by the number of observations. For example, if the data set of heights from 11 patients was 1.55, 1.62, 1.65, 1.72, 1.75, 1.76, 1.80, 1.80, 1.80, 1.85, 1.86 m, then the mean will be the sum of these heights divided by 11 (the number of patients), which is 1.74 m. This mean does not give any indication of the spread or *variance* of the values or whether the values are spread equally along the number axis or clustered together. This will be explored later.

Other measures to describe averages are the *mode*, the *median* and the *geometric mean*. The mode is the most frequently occurring observation in a data set. In the height data set used above, the sample mode is 1.80 m, as this height has three observations and the other height values have only one observation each. The median is the middle value of the set of measurements when they are placed in order according to their numerical value, or *ranked* in order. In the above example of heights, the numbers are ranked in ascending height values, and so the middle value, or median, will be the 6th value, which is 1.76 m. If there were an even number of observations, then the median would be the average of the middle two values. The geometric mean is used when a set of data is *transformed* by, for example, taking logarithms of each of the data points. The arithmetic mean of this new set can be calculated as described above. The transformation process will be discussed later. Transformed data must be transformed back again so, in the case of logarithm transformation, antilogarithms have to be made. The antilog of the mean is the geometric mean.

Variability

Graphical methods are important for examining the variability of data, but it is necessary also to be able to quantify the amount of variability in the data. This numerical value, in conjunction with the mean, would provide an informative, but brief, summary of a set of data. There are three main approaches to giving the variability a numerical value: the *range*, the *centiles*, and a numerical measure of the dispersion around the mean.

The range, which is the simplest variability measure, is quoted as the lowest and highest values of a series of numbers. Obviously outliers, or unusual values, will confuse the issue and make the spread of values appear greater than it really is. The situation can be improved by giving two numbers that encompass most, rather than all, of the data values. For example, the range between which 90% of the observations lie can be calculated – this would then omit the outliers. This is called the *central range*. The data set can be further split into *centiles*, or *percentiles*. If the data set is ranked from lowest value to highest value, then half way along is the 50% centile (or median). For example, if there are 100 data points, set out and ranked in ascending order, then the 5% centile will be the 5th data value. The 95% will be the 95th and so on. For other data sets, interpolation may be needed. The central range discussed above is the range between 5% and 95% of the data values. As an example, using the data shown in the histogram of Figure 2.3, then this data could be ranked as follows:

$$\{0.1, 0.1, 0.1, 0.2, 0.2, 0.2, 0.2, 0.2, 0.2, 0.2, \ldots\}.$$

As shown in Figure 2.3, to finish the data set, nineteen 0.3 values would be added, then twenty-seven 0.4 values, and then finally thirty-two 0.5 values. The 5% value (of the dataset) would be worked out by:

$$0.05 \text{ (the 5\%)} \times 89 \text{ (the total number of concentrations plus}$$
$$1 = 3 + 7 + 19 + 27 + 32 + 1) = 4.45.$$

So the 5% value is the 4.45th value, but there is not a 4.45 value, so the 5% value will be half way between the 4th and the 5th values. In this set both of these values are the same and so the 5% value is 0.2. In a similar calculation, the 95% value is 0.5. Therefore, in this dataset the 5% and 95% centiles are 0.2 and 0.5 g l^{-1}. All other centiles can be calculated in the same manner. The difference between the 25% and the 75% values is called the *inter-quartile range*, which again can be used to describe variability in a more useful way than just the total range.

Variance and Standard deviation

A mean ($\bar{x}$) has been calculated for the dataset, and all the data points (x_i) may have values similar to the mean value, i.e. their differences may be small. In this situation, there is a low spread or *variance*. All the data points may have large differences between the mean and the points and then there will be high variance. A numerical value can be worked out for this variance by first working out the difference from each of the data points to the mean. If all these values were added together, to get some 'feel' for the variance, then this would not be useful as the sum might be zero, as some differences would be negative and some equally positive. If the differences are squared (to remove the negatives) and then summed, then this provides an indication of the spread. The average of these 'squared differences' gives a measure of individual deviation from the means. The result can be divided by ($n - 1$), where n is the number of data points, to get an estimate of the population variance. (If the result is divided by only n, then the answer is underestimated.) The sample variance can be written as:

$$\frac{\sum_{i=1}^{n}(x_i - \bar{x})^2}{n-1}.$$

To put the *variance* in the same units as the mean, the equation needs to have its square root taken, and this result is then called the *standard deviation*. An easier form (which is mathematically the same, only rearranged) for calculation purposes is:

$$s = \sqrt{\frac{\sum x_i^2 - \frac{1}{n}(\sum x_i)^2}{n-1}}.$$

For most circumstances, the large majority (95%) of a set of observations will be within ± two standard deviations of the mean. This statement depends on the *shape* of the distribution. If a histogram is drawn of some data (see Figure 2.5), then it is symmetrical if each side of the mode (the most commonly occurring value, in this example 0.7) has a similar shape. If it is not symmetrical, it is

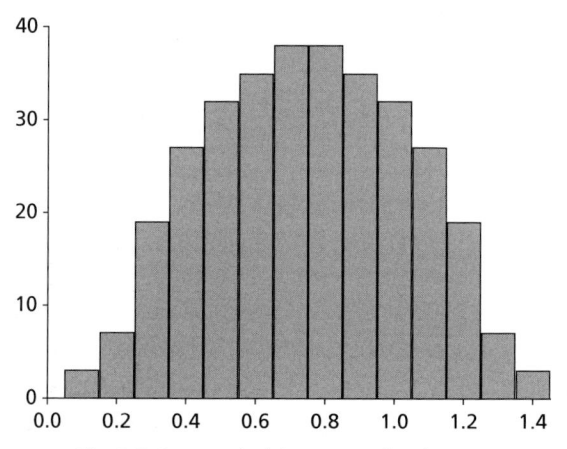

Fig. 2.5 Symmetrical frequency distribution.

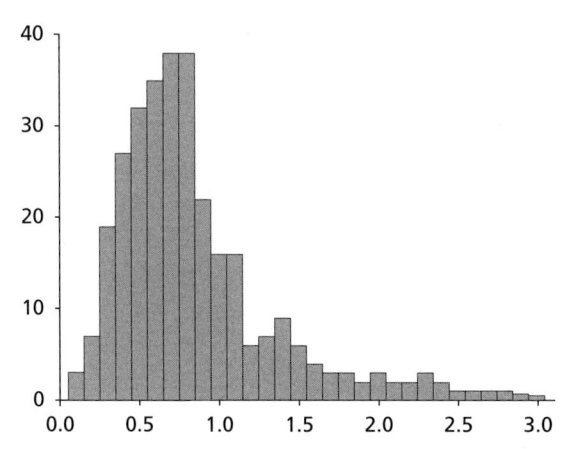

Fig. 2.6 Unsymmetrical frequency distribution with positive skew.

called a skewed distribution and an example is shown in Figure 2.6. This figure shows positive skewness where there is a long right hand tail, and this is most common skewed distribution. Negative skewness is where there is a long left hand tail. The data may be able to be transformed mathematically by using, for example, logarithms of each data point, and this may give a more symmetrical picture. The shape and properties of distributions will be introduced further in the next section on probability.

Probability

The probability of some outcome is the proportion of times that a certain outcome would occur if the experiment were repeated a large number of times. The probability that a baby will be a girl can be estimated by observing what proportion of a large population of babies are girls. Probability lies between 0 and 1. Something that cannot happen has the probability of 0 and something that is certain has the value of 1. There are rules about probabilities: for a given event, for any two outcomes that might happen, the probability of either occurring is the *sum* of the individual probabilities, as long as they cannot occur at the same time (they are exclusive). For example if the probability of an individual being an anaesthetist is 0.01 and that of being a physicist is 0.02, then the probability of being either an anaesthetist or a physicist is 0.03. It follows that the sum of *all* possible occupations must add up to 1, since all of the occupations must occur. If two or more different events happen that are independent of each other (the outcome of one tells us nothing about the other event), then to obtain the probability of a combination of specific outcomes, the individual probabilities are multiplied. For example, if there are three people in a library, the probability that they are all anaesthetists is 0.01 × 0.01 × 0.01, i.e. 0.000 001. If the events are not independent then the multiplication rule does not apply, and the calculations are more complex.

Probability distributions

Many statistical methods are based on the assumption that the observed data are a sample from a distribution that has a known shape. If this assumption is reasonable then statistical methods of analysis are called parametric analyses and are simpler to use than those from a sample without such a distribution. Those methods that make no assumptions about distribution shape are called non-parametric methods. The most commonly used distribution for continuous variables is called the Normal distribution (or Gaussian).

The Normal distribution

Histograms (an example was shown in Figure 2.3) have already been discussed as a graphical method of showing the frequency with which different values of a variable occur in the dataset. If the midpoint of each band in Figure 2.5 is joined up then the resultant curve will be a frequency polygon, as shown in Figure 2.7. The shape of the curve is bell shaped, symmetrical and has one peak. Its shape is loosely similar to the Normal distribution, which is a probability distribution that has one peak and is symmetrical. This distribution (unlike the one in Figure 2.7) has no upper and no lower limit – it is defined for all values to infinity. The height of the frequency curve is the probability density. As with the histogram of observed data, the Normal distribution is used by considering areas. The total area under the curve is always 1.0. A restricted area gives the probability between the values within the restricted area. For example, the probability of the whole range or area occurring is certain or 1, and the area to the left of the median is 0.5. Therefore the probability of being below the median is 0.5 or 50%.

The Normal distribution is completely described by two parameters – the mean (μ) and the standard deviation (σ). For example the distribution in Figure 2.8 has a mean of 10 and a standard deviation of 2.

If we assume that in the population, some variable (for example blood pressure) has a Normal distribution, then the probability of a certain value (or pressure) occurring can be estimated from the area under the curve. For example, the probability of the pressure being between one standard deviation on one side of the mean (12 in the figure) to one standard deviation the other

THE NORMAL DISTRIBUTION | 31

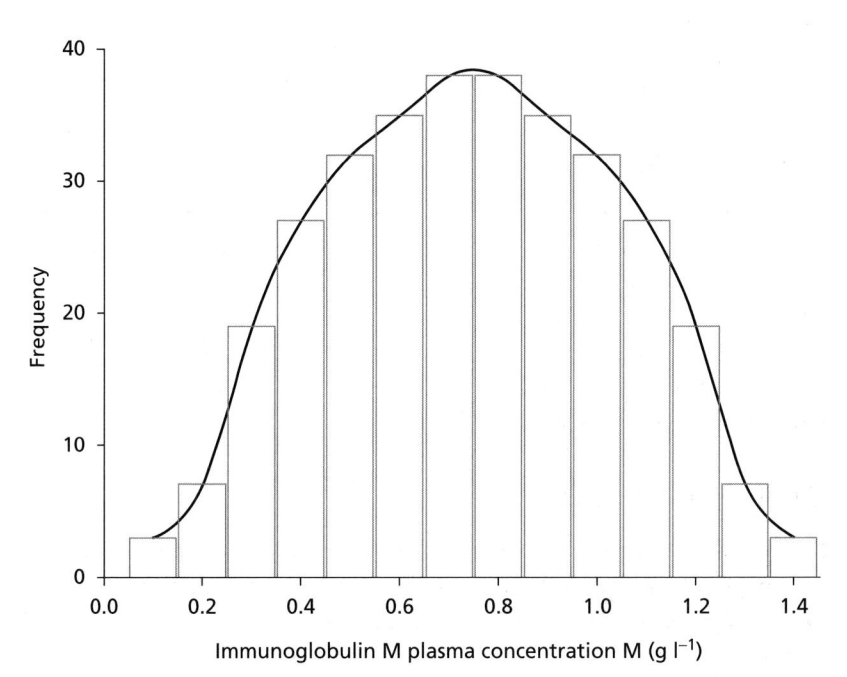

Fig. 2.7 Frequency distribution, showing bell shaped curve.

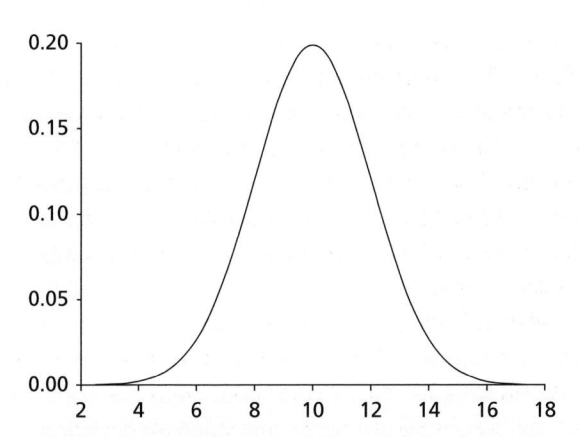

Fig. 2.8 Normal distribution with a mean of 10 and a standard deviation of 2.

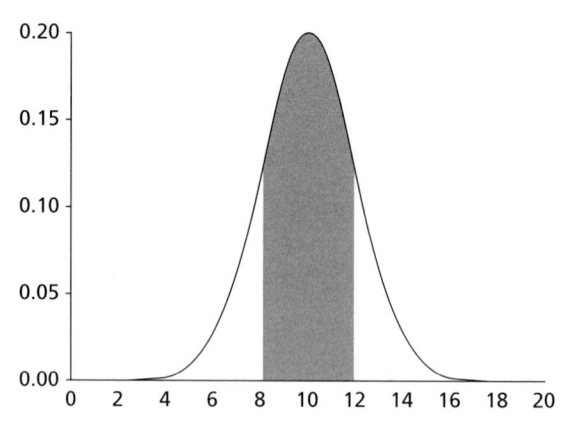

Fig. 2.9 Normal distribution (mean of 10, standard deviation of 2) showing area between mean plus one standard deviation (12) and mean minus one standard deviation (8).

side (8 in the figure) is 68% of the area under the curve, or 0.68. This is illustrated in Figure 2.9. Conversely being outside it (i.e. greater or less than one standard deviation) is the difference between the total area (1) and this value, i.e. 1 − 0.68, or 0.32.

Sample and population means

When a mean of a sample of data points is used to estimate the mean of a certain variable, it is not known how good this estimate of the mean is and how near in value it is to the true mean of the complete population of that variable. If, for example, the complete population of the UK has its resting blood pressure taken, then this extremely large data set will have a true mean μ and a true standard deviation σ. A subset of the population (for example 10 patients) can have their blood pressure taken and this sample will have its own mean ($\overline{X_i}$) and standard deviation (s_i), which may well be very different from the population values. Figure 2.10 shows this in schematic form, with the top box representing the complete population and the smaller boxes below representing small samples, taken at random from the population. Each of these samples will have their own mean and standard deviation.

The mean of Sample 1 (a small number of patients) will be a certain value and is probably different from the mean of Sample 2 and so on. If the number of patients in each sample is very large, then the sample mean will become

```
                    ┌─────────────────────────┐
                    │ Population              │
                    │ Blood pressure resting  │
                    │ μ mean,                 │
                    │ σ standard deviation    │
                    └─────────────────────────┘
           ┌────────────────┬─────────────────┐
           ▼                ▼                 ▼
   ┌──────────────┐ ┌──────────────┐ ┌──────────────┐
   │ Sample 1     │ │ Sample 2     │ │ Sample 3     │
   │ $\bar{X}_1, s_1$ │ $\bar{X}_2, s_2$ │ $\bar{X}_3, s_3$ │
   └──────────────┘ └──────────────┘ └──────────────┘
```

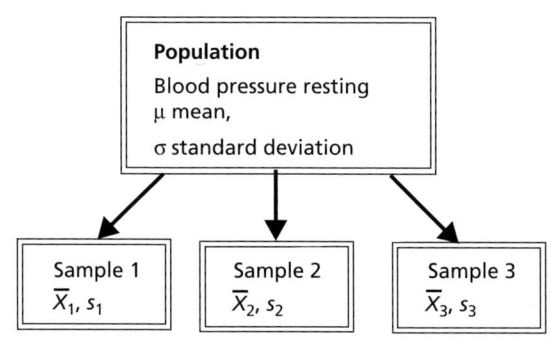

Fig. 2.10 Diagram showing the mean and standard deviation of the whole population with random sample sets drawn from this, each having a mean ($\bar{X}$) and standard deviation (s).

closer to the population mean. If the number of patients in each sample is small, then it is likely that this sample could be less representative of the population mean. If all the means of Sample 1, Sample 2, Sample 3 and so on are calculated and the *grand* mean of these sample means is calculated, it will be found that this grand mean is in fact the population mean, μ, with a standard deviation of:

σ (*the population standard deviation*)$/\sqrt{}$(*the number of patients in each sample*).

The standard deviation of the means of the samples (of size n) is termed the *standard error of the mean* (SEM), and the best estimate of it is obtained using the patients drawn from one of the single sample sets of data:

s (*the sample standard deviation*)$/\sqrt{}$(*the number of patients in the sample*).

The standard deviation quantifies variability in the population and the SEM quantifies uncertainty in the estimate of the mean. In the whole population 95% of the time, the mean obtained from each sample is within two (actually 1.96) standard errors (SE) of the true actual mean. The interval from the actual mean −2 SE to the actual mean +2 SE is known as the 95% confidence interval. The 99% confidence interval would be obtained by using ±3 SE (actually 2.58). However, these multiples of the SE are only true if the number of patients in each sample is large (large is normally taken to be greater than 30). If the number is small, then the estimates of the standard deviation of the population obtained from the small sample become very different from the actual value. For the small sample size, a multiplier of more than two is needed to calculate the 95% confidence intervals for the mean, because there is a greater imprecision involved. The 95% confidence interval of the total population can be obtained from a Normal distribution table found in statistical textbooks and this table quotes 1.96 SE. Because of this imprecision, the confidence intervals will be wider for samples containing small numbers of data points, and a different distribution is needed to calculate the intervals. This distribution is called the *Student's t-distribution*. For example, for a sample containing five patients, the 95% confidence interval will be, using the t-distribution, ± 2.57 SE. The Student's t-test makes use of comparing the confidence intervals of two sets of data using the t-distribution, and this will be discussed in the next section. The confidence intervals of two sets of data can be compared to see if they overlap or a hypothesis test can be used. An important parameter of the t-test is the number of *degrees of freedom*. This is a strange concept, but is defined as the *size of the sample – the number of estimation parameters (here, standard deviation)*. So for a sample of 10 patients, the degrees of freedom will be $n - 1 = 9$.

Hypothesis testing

The null hypothesis is that there is no difference between two samples and that both samples come from the same population. The P value is the probability of getting a difference between the two samples as large or larger as is observed, if the null hypothesis is true (i.e. the two samples are the same). The cut-off values for P can be chosen to be any value, but are normally taken to be 0.05 or 0.01. If the P value is higher than the cut-off value, then there is no evidence to reject the null hypothesis, if it is below then the null hypothesis is unlikely to be true and is rejected. A value of $P = 0.01$ means there is a 1% probability of getting the data if the null hypothesis is true, i.e. it is much more likely that the two samples are from different populations.

Before the statistical test is carried out, a cut-off value for P must be set. When the P value is generated, then it is either below the cut-off or above it. If it is below this value, then this is called the critical region and the null hypothesis is rejected. This of course can be the correct interpretation and the sample sets are from different populations. Or this rejection might be incorrect when the null hypothesis is in fact true. This is a type 1 error, an α error, or false positive (i.e. reject when it is true). If the value is greater or above the cut-off value, then this is the acceptable region and the null hypothesis is accepted. This can be the correct interpretation where the sample sets are from the same population, or the sample sets can be different and then the null hypothesis is not true. In this case there is a type 2 error, a β error, or false negative (i.e. don't reject it but

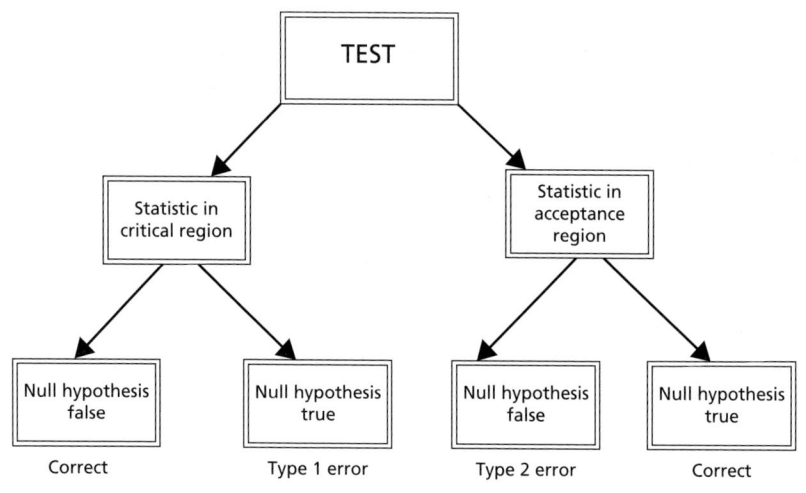

Fig. 2.11 Test flow diagram.

it is false). A summary of hypothesis testing, giving all the possible outcomes from a test is shown in Figure 2.11.

Parametric methods

Student's t-tests

A test can be carried out of the null hypothesis, that the mean from a sample is from a population with a known mean. If the data sets are large (>100 for example) then the Normal distribution is used, otherwise a distribution called the Student's t-distribution is used (as discussed previously). The test, called a *one sample t-test* uses the test statistic:

$$t = (sample\ mean - known\ mean)/(SE\ of\ sample\ mean).$$

This test uses the t-distribution, which becomes identical to the Normal distribution when the numbers in the data set are large. The t value will be 0 when the means are the same: it is likely that the sample is from the same population. As the difference between the sample mean and the known mean increases, then the t value increases and it becomes more likely that the null hypothesis is not true, and the sample is from a different population. Normally it does not matter whether the difference between the means is positive or negative, and then a *two sided* t-test is used. The t statistic can be translated into a P (probability) value, using a computer or statistical tables. For example, a population has a mean systolic blood pressure of 120 mmHg, and in our sample of nine patients having a certain anaesthetic, the blood pressure is 110 mmHg with a standard deviation of 10.6. This gives a test statistic of:

$$t = (110 - 120)/(10.6/3), \text{ which gives } t = 2.8.$$

In this example, the P value is *calculated* (from statistical tables) to be 0.05. If the null hypothesis cut-off is 0.05 (i.e. $P < 0.05$), then the calculated P value is not less than the cut-off so the null hypothesis is not rejected, but only just! If the cut-off was set at 0.01, then the null hypothesis would be accepted.

The *paired t-test* is an extension to the one sample test, and is a special case of the *two sample* t-test. Paired data can result when the same individuals have been studied more than once and also when each member of the sample has been individually matched to another in the other sample. An example could be women who take their blood pressure before and after their menstrual periods. One sample would consist of all the women before and the other sample would consist of all the women after. Each woman would be paired with herself, before and after. With paired data, the interest is in the average difference between observations and the variability of these differences. The *within-subject* differences are important. The test statistic value is now calculated as, (with $n - 1$ degrees of freedom):

$$t = (\textit{mean of differences between each before and after subject})/(\textit{SE difference}).$$

The other two sample t-test is where the two samples are truly independent, i.e. there is no pairing and the number of samples in each group can be different. The mean difference between the two groups and the variability between subjects is important. The test statistic is calculated (with $n_1 + n_2 - 2$ degrees of freedom), where Sample 1 contains n_1 patients, and Sample 2 contains n_2:

$$t = (\textit{mean of first group} - \textit{mean of second group})/(\textit{SE}(\textit{mean 1} - \textit{mean 2}))$$

Non-parametric methods

Wilcoxon signed rank sum test

This test is used for non-parametric data, that is when the data cannot be assumed to be from a Normal distribution. This test is the non-parametric equivalent of the one sample t-test.

If the sample values and the set value (or known mean) were from the same population, then, on average, there would be an equal number of observations greater and less than the set value. This signed rank test calculates how likely it would be to observe the data values if the null hypothesis is true. The test method calculates the difference from each data point to the set value. In the example in Table 2.1 the set value is 100 mmHg. Once each difference is

Table 2.1 Blood pressure values of ten patients with rank order of differences (ignoring their signs) from a set value of 100 (mm Hg)

Patient	Blood pressure (mm Hg)	Difference from 100 (mm Hg)	Ranks
1	80	20	10
2	85	15	9
3	87	13	8
4	89	11	7
5	90	10	5.5
6	92	8	4
7	95	5	1.5
8	95	5	1.5
9	107	−7	3
10	110	−10	5.5

calculated then the differences are ranked in order (i.e. the least magnitude is position 1, the next position 2, etc) but the signs are ignored for the ranking. Any differences that are the same will have their ranks averaged, e.g. 1.5 and 5.5 in the example. Once the ranks have been worked out, the 'negative' rank numbers are added together. This is differences −7 and −10 in the example, which give ranks 3, and 5.5. These add up to 8.5, and this value is looked up in a distribution table (similar to the Normal tables) to give a P value of 0.05. The probability of getting our data if the null hypothesis is true is 0.05, that is, it is likely to be rejected.

Mann–Whitney U test

This is another non-parametric test, an alternative to the t-test for comparing data from two independent groups (the test is also called the Mann–Whitney–Wilcoxon). The test is also used when the variances of the two populations of data are significantly different (the F test will determine this, which is the ratio of the variances), but alternatively a modified t-test called the Welch test could be used in this situation. The Mann–Whitney ranks both groups together as illustrated in Table 2.2, and then the sum of the ranks of each group are compared (comparing columns 1 and 3, which give totals of 19.5 and 35.5). If the null hypothesis is true then the sum of each rank of the separate groups will be similar. The more different they become, the more likely the null hypothesis is false. Again distribution tables are used to find the P values.

Table 2.2 Blood pressure values from two groups of patients with difference treatments

Group 1 rank	Group 1 values (mm Hg)	Group 2 rank	Group 2 values (mm Hg)
1	80		
2	85		
3	87		
		4	89
		5	90
6	92		
7.5	95		
		7.5	95
		9	107
		10	110
Sum = 19.5		Sum = 35.5	

Diagnostic tests

In a diagnostic test, patients can be classified into two groups, according to the results of an investigation or test. For example, two groups could be if a patient is awake or asleep. A measurement often used to detect whether the patient is awake or asleep during anaesthesia is a parameter obtained from the *electroencephagram* (EEG). A *threshold* value is set so that if the EEG value is above this, then the patient is assumed to be awake, and below this the patient is assumed to be asleep. Ideally this value of the threshold would discriminate completely between the two groups but usually perfect discrimination is not possible. Different values of the threshold would give better or worse discrimination. *Sensitivity* is the proportion of true positives (i.e. correct diagnosis of awake) that are correctly identified by the test. *Specificity* is the proportion of true negatives that are correctly identified by the test (i.e. the other condition, correct diagnosis of asleep). Table 2.3 shows a useful summary of this.

Receiver operating characteristics curves

Receiver operating characteristics (ROC, first used in testing radar receiver installations) curves are a graphical representation of how robust a diagnostic test (for example the EEG test described above) is and is a plot of the true positive rate (sensitivity) against the false positive rate (1-specificity). The ROC

Table 2.3 Representation of a diagnostic test

Test result	Actual condition		
	Positive (patient is awake)	Negative (patient is asleep)	
Positive (EEG = awake)	True positive (correct) = a	False positive = b	total positive tests a + b
Negative (EEG = asleep)	False negative = c	True negative (correct) = d	total negative tests c + d
	total actual positives a + c	total actual negatives b + d	

Sensitivity is defined as: true positives/total actual positives. Specificity is defined as: true negatives/total actual negatives.

curve is a plot of the sensitivities and specificities at different thresholds (for example, different values of the EEG for determining awake and asleep). An ROC curve shows the trade-off between sensitivity and specificity (any increase in sensitivity will be accompanied by a decrease in specificity). The closer the curve follows the left hand border and then the top border of the ROC space, the more accurate the test. The closer the curve comes to the 45-degree diagonal of the ROC space, the less accurate the test. The area under the curve is a measure of test accuracy or the *goodness* of the test. Three ROC curves are shown in Figure 2.12, and show worthless, good and excellent tests. Different diagnostic tests (for example, different EEG parameters) will have different curves. The area under the curves will be different and allow some comparison between methods.

Analysis of variance

It is not a good idea to compare multiple groups of data using multiple t-tests, but to use *analysis of variance* (ANOVA). This is an extension of using the ideas discussed using two groups of data into three groups or more. The simplest form is *one way analysis of variance* where there is a single way of classifying individuals (e.g. types of ventilation). When there are two classifiers (or factors) (e.g. where each patient also receives more than one treatment), then *two way* analysis of variance is used and so on. This discussion will deal only with one way analysis of variance. The principle behind ANOVA is to put the total variability of a set of data into components according to different sources of variation. The variation between the samples within each group is compared with the variation between the groups. If the groups of data are from the same parent population, then the ratios of these two variances should be near 1.0,

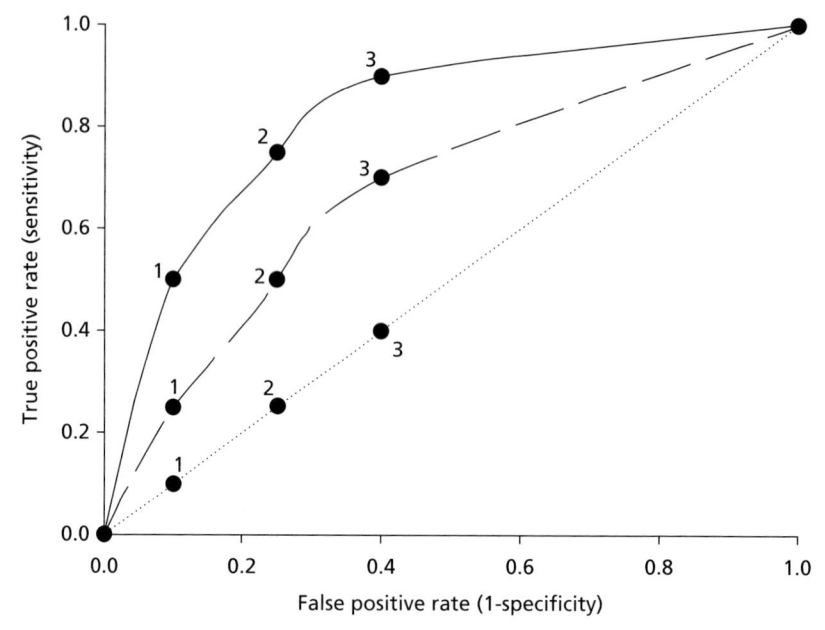

Fig. 2.12 Three receiver operating characteristics curves. The dotted is a worthless test, the dashed a good test, and the solid line an excellent test. The 1, 2 and 3 represent different values of the thresholds.

and the larger this number (or 'F' ratio) is, then the more unlikely that the null hypothesis is true. The assumption for this test is that the data are normally distributed.

As an example, if three groups of patients are on different types of ventilation, then a measure could be their oxygen saturation after 5 minutes. If the groups contain 8, 9 and 7 patients, then the total number of patients would be 24. Initially the total *sum of squares* is calculated by summing the 24 deviations from the overall mean (of the 24 patients). This result would be broken down into two parts: the *within-groups* and the *between-groups*. For the within-groups calculation, the sum of squares would be made by summing the 24 observations from the mean in its own group (group 1 patients have their deviation from the mean of group 1, group 2 from the mean of group 2 and so on). The in-between groups would be calculated by the sum of the squares of the mean of each group from the overall mean. If the complete data were from the same population then these two variances would be similar, and the null hypothesis would be accepted. The greater the difference between the two variances, the more likely that one or more of the groups are from different populations.

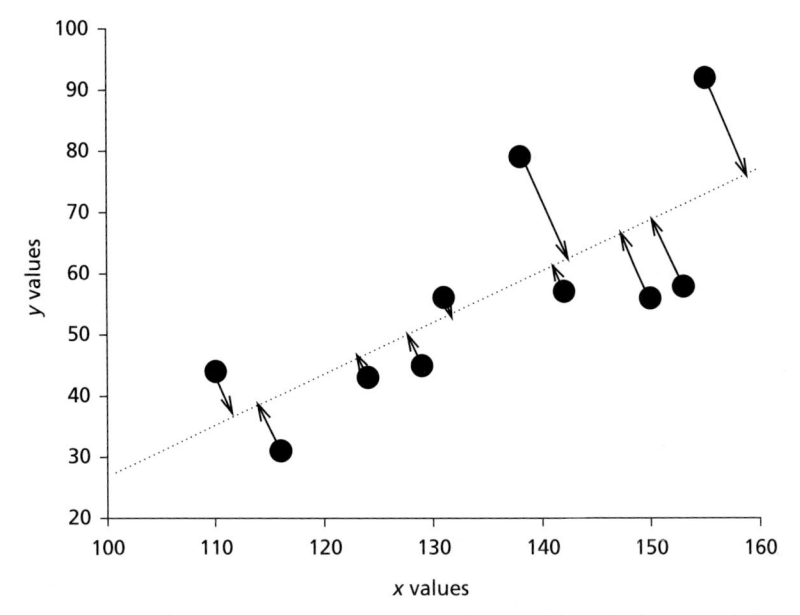

Fig. 2.13 Graph showing ten observations with line of best fit drawn and also illustrating the perpendicular distance from each data point to the line.

Meta-analysis

Meta-analysis is a statistical procedure that integrates the results of several independent studies that are considered possible to be combined. Well conducted meta-analysis can provide a more precise estimate of treatment effect, but badly conducted analysis may be biased owing to exclusion of relevant studies or inclusion of inadequate studies. Misleading analysis can generally be avoided if a few basic principles are observed. The analysis can be considered as an observational study of all the evidence from all previous studies. The steps involved in the process are really similar to general research studies, where the problem to be addressed is set out, the data is collected and analysed, and the results are reported.

For any meta-analysis, a detailed research protocol should be carried out in advance, that states the objectives, the hypothesis to be tested, and the criteria for selecting (or rejecting) relevant studies. The criteria for studies to be included are very important, and relate to the quality and design of the trials. Ideally only proper controlled, blinded trials with proper randomisation should be included. It can sometimes be difficult to assess the quality of a study to be included, as occasionally the published information on the quality of the study can be sparse. It has also to be decided whether to include only published studies in the analysis. Either way (to include or not to include) can produce difficulties, as using only published studies can result in distorted results due to publication bias, but also unpublished results may distort the results due to the lack of peer-review. Unpublished results can be included, however, as long as care is taken in assessing the quality of the study.

Before studies are combined together analytically, it can be useful to combine the various studies graphically using a common scale, to assess the degree of similarity between the studies.

There can be difficulties in meta-analysis in combining the data from all the different studies. A simple arithmetic average of the results from all the trials would give misleading results, as the results from smaller trials are more subject to chance variations and also these studies should be given less weight. Methods used to combine trials use a weighted average of the results, in which the larger trials have more influence than the smaller ones. The variability between the various studies is also taken into account in the combining process, and this can be a complicated.

Correlation and regression

Correlation

Correlation looks how associated two variables are with each other. Figure 2.13 shows a scatter diagram of two variables, x and y, and 10 observations. The degree of association between the points is measured by the correlation coefficient (r), which is how much the data points deviate (their perpendicular distance) from a line of best fit. This is illustrated in the Figure 2.13, and the equation is:

$$r = \frac{\sum (x - \bar{x})(y - \bar{y})}{\sqrt{[\sum ((x - \bar{x})^2 \sum (y - \bar{y})^2]}};$$

r can have values between -1 and 1, and is zero if the variables are not associated at all. Perfect positive correlation, $r = 1$, is shown in Fig. 2.14(a) and this is where the increase in one variable gives an exact linear increase in the other variable. Negative association (r is negative) is when lower values of one variable are associated with lower values of the other. $r = -1$ is a perfect negative association and is shown in Fig 2.14(b). A positive correlation of 0.8 is shown in Figure 2.14(c), and -0.8 is shown in Figure 2.14(d). The coefficient is based in a linear trend and the fit is normally by the least squares method.

A significance test can be used to determine whether the observed correlation could be simple due to chance (based on the t-test).

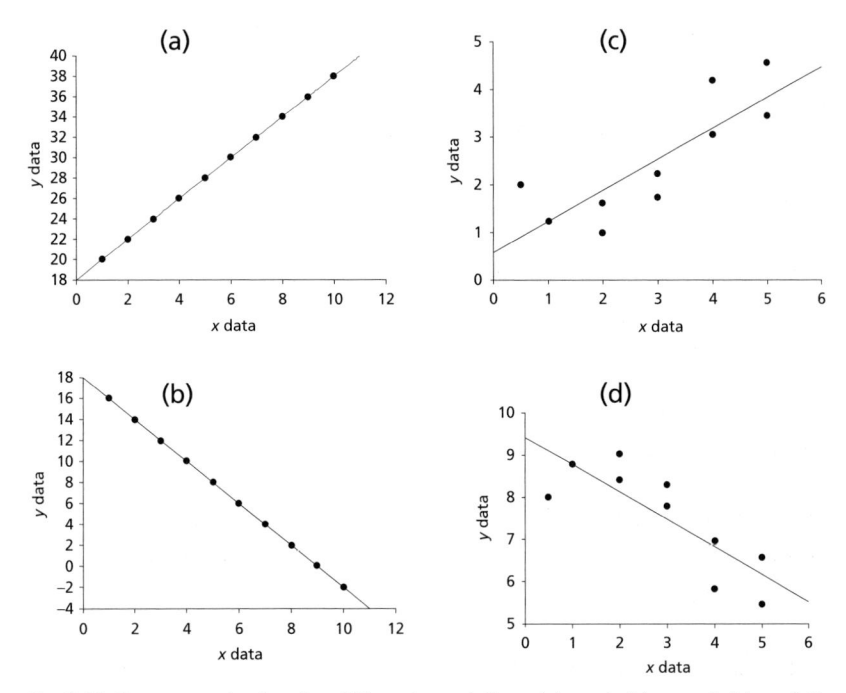

Fig. 2.14 Scatter graphs showing different correlations. (a) $r = 1$; (b) $r = -1$; (c) $r = 0.8$; (d) $r = -0.8$.

Linear regression

This both describes the relationship between the two variables and gives the equation of the straight line that predicts how the y variable (the dependant or response variable) alters with a change in the x variable (the independent or predictor variable). The equation of the regression line is:

$$y = mx + c,$$

where c is the intercept, and m is the slope of the line (see Chapter 1). The fit is using the least squares fit, which is to minimise the sum of the squared vertical distances of the points from the line. A test statistic can be used to determine whether the gradient is statistically different from zero. The distance of each sample from the line is called the *residue* and the residual variation is the amount of variation of the points from the line and is a measure of the *goodness of fit* of the line. A measure used in regression is R^2 (square of the correlation coefficient), which is the sum of squares explained by regression as a percentage of the total sum of squares. A high value of R^2 indicates that the majority of the variability in one variable is explained in the other.

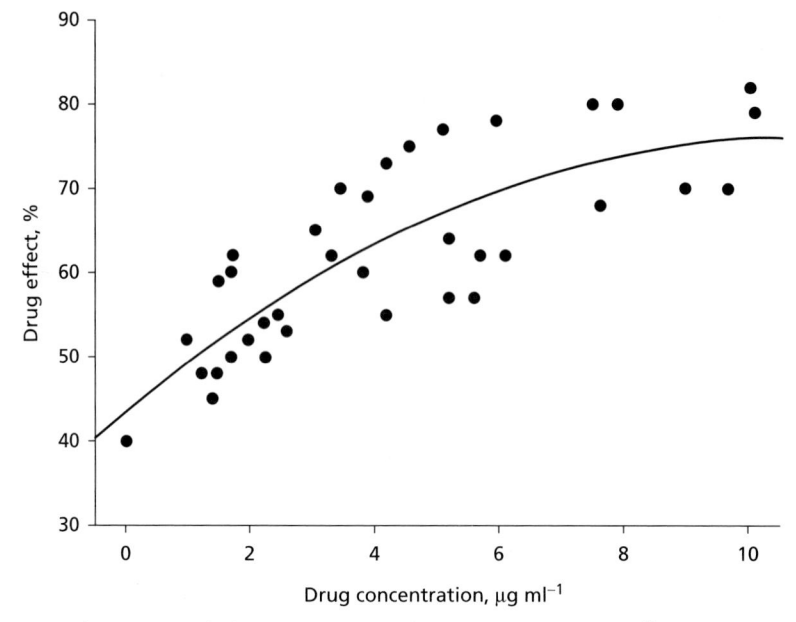

Fig. 2.15 graph showing example of non-linear regression, $R^2 = 0.65$.

There are non-linear relationships as well, such as polynomials. The simplest one being a quadratic, and an example of this is shown in Figure 2.15, which shows drug effect against drug concentration. Instead of being modelled with the equation for a straight line, the regression equation is:

$$drug\ effect = a\ (concentration)^2 + b\ (concentration) + c.$$

Comparing groups of categorical data
The analysis of frequency tables

A general way of showing frequency of events is to use a table (contingency table) where each cell of the table corresponds to a particular combination of characteristics relating to two or more classifications. A two way contingency table, which relates to two categorical variables, will be discussed. An example of a two way frequency table is shown in Table 2.4, which shows coffee consumed by grade of anaesthetist. The analysis of the frequency tables is (largely) based on hypothesis testing: the null hypothesis is that the there is no association between the two classifications (coffee and medical grade) in the relevant population (anaesthetists). The observed frequencies are compared with what is expected if the null hypothesis is true. The calculations are based

Table 2.4 Two way frequency table showing coffee consumed by grade of anaesthetist

	Cups of coffee per day						
Grade	0	1	2	3	4	5	Total
Medical student	0	0	1	3	1	0	5
Foundation year 1	0	0	0	1	4	10	15
Speciality doctor	0	2	3	5	20	10	40
Consultant	5	10	10	20	5	0	50
Total	5	12	14	29	30	20	110

on the expected frequencies on the distribution of the variables of the whole sample, as indicated by the row and column totals. The combination of row and column categories are called cells. The test statistic is calculated:

$$\sum (observed - expected)^2 / (expected).$$

Obviously the larger this value, the more likely that the null hypothesis is not true.

Expected frequencies

If the null hypothesis is true and the two variables are unrelated (independent), then the probability of an individual being in the particular row (i.e. specialty doctor) is independent of which column they are in (i.e. coffee consumption). The probability of being in a particular cell of the table is the product of (being in the row) and (the column) containing that cell. These probabilities are estimated using the observed proportions. For example there were five cups of coffee drunk by medical students in the department where 110 cups were drunk altogether, so the proportion drunk by medical students was 0.045. Also the proportion of staff consuming no coffee was the same, 5/110 or 0.045. If the grade of staff and coffee consumption are independent, then the expected proportion of the whole sample that are medical students and drink no coffee is the product of these proportions: $5/110 \times 5/110 = 0.002$. To get the expected frequency in that cell of the table this is multiplied by the total drunk in the department to get: $0.002 \times 110 = 0.23$. This procedure is done for each cell and the expected table of grade and coffee is shown in Table 2.5.

When the null hypothesis is true the statistic X^2 has a Chi squared distribution (the X would have Normal distribution). The data will have a cut-off where the null hypothesis is rejected, that is where there is evidence that the amount of coffee is related to career grade.

Table 2.5 Expected table of coffee consumed by grade of anaesthetist

GRADE	Cups of coffee per day						Total
	0	1	2	3	4	5	
Medical student	0.22	0.54	0.63	1.3	1.35	0.9	5
Foundation year 1	0.68	1.6	1.9	3.9	4	2.7	15
Speciality doctor	1.8	4.4	5	11	11	7	40
Consultant	2.3	5.5	6.4	13	13.6	9	50
Total	5	12	14	29	30	20	110

The hypothesis test is based on how alike these tables are. The test statistic X^2, is calculated as:

$$\sum \frac{(O-E)^2}{E}, \quad \text{where } O \text{ is observed and } E \text{ is expected.}$$

Evidence based medicine

Evidence based medicine is where individual clinical expertise is integrated with the best external evidence. It is about the conscientious, explicit and judicious use of current best evidence in making decisions about the care of individual patients. This best evidence can come from both clinically relevant basic sciences and patient centred research. External clinical evidence can make previously accepted diagnostic tests or anaesthetic procedures invalid and replace them with new ones that are more accurate, safer and more effective. However, evidence based medicine is not restricted to randomised trials and the meta-analysis discussed earlier. It involves tracking down the best external evidence to answer the various clinical questions and combining this with the best clinical care.

References

Isaacs, D., Altman, D. G., Tidmarsh, C. E., Valman, H. B. and Webster, A. D. (1983) Serum immunoglobulin concentrations in preschool children measured by laser nephelometry: reference ranges for IgG, IgA, IgM, *J. Clin. Pathol.*, **36**(10), pp. 1193–6.

Further reading

Altman, D. G. (1991) *Practical Statistics for Medical Research*, Chapman and Hall, London.
Bland, M. (2000) *An Introduction to Medical Statistics*, 3rd edn, Oxford University Press.
Egger, M., Davey Smith, G. and Phillips, A. N. (1997) Meta-analysis: principles and procedures, *BMJ*, 315, pp. 1533–37.
Cruickshank, S. (1998) *Mathematics and Statistics in Anaesthesia*, Oxford University Press.

Chapter 3

Background physics

This chapter contains: dimensions and units; input, output, linearity, drift and hysteresis; frequency response; calibration of transducers; atomic structure; basic mechanics.
The chapter links with: Chapters 1, 4, 5, 7 and 29.

Introduction

This chapter covers the background physics that is not otherwise covered in the relevant chapters associated with the equipment discussed. It is a useful introduction to some loosely related, but widely applicable concepts.

Dimensions and units

Any measurement made by the anaesthetist can be simplified and made easier to understand if it is represented by its basic dimensions. Dimensions are the basic components of equations and are independent of the units used. For example a common term in physics and medicine is velocity, which has the dimensions of length per unit time, written dimensionally as $[L][T]^{-1}$. This means that, independent of the measurement system used, the measurement of velocity requires that the numerical value of a length be divided by the numerical value of a time. Equations or graphical axes can be predicted and their validity can be checked by dimensional analysis. Each side of an equation can be represented in basic dimensions and both sides should balance. The dimensions needed to describe most events are mass $[M]$, length $[L]$, time $[T]$, and temperature $[\theta]$. An example of dimensional analysis is described later in this chapter but first units must be discussed. All measurements need to have their correct unit and symbol attached to them. Equations have a unique language and the syntax must be correct for communication within the international community. In the past many different systems of units were used; one was Imperial, another c.g.s. The Système International (SI) of units was established in 1960 and is now the recognised system of measurement communication. For completeness the base quantities are shown and some useful derived physical quantities are shown in Table 3.1

Table 3.1 SI base units and derivations

Physical quantity	Name	Symbol	SI derived unit
length	metre	m	base
mass	kilogramme	kg	base
time	second	s	base
electric current	ampere	A	base
temperature	kelvin	K	base
amount of substance	mole	mol	base
luminous intensity	candela	cd	base
frequency	hertz	Hz	s^{-1}
force	newton	N	$m\ kg\ s^{-2}$
pressure, stress	pascal	Pa	$N\ m^{-2}$
energy, work	joule	J	$N\ m$
power	watt	W	$J\ s^{-1}$
electric charge	coulomb	C	$s\ A$
electric potential	volt	V	$W\ A^{-1}$
absorbed dose	gray	Gy	$J\ kg^{-1}$
dose equivalent	sievert	Sv	$J\ kg^{-1}$
temperature	degree Celsius	°C	K

In the SI system the combination of basic units involves multiplication and division but multiplication is shown as a space and division is shown as a negative superscript. For example, velocity, in metres per second, is m s^{-1}. Prefixes to the name of each unit are usually in multiples of 10^3 and 10^{-3}. There are a few non-SI units that are still used in medicine, (not just anaesthesia) and these seem resilient to change. One is the millimetre of mercury (mmHg) for intravascular pressures (100 mmHg = 13.3 kPa), and total and partial gas pressures. The standard atmosphere (1 atm = 101 kPa) and the wavelength unit, the ångstrom (Å = 10^{-10} m) are also still used.

As an example of dimensional analysis, Poiseuille's equation will be examined, which is applicable to blood flow analysis. This states that flow, the rate of change of volume with respect to time, $\Delta V/\Delta t$, is dependent on the fourth power of the radius of the vessel. Resistive forces act against the flow of the liquid and this term is called dynamic viscosity η and its units are kg m^{-1} s^{-1}. Its dimensions are:

$$[\eta] = [M]\,[L]^{-1}\,[T]^{-1}.$$

A thought experiment suggests that the rate of flow of volume depends in some way on the pressure gradient along the tube $\Delta p/l$, on r, the radius of the tube and on η. This can be written as:

$$\frac{\Delta V}{\Delta t} = k \left(\frac{\Delta p}{l}\right)^x \eta^y\, r^z.$$

The dimensions of the left side of the equation are written ($dV = [L]^3, 1/dt = [T]^{-1}$):

$$\frac{\Delta V}{\Delta t} = [L]^3\, [T]^{-1}.$$

For part of the right side of the equation, pressure is force (F), per unit area (A) and force is mass (m) multiplied by acceleration (a):

$$\frac{\Delta p}{l} = \frac{F}{(A\,l)} = \frac{ma}{(A\,l)}.$$

This can be written with dimensions:

$$[M][L][T]^{-2}\, [L]^{-2}[L]^{-1}$$
$$= [M][L]^{-2}[T]^{-2}.$$

Since the dimension of r is [L], the right side becomes:

$$= ([M]\,[L]^{-2}\,[T]^{-2})^x\ ([M]\,[L]^{-1}[T]^{-1})^y\,(L)^z.$$

Equating both sides using powers of [L] the equation becomes:

$$3 = -2x - y + z.$$

Equating powers of [M], the equation becomes:

$$0 = x + y.$$

Equating powers of [T], the equation is:

$$-1 = -2x - y,$$

therefore $x = 1;\ y = -1$ and $z = 4$.

These were three equations to solve for three unknowns. This gives the original equation as:

$$\frac{\Delta V}{\Delta t} = k \left(\frac{\Delta p}{l}\right) \eta^{-1} r^4,$$

which is a form of the Hagen–Poiseuille equation.

Input/output, linearity, drift, hysteresis

Many measurements in anaesthesia can be simplified to a black box in which an input is applied to a system, a transfer function acts on the input, and a modified output is produced which is the product of the transfer function and the input. This is illustrated in Figure 3.1.

For clinicians, it is less important to know how the transfer function is carried out, than to know its function. One aspect of a system that is important, especially in such a system as converting a transducer output to a useful value, is its linearity (and non-linearity). A linear system is one where the same function (e.g. amplification of a certain gain) is applied to the whole range of the input values, whereas in a non-linear system the function applied to the input changes throughout the range. If the relationship between the input and output obeys the following three rules the system is said to be linear.

- If the input increases N times, the output must also increase N times.
- A sinusoidal input gives a sinusoidal output of the same frequency.
- The output does not depend on the previous history of inputs; this can be expressed as no memory or no *hysteresis*.

Hysteresis causes distortion (non-linearity) and can occur in measurement in medicine, especially using certain transducers. If hysteresis is present, the signal output from a transducer, produced by a physiological condition such as temperature, can differ depending on whether the temperature is increasing or decreasing. Also hysteresis can occur naturally, for example in the lung. Figure 3.2 shows the curve of pressure against volume for the lung, the difference between the curves of expansion and deflation shows hysteresis, and this area represents the energy spent overcoming airway and tissue resistance.

Drift is where there is slow change in the output signal where there is no actual change in the variable being measured. This effect can appear to be at random but there is normally some underlying cause. This can be, for example, in a strain gauge pressure transducer (see Chapters 5 and 12). When this is first

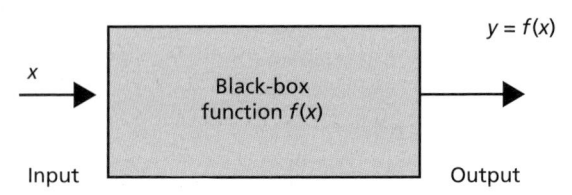

Fig. 3.1 Diagram showing black-box concept.

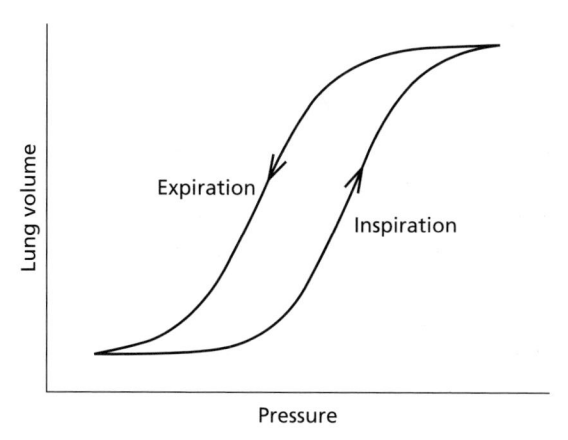

Fig. 3.2 Inspiration-expiration loop showing hysteresis.

switched on, there will be some slight heating of the resistors in the transducer, due to current passing through them. This will alter the balance and affect the output. Normally this drift will stabilise as the tiny heating effects equilibrate.

Frequency response

Another important aspect in a system is the frequency response. Signals described with respect to time (the time domain) are typically those where a physiological signal such as blood pressure or the electroencephalogram (EEG) are represented as a voltage signal, either continuous or sampled, against time. Time domain signals can also be transformed or represented in the frequency domain. The frequency domain is where the amplitudes of the frequency components are plotted against frequency. This can have advantages with the analysis of the signals, by presenting the information in a different manner and this can unmask hidden details. An example of a simple spectrum of a square wave is shown in Figure 3.3.

Any complex waveform may be represented as the algebraic sum of a number of sine or cosine waves (obtained using Fourier analysis, discussed in Chapter 1). This is usually a fundamental frequency and a number of *harmonics* that are multiples of this frequency. For example a square wave, as shown in the centre of Figure 3.3 has a fundamental frequency, which has the same frequency as the fundamental sine wave, and an infinite number of odd harmonics. In reality a reasonable version of a square wave can be made up of several harmonics, as shown in the left hand diagram of Figure 3.3. The diagram shows the harmonic make up of the square wave with three

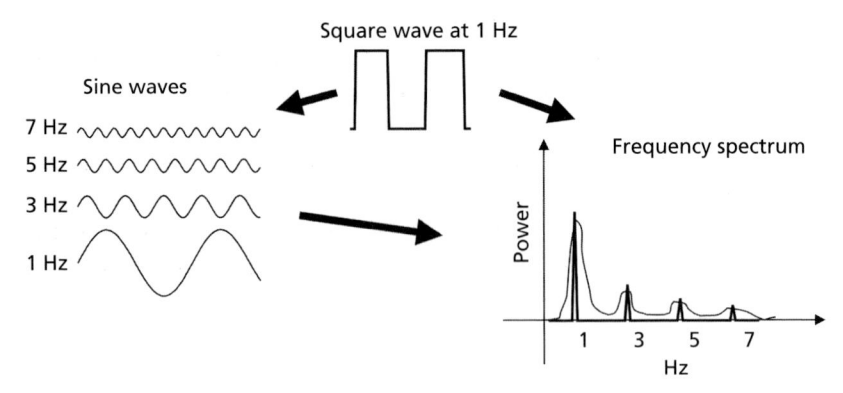

Fig. 3.3 Figure showing a square wave and its frequency components. See text for details.

odd harmonics included and the *spectrum* (the plot of the amplitude of the frequency components against the frequency in Hz) of the square wave shown in the right hand diagram. The amplitude of each frequency component in the spectrum is related to the amplitude of each sine wave component in the left hand diagram.

A physiological example of the harmonic make up of a signal is shown in Figure 3.4. This figure shows a blood flow waveform plotted in the time domain, obtained from a common femoral artery. Eight harmonics can be used to represent a good approximation of the femoral arterial flow waveform.

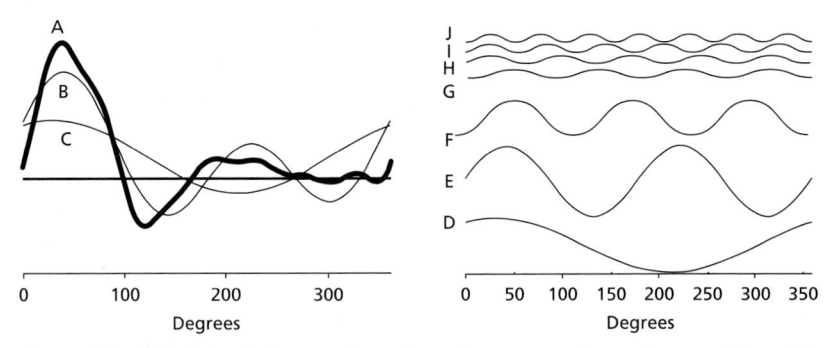

Fig. 3.4 Physiological velocity waveform from the common femoral artery. Wave A is the waveform produced with eight harmonics (sine waves), B is the wave with the fundamental and 2nd harmonic. C is the fundamental frequency. On the right of the figure, D–J are the fundamental and 2–7th harmonics respectively. Data are from Evans et al. (1989).

In the square wave example, each odd harmonic wave starts at the same point, or *phase*, in time (zero degrees) and these are added to produce the square wave. In the femoral artery example, the harmonics are consecutive and each one has a different starting point, or phase, in the sine wave cycle (0–360 degrees). These are again added together to produce the desired waveform. The figure illustrates each harmonic with its correct phase shift, apart from the 8th harmonic, which is too small to visualise. It should be noticed that if the number of harmonics is reduced to two (waveform B), then there is considerable distortion in the resultant waveform. Any amplifier processing this signal must have sufficient bandwidth to include at least the eight harmonics. This will be discussed in more detail later.

A waveform such as the EEG is very complicated but still each time period can be broken down into different frequencies, with each frequency having a different amplitude and phase. This process is carried out by a mathematical method using *Fourier* series. This is also used in the *digital domain* by using Fourier transforms, which is just a digital version of the Fourier series. These topics will be discussed again, in context, in the section on signal processing in Chapter 5.

Calibration of transducers

It is highly desirable for the electrical output of a transducer to be linearly proportional to the amplitude of a physiological event. The act of applying a physiological event (real or simulated) of known amplitude and measuring the transducer output is known as calibration. The transducer output may be voltage, current, or a digital value. A correct calibration procedure consists of both static and dynamic calibrations. As an example of the procedure, the calibration of an arterial blood pressure measuring system will be considered. Using a column of mercury to correlate mmHg to voltage carries out static calibration. The blood pressure transducer is assumed to be linear throughout its working range and so a single point calibration could be used along with *zeroing*. It is prudent to check the linearity by carrying out several point calibrations, at least one at each end of the working range. Dynamic calibration tests the frequency response of the system. What is wanted is a perfect electrical equivalent of the arterial pressure at the output of the amplifier. As discussed above, any arterial waveform is a complex wave consisting of many frequency components. The system under calibration must be capable of accurately passing all frequency components from the arterial site to the output of the amplifier, with no amplitude or phase (time) distortion. To give an example of the frequency response required, the highest heart rate that is usually

54 | BACKGROUND PHYSICS

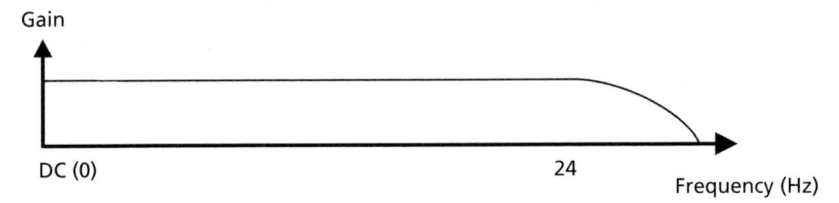

Fig. 3.5 Graph of the required amplitude frequency response for measurement in the arterial system.

encountered is 180 beats per minute, which is equivalent to a frequency of 3 Hz. Fourier analysis shows that the arterial waveform is sufficiently constructed using eight harmonics and so the system must have a flat response (i.e. all harmonics have the same amplitude and phase shift) from DC to 24 Hz (i.e. 3 × 8 harmonics). This is illustrated in Figure 3.5, showing the constant gain of the system from DC to 24 Hz.

A mechanical sine-wave generator applied to the transducer can carry out this dynamic calibration from the transducer to the amplifier. This is shown in Figure 3.6. The sine-wave generator will sweep (either manually or automatically) from DC until the highest frequency of interest, typically a few hundred Hz. The output from the reference amplifier should show a flat response throughout the sweep range (as demonstrated in the output from the reference amplifier), and the calibration side should ideally show a flat response in the working frequency range (typically 0–24 Hz), and any resonant peak (see Chapter 12) should appear beyond this, as demonstrated in the diagram. At higher frequencies, the response will fade.

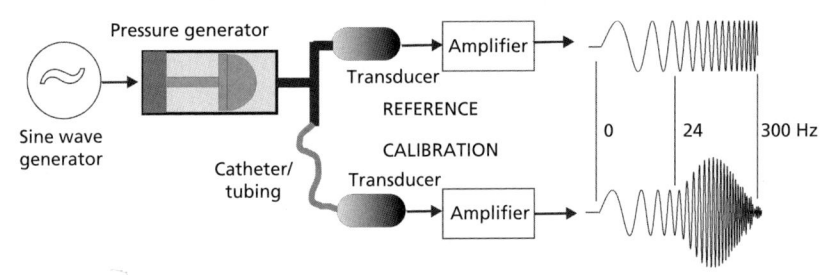

Fig. 3.6 The dynamic calibration of a transducer and catheter system. The reference transducer records directly from the pressure generator and the calibration side is connected to the catheter measuring site. The waveforms show examples of the outputs from the amplifiers, showing flat waveforms from 0 to 24 Hz, and resonance between 24 and 300 Hz on the calibration waveform.

A transducer can be calibrated perfectly and the amplification following it can be perfect. There can be problems with how the transducer is connected to the physiological event, and this must be given some thought. If the transducer, or connecting amplifiers, are non-linear, then multiple calibration points throughout the working range must be made to construct a calibration curve.

Atomic structure

All matter can be broken into a limited number of substances, the atomic elements. There are just over a hundred elements and the basic building block in all of these is the atom.

A simplified model of an atom consists of a nucleus of protons and neutrons, with electrons in orbit around the nucleus. The characterising difference between elements is the number of protons, called atomic number, Z, that the atoms contain. The nucleus will be positively charged due to the protons, creating an attractive electrostatic force between the nucleus and the electrons. The electrons occupy discrete orbits or *shells* around the nucleus, which occur at different discrete energy levels. By convention, the binding energy of electrons in an atom is negative. Electrons in the shell nearest the nucleus have the lowest energy (the ground state) and successive shells have electrons of progressively increasing energy as the distance from the nucleus increases. The nucleus has a diameter of approximately 10^{-14} m, and the whole atom a diameter of 10^{-10} m. There are a limited number of energy shells and the total number of electrons in a shell is restricted. The shells are known by the letters K, L, M, etc. as shown in Figure 3.7, and the maximum number of electrons in these shells are 2, 8 and 18 respectively. The atom is electrically neutral, since the electrons will balance the protons. The atomic number determines the electron structure of the atom and therefore its chemical properties. The combined number of protons and neutrons is the atomic mass number, A. The normal terminology is that A is written as superscript and the proton number Z is written as a subscript. So, for example, helium is $^4_2$He, meaning 2 protons and 2 neutrons, and therefore 2 electrons.

All the nuclei of the atoms of one particular element have the same number of protons, but the number of neutrons can vary. Isotopes are atoms of the same element that have different numbers of neutrons and many are artificially produced. For example, $^{12}_6$C refers to a stable carbon atom with $A = 12$ and $Z = 6$. Carbon 15 still has $Z = 6$, but has three more neutrons, which make it unstable and radioactive. It is called a radionuclide.

In each atom the outermost or valence shell is concerned with the chemical, thermal, optical and electrical properties of the element. If an atom gains or

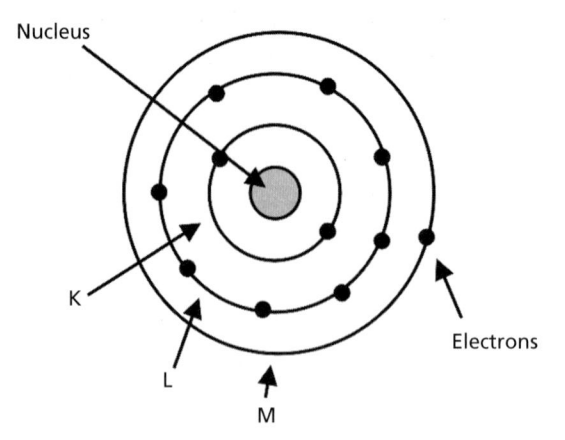

Fig. 3.7 Electron shells in an atom.

loses an electron, the electric charge no longer balances. An atom that loses or gains an electron becomes an ion and the process is known as ionisation.

Basic mechanics

Mass, weight and force

The mass of a substance is the amount of it, and the SI unit is the kilogram (kg), which is often how mass is described. Units such as pounds (lb) or stone are archaic in the UK and EU, and should not be used. Another way of thinking of the amount of the substance is its volume, the SI unit of which is m^3, or litre where 1000 L = 1 m^3. The relationship between mass and volume is density, where density is mass per unit volume in kg m^{-3}. A very dense substance such as mercury weighs 13 600 kg m^{-3}, while air weighs 10 000 times less at 1.3 kg m^{-3}.

So what is the difference between mass and weight? On the surface of the Earth, the force of attraction between an object of mass m kg, and the centre of the Earth is $m \cdot g$ newtons, where g is the acceleration due to gravity at the Earth's surface. Colloquially, this force of attraction is often referred to as *weight*, whose value is m kg wt or $m.g$ newtons. In a 'weightless' or microgravitational environment, where the effect of g has been eliminated by the laws of orbital mechanics, the same object would have the same mass m, but its weight would be zero. Astronauts walking on the Moon had the same mass as their Earth mass, but their weight was one sixth of their Earth weight due to the smaller gravitational acceleration on the moon.

As mentioned above, weight is a special sort of force. Force F is a term whose units are newtons and, where it causes a given mass m to move with acceleration a, in the absence of other forces, such as friction, then:

$$F = m.a$$

This is Newton's second law of motion; where $a = g$, force F becomes identical to the weight. Another way of expressing Newton's second law of motion is that

$$force = rate\ of\ change\ of\ momentum,\ or\ F = d(m.v)/dt.$$

Newton's first law of motion states that every body stays at rest or in uniform motion (constant velocity) unless acted on by such a force. The third law is that every action (or force) must have an equal and opposite reaction. Thus your weight on the ground is balanced by the reactive force from the ground. A force of action from a missile propulsion system has the equal and opposite reaction of acceleration of the missile.

Pressure is force per unit area with units of $[kg].[m\ s^{-2}]/[m^2]$, or $N\ m^{-2}$, or Pa. Pressure will be discussed in detail in Chapter 7 on Behaviour of Fluids. The pressure generated by a ventilator can be a force or weight acting on a bellows over a cross-sectional area. The same concept applies to car brakes, where a force from the driver's foot on the pedal acts over a small piston surface area, causing a high pressure on the brake. A classical anaesthetic example is the fixed force from a thumb on a syringe plunger will create a higher pressure at the outlet of a 2 ml syringe than a 20 ml syringe.

Energy and work

When a force F moves an object a distance d in the same direction as F, then an amount of work $F.d$ N m (joule) is done. This is shown schematically in Figure 3.8 as the area under the F–d curve. F might be the result of a ventilator bellows pressure P acting over the bellows cross-sectional area A. Hence $F = P.A$ and work $= P.A.d$ joules.

Work is transferred directly into energy. This may be potential energy, when an object is given the potential to do further work, or to kinetic energy, when movement takes place. In a process involving interchange between work and energy, some work is converted into thermal energy, and is usually lost to the surroundings.

Energy can also be lost by being absorbed by interacting surfaces. However, if the distance d that the object of mass m is moved is a vertical height h, the potential energy of the object is $m.g.h$ joule. If the object were then allowed to fall freely under gravity, and most of the potential energy was recovered as

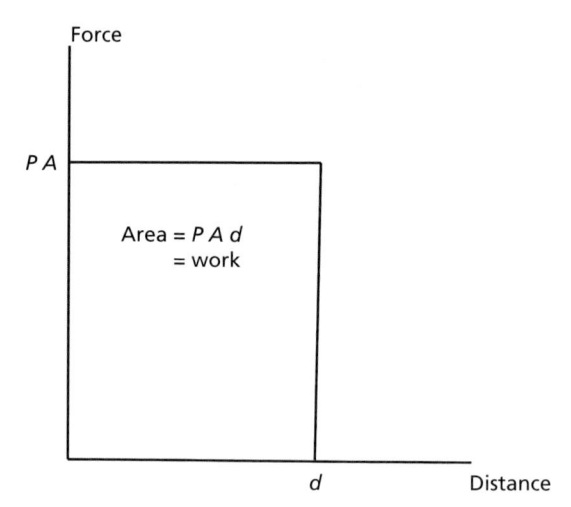

Fig. 3.8 The work done in moving a ventilator bellows of cross-sectional area A through a distance d, by applying a pressure P (see text).

kinetic energy $\frac{1}{2} m v^2$, with some energy loss from air resistance, where v is the final velocity, then

$$\tfrac{1}{2}.m v^2 \leq m g h.$$

Figure 3.9 attempts to show the relationship between this form of work and energy.

The ventilator bellows that produced work PAd, converts this to energy in inflating the lung to a volume V, through a pressure change ΔP. At the end of inspiration the potential energy of the lung would be $V \Delta P$, although some losses will have occurred in overcoming airway resistance and elastic forces in the lung. This expression would also represent the respiratory work done by the patient if he or she had inflated her own lung. A short exercise will reveal that the units of pressure × volume are the same as those of work or energy. Notice from Figure 3.2 that the expiratory pathway on this loop is not the same as the inspiratory pathway. The difference is called *hysteresis* and represents heat dissipation.

Power

Power is work or energy per unit time, and therefore a 'powerful' device is one that is capable of giving high energy or work per unit time. The units are N m s^{-1}, or J s^{-1} or W. Electrical power is measured as voltage × current, and

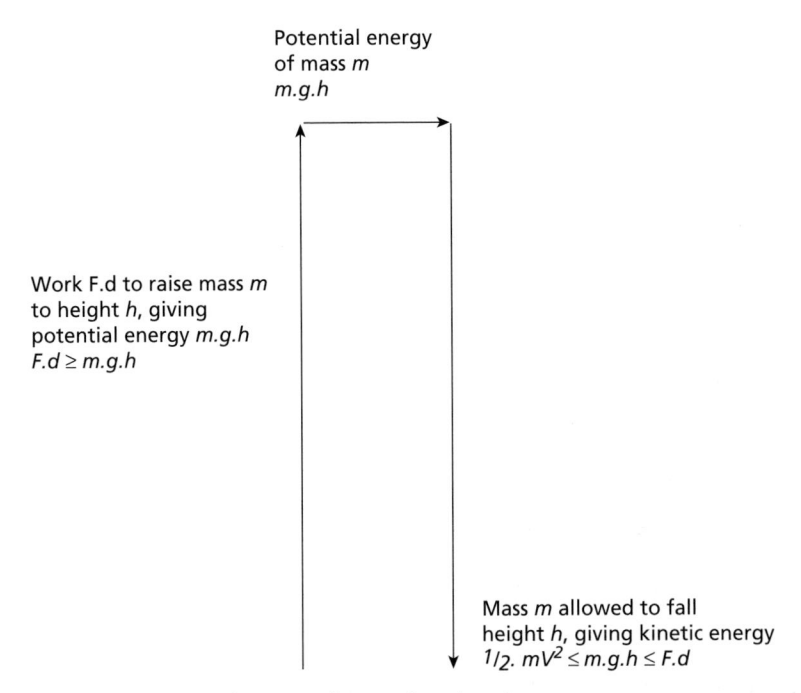

```
                    Potential energy
                    of mass m
                    m.g.h
                    ▲
                    │
                    │
                    │
Work F.d to raise mass m
to height h, giving
potential energy m.g.h
F.d ≥ m.g.h
                    │
                    │
                    │
                    │
                    │   Mass m allowed to fall
                    │   height h, giving kinetic energy
                    ▼   1/2. mV² ≤ m.g.h ≤ F.d
```

Fig. 3.9 Exchange of work and energy.

is converted into light, heat, power transmission or other electro-mechanically useful activity. For example an electric motor converts electrical power or energy into mechanical power or work, and a generator does the converse.

References
Evans, D. H., McDicken, W. N., Skidmore, R. and Woodcock, J. P. (eds) (1989) *Doppler Ultrasound*, JohnWiley & Sons Ltd, Chichester.

Further reading
Dorrington, K. L. (1989) *Anaesthetic and Extracorporeal Gas Transfer*, Oxford Medical Engineering Series, Oxford Science Publications, Oxford, pp. 4–7.

Hill, D. W. (1976) *Physics Applied to Anaesthesia*, Butterworths, 3rd edn, pp. 226–28.

West, J. B. (1979) *Respiratory Physiology, the Essentials*, 2nd edn, Williams and Wilkins, p. 67.

Chapter 4

Electricity, magnetism and circuits

This chapter contains: electric charge, current, voltage, power, analogies, magnetism, circuit rules, bridge circuits, alternating and direct current and voltage, capacitors, inductors, transformers.
This chapter links with: Chapters 3, 5, 6, 10, 12, 18, 20 and 21.

Introduction

Electricity is a broad term that includes a variety of phenomena resulting from the flow (or presence) of electric charge. It is a complex subject, but this chapter will provide the necessary background for simple electric circuits here, and the following chapters on electronics, biological signal processing, and electrical safety.

Electrical properties

Electric charge

Electric charge is a property of certain subatomic particles. Charge originates in the atom (see Figure 3.7 in Chapter 3), in which its most familiar carriers are the proton and electron. By convention, electrons carry a negative charge while protons carry a positive charge. The protons are located in the centre of the atom and the electrons are in motion outside of the nucleus in orbits. In simplistic terms, the protons are trapped inside the nucleus making it difficult for them to 'escape'. Under certain conditions, however, electrons can escape, and the movement of them is 'electricity'. Charge may be transferred between bodies, either by direct contact, or by using a conducting material such as copper or wire. *Static* electricity refers to the imbalance of charge on a body, usually caused by friction when two dissimilar materials are rubbed together, transferring charge from one to the other. Charge is measured in coulombs (C) and one coulomb contains 6.24×10^{18} electrons.

Electric current

The movement of this electric charge is known as an electric current, and the magnitude of this is measured in amperes. One ampere flowing for one second passes a coulomb of charge. So the current $(I) = C\,s^{-1}$. By historical convention, conventional current flows from the most positive part of a circuit to the negative and this concept is continued in this text and most others (but in fact the actual flow of the electrons is in the opposite direction). Current flowing through a wire causes many observable effects, for example heating and *magnetism*. Magnetism will be discussed later.

Voltage or volts

Voltage was historically called *tension* or *pressure*, which captured the concept well, and a water analogy can further help with this understanding and is discussed later. Voltage is the electric potential energy per unit charge, measured in joules per coulomb (V, volts = $J\,C^{-1}$). It can also be referred to as the electric potential, noting that the potential is a per unit charge quantity. It is the difference in voltage that is meaningful. The difference in voltage measured when moving from point A to point B is equal to the work that would have to be done, per unit charge, to move the charge from A to B.

When a voltage is generated by a battery, or by a *magnetic force* (see later in this chapter), this generated voltage has been traditionally called an *electromotive force* (emf). The emf represents energy per unit charge (voltage) that has been made available by the generating mechanism but it is not a force as such. The term emf is retained for historical reasons. It is useful to distinguish voltages that are generated from the voltage changes that occur in a circuit as a result of energy dissipation, e.g. in a light bulb (or resistor).

Power

The power P (work done per second, in watts, $J\,s^{-1}$) consumed by a circuit is:

$$V\,(J\,C^{-1}) \times I(C\,s^{-1})$$

Water analogy

A simple analogy to help understand voltage is that of water flowing in a closed circuit of pipework, driven by a mechanical pump. The voltage difference between two points corresponds to the water pressure difference between the two points. If there is a water pressure difference between two points, then the water flow (due to the pump) from the first point to the second will be able to do work, such as driving a *turbine*. In a similar way, work can be done by the electric current driven by the voltage difference due to the *battery*: for example, the current generated by the battery can light a bulb in a circuit. If the pump isn't working, it produces no pressure difference, and the turbine will not rotate. Equally, if the battery is flat, then it will not light the bulb.

This water flow analogy is a useful way of understanding many electrical concepts and is also further discussed later. In such a system, the work done to move water is equal to the pressure multiplied by the volume of water moved. Similarly, in an electrical circuit, the work done to move electrons or other charge-carriers is equal to the electrical pressure multiplied by the quantity of electrical charge moved. Voltage is a convenient way of measuring the ability to do work. In relation to flow, the larger the pressure difference between two points (voltage difference or water pressure difference) the greater the flow between them (either current or water flow).

As discussed earlier voltage usually means voltage difference. Obviously, when using the term voltage in the shorthand sense, it must be clear about the two points between which the voltage is specified or measured. When using a *voltmeter* (a special device that gives a voltage reading on its numerical display) to measure voltage difference, one lead of the voltmeter must be connected to the first point and the other to the second point.

A common use of the term voltage is in specifying how many volts are dropped across a device such as a light bulb. In this case, the voltage or, more accurately, the voltage drop across the device, can usefully be understood as the difference between two measurements.

Magnetism

Magnetism is the effect that is observed when the north and south ends (poles) of a bar magnet attract or repel neighbouring ferromagnetic items. Two bar magnets brought into close proximity to each other show that opposite poles attract each other and like poles repel.

A wire carrying an electric current also behaves like a magnet by inducing a *magnetic field* around it. Therefore magnetic fields can be created using electrical power, and can be used to provide mechanical force, as in an electric motor. Conversely, if a wire is moved through a magnetic field, an electric current is induced in the wire proportional to the wire's speed of movement. Therefore mechanical force, or movement, can create electric current, as in a dynamo, or in a power station.

If the current-carrying wire is coiled, a more intense magnetic field is created, because each wire in the coil that carries current induces a magnetic field around it. The close proximity of the coils to each other magnifies the flux density as many times as there are coils. If the single current-carrying wire, which induces flux Φ (in webers), is coiled into N coil turns, then the flux thus induced is $N.\Phi$ (weber). An elongated coil ensures uniformity of the magnetic field along its length. The coil can have an air core or an iron core. The high

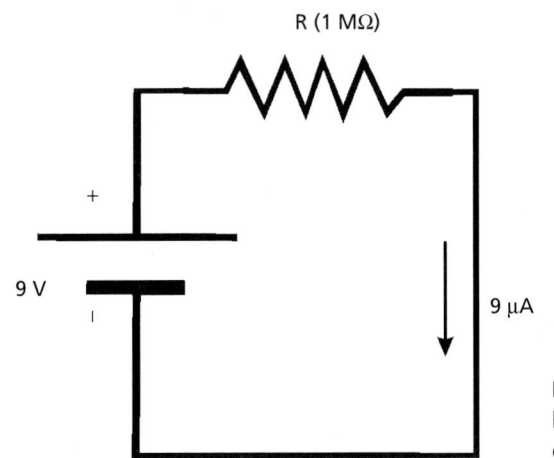

Fig. 4.1 Simple circuit with battery of 9 V and resistor of 1 MΩ.

magnetic permeability, μ, of iron compared with that of air means that the flux density B of such a coil will be greatly enhanced for a given magnetic field strength H, because $B = \mu H$. If the wire is coiled around an iron core in a toroidal form, the flux density is further enhanced because of the high permeability of iron.

A changing magnetic field induces a voltage in the coil proportional to the rate of change of the magnetic field, and the induced voltage opposes the change of magnetic flux that produced it. The transformer uses this property and transfers electrical energy from one circuit to another by means of a magnetic field linking both circuits, and this is discussed later in this chapter.

Basic electrical circuit rules

Ohm's law states that the current (I) in *amperes* flowing through a conductor of *resistance* (R) measured in *ohms* (Ω), is proportional to the electromotive force (*voltage, V*) in volts. For a constant voltage, $V = IR$. This law gives the basis of solving most basic electrical circuits. Current flows through a component when a voltage is supplied across it. A simple circuit showing a battery (voltage) across a resistor, is shown in Figure 4.1. If the battery voltage is 9 V, and the resistance is 1 M (10^6)Ω, then the current through the circuit will be $9/(1 \times 10^6)$ A, which is 9 μA.

There are some basic rules about voltage and current in simple circuits:

1. The sum of currents into a point equals the sum of currents out of it.
2. Components connected in series have the same current going through each of them.

3. Components connected in parallel have the same voltage across each of them. Each parallel limb can consist of a number of components in series.
4. When two resistors R_1 and R_2 are connected in series the total resistance (R_t) is the sum of R_1 and R_2. This can be shown by the following: The voltage across both will be V, and the current will be the same through both resistors. Therefore V_1 will be IR_1 and V_2 will be IR_2. The two voltage drops across the resistors will be equal to the total V, so:

$$V = IR_1 + IR_2 = I(R_1 + R_2) = IR_t.$$

Therefore $R_t = R_1 + R_2$.

5. When the two resistors are connected in parallel the total resistance (R_t) is the reciprocal of $(1/R_1 + 1/R_2)$. This can be shown by the following. The voltage across both resistors will be V, and the total current flowing before the branch is I. The current through R_1 will be V/R_1 and the current through R_2 will be V/R_2. The sum of the current flowing through the two resistors will be equal to the total current I, so

$$V/R_1 + V/R_2 = V(1/R_1 + 1/R_2) = I.$$

I is also equal to V/R_t. Therefore $1/R_1 + 1/R_2 = 1/R_t$. The effect of putting resistors in parallel is to decrease the total resistance to below the value of the lowest resistor.

Voltage dividers

Resistors can be usefully used as voltage dividers. This is illustrated in Figure 4.2. Here: $V_{out} = V_{in} R_2/(R_1 + R_2)$. This is solved as follows: $V_{out} = IR_2$ and $V_{in} = I(R_1 + R_2)$, therefore $V_{out} = V_{in} R_2/(R_1 + R_2)$.

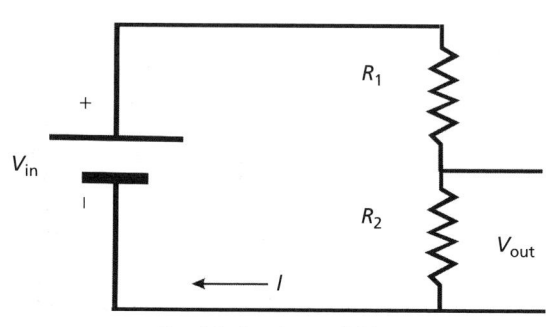

Fig. 4.2 A voltage divider.

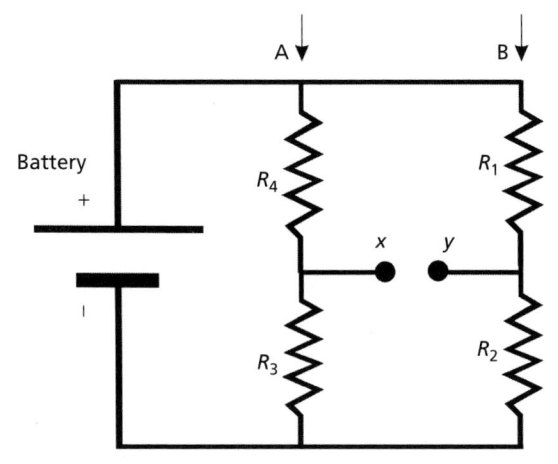

Fig. 4.3 A bridge circuit.

Bridge circuits

Circuits based on voltage dividers can be expanded to form bridge circuits, typically the Wheatstone bridge. An example is shown in Figure 4.3.

These circuits can be found in many medical devices. The circuit shown can be thought as two voltage dividers in parallel (A and B in the diagram). If $R_1 = R_4$ and $R_3 = R_2$, then both points x and y will be at the same potential. Both voltage dividers A and B will divide the battery voltage by the same amount. Therefore the voltage between x and y will be zero, i.e. the bridge is in balance. If R_3 is replaced with an unknown resistance such as a thermistor or a strain gauge as shown in Figure 4.4, then R_2 can be substituted with a variable

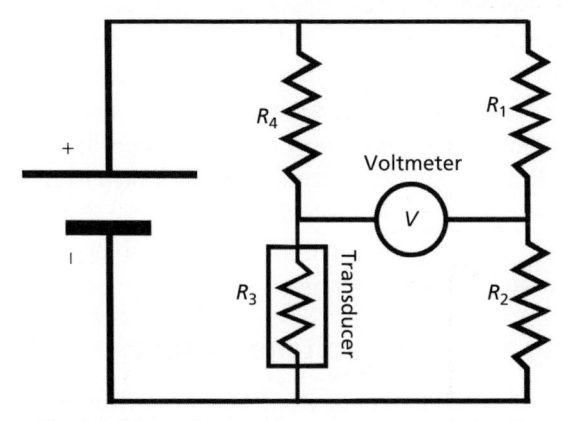

Fig. 4.4 Bridge circuit with strain gauge replacing R_3.

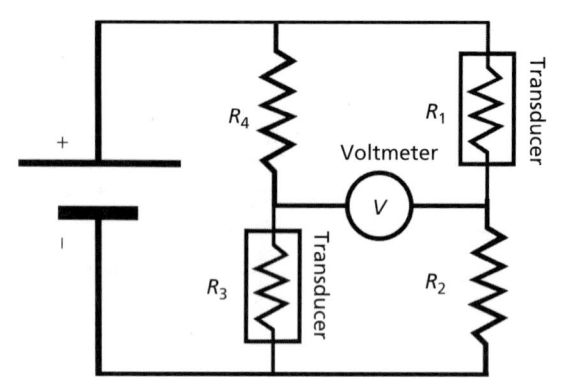

Fig. 4.5 Bridge circuit with two strain gauges replacing R_3 and R_1.

resistance, similar in value to the R_3. Again the bridge can be balanced by adjusting R_2 so that the voltage between x and y is zero. If a sensitive voltmeter is connected between x and y, then a very small change in resistance R_3 can be measured. In this way the tiny change in, for example, pressure transducers that employ strain gauges can be detected. The bridge in effect balances out the resting potential (which can be a large voltage), so that tiny pressure signals can be further amplified by other components, as discussed later.

The bridge can be made doubly sensitive by having a second strain gauge added on the opposite side of the bridge, as shown in Figure 4.5. In the pressure transducer example, both strain gauges are under strain under resting conditions. As pressure is put on the transducer one strain gauge will compress whilst the other will expand, making one increase its resistance and the other decrease.

If the supply potential to the bridge changes, or there is noise on the supply, then this does not affect the balance of the bridge (both voltages on either limb go up or down the same amount). In a similar manner, temperature changes will have little effect on the bridge, as the temperature change will affect each arm of the bridge equally. This property is the reason why the bridge circuit is used in medicine, to maintain accuracy in a measured signal in the face of potentially unwanted changes elsewhere. The bridge can also remove the level of the resting baseline pressure, and therefore make it incredibly sensitive to tiny changes from this baseline pressure.

Alternating and direct voltage/current

Two types of current flow exist in electric circuits: when the voltage source (or current) is at a constant level, this term is defined as *direct current* (DC).

The other is *alternating current* (AC), where the signal oscillates, either at its simplest as a sine wave as in the domestic mains, or it may be a complex random oscillation such as an electroencephalogram (EEG) signal. Part of Figure 4.6 shows a 50 Hz domestic UK mains sine wave (the solid waveform). The AC wave needs to be quantified, and there are several possible waves of doing this. The mean of the wave is not useful as this is zero. The peak values could be used (+325 V, −325 V) but these are seldom used. The normal way that the 50 Hz wave is quantified is by giving the *root mean square* (RMS) of the waveform. This is the value that would give the same heating effect (a resistor or electric fire) if the same value of DC voltage were supplied. For example, if 325 V AC peak, (230 RMS) were applied to an electric fire, then the same amount of heat would be obtained from if 230 V DC were to be applied to the same electric fire. The RMS value is really the average of the heat producing parts of the waveform and since the negative and positive parts of the wave both produce the same heating effects, they can be taken as going in the same direction, as described by the dashed waveform. The AC voltage is converted to RMS by first squaring the waveform. This has the effect of *converting* the negative parts of the waveform to positive values (the dashed waveform). The mean is taken (the average of the waveform, the 'heating' effect) and it is then square rooted to take account of the squaring at the first stage of the procedure.

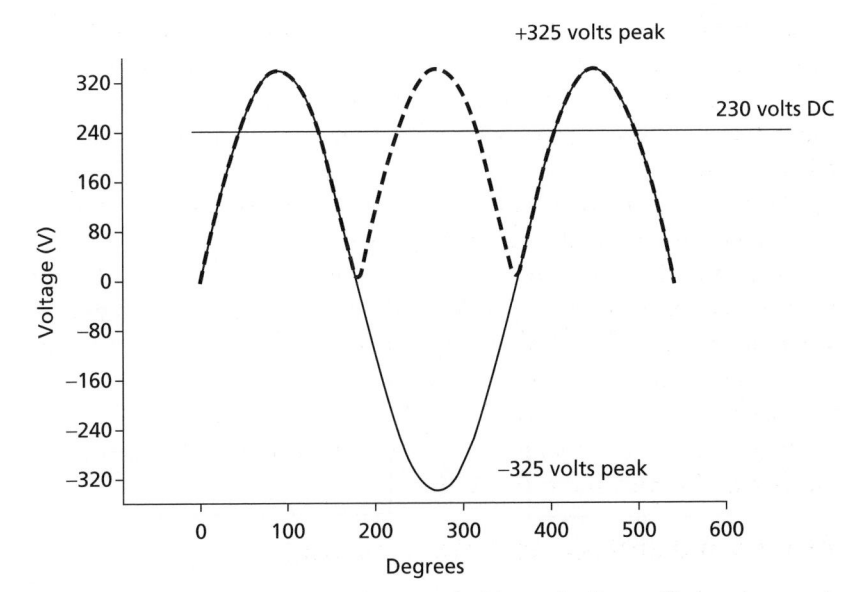

Fig. 4.6 Domestic UK mains supply wave (solid curve) plus rectified mains supply (dashed wave). The 230 V DC level is shown.

The domestic electricity supply is a current, which alternates 50 Hz in the UK and 60 Hz in the US. (The voltage in the US is 110 V RMS.) These frequencies are used as they are the most efficient for power transmission, but by coincidence these frequencies fall into the range that is the most dangerous for the human body. This is discussed further in the Chapter 6. Other forms of alternating current oscillate above a DC level, e.g. a blood pressure signal obtained from a pressure transducer, which was discussed in the last section.

Capacitors and inductors

Two important components in electrical and all electronic circuits are the capacitor and the inductor. The capacitor consists of two conductor plates separated by an insulator (which could be air or a solid material). An inductor consists of a number of turns of a wire conductor wound around a core of normally magnetic material. The capacitor and inductor behave very differently depending whether AC or DC is applied across them. If a constant (DC) voltage is applied (by closing the switch) across a capacitor, via a source resistance as shown in Figure 4.7, the capacitor will charge up exponentially from zero volts to the DC value across it and the current will decrease exponentially with a time constant of RC seconds (in Figure 4.7, $RC = 1$ s).

The steady state current (i.e. DC) through an inductor, as connected as shown in Figure 4.8, will be limited only by the resistance of its coil but a change of current produces a change in the magnetic field around the coil, which in turn generates a voltage in the coil with the opposite polarity to the driving voltage, which opposes the change in current. It therefore *slows down* any change of current. Figure 4.8 shows that when DC voltage is applied to the

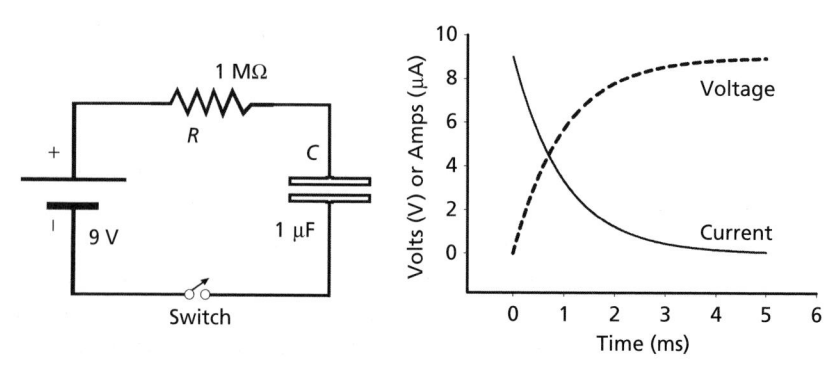

Fig. 4.7 Capacitor charging circuit showing components and the corresponding current and voltage waveforms across the capacitor C.

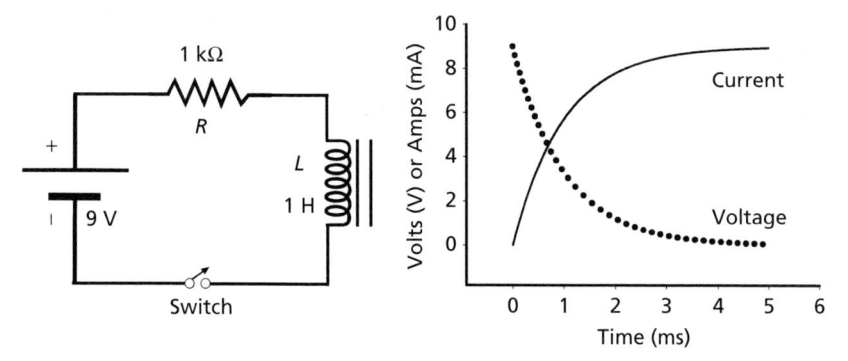

Fig. 4.8 Inductor circuit showing components and the corresponding voltage and current waveforms.

inductor (by closing the switch), the effect is the opposite to that which occurs with a capacitor. The time constant in this case is L/R seconds.

In an AC circuit (a similar circuit to Figure 4.7, but the battery is replaced by the AC source) the voltage across the capacitor is constantly changing direction and the current and voltage never reach steady state conditions as in the DC case. The current through a capacitor is proportional to the rate of change of voltage and so when a sinusoidal voltage waveform is applied (starting at zero), the current is at a maximum at the start of the cycle. As the capacitor charges up, the current decreases and the voltage increases. The minimum of the voltage cycle is the peak of the current waveform where the rate of change approaches zero. The current waveform peaks before that of the voltage, and therefore the voltage *lags* the current by a quarter of the whole cycle (360 degrees), or 90 degrees. This is illustrated in Figure 4.9.

The current *flow* increases as the frequency increases. The resistance to flow is inversely proportional to frequency and is called *reactance*. It is calculated as $X_c = 1/(2\pi fC) \Omega$, where f is the frequency of the supply waveform in Hz.

The current and voltage relationships with an inductor are different and in this case the voltage peak occurs before the current peak and so the voltage *leads* the current by 90 degrees as illustrated in Figure 4.9. The inductive resistance or reactance is, $X_L = 2\pi fL \Omega$, where L is the inductance in Hz.

Although reactance in both the C and L circuits is measured in Ω, it cannot be used directly in Ohm's law. A term called *impedance* is used and this is resistance plus reactance. Both the reactive ohms and resistive ohms present in the circuit must be considered and the overall opposition to current flow (the impedance) calculated at the frequency required, taking the phase difference

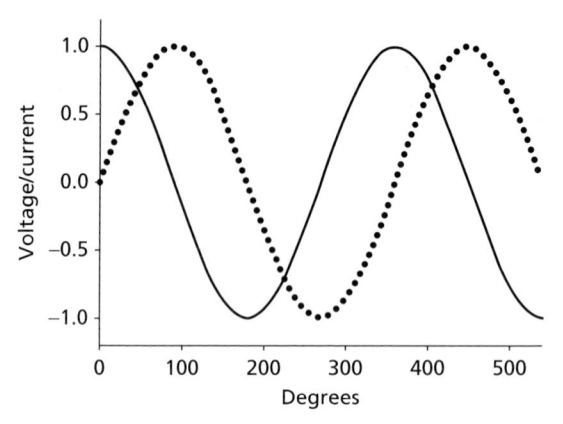

Fig. 4.9 Voltage waveforms across circuit components and the corresponding current waveforms, showing the relative phase relationships. Across a capacitor, the voltage waveform is the dotted wave and the current waveform is the solid wave. The inductor is the other way round: the voltage is the solid wave and the current is the dotted one.

into account. This can be dealt with more rigorously using a mathematical technique called *complex numbers*. Resistors in AC circuits can still use Ohm's law as in the DC case.

Transformers

A transformer consist of two separate (or more) coils (inductors) normally placed around a iron, or ferrite core. The transformer diagram and circuit diagram is shown in Figure 4.10. The voltage in the *primary* circuit induces magnetic flux in the iron core. The magnetic field to which this flux belongs induces a voltage in the *secondary* coil.

The changing voltage applied to the primary coil causes a changing current flow, which causes a changing magnetic flux in the core. Because of the changing flux, a voltage is induced in the secondary coil, which opposes the direction of the flux change in the core. No electrical connection between input and output exists or is necessary. Because only changing currents (AC) are involved, the output circuit can be isolated from a DC component at the input.

The coils are usually made of copper to reduce losses and the core is made of steel with a high permeability to minimise hysteresis loss and laminated to minimise other current losses.

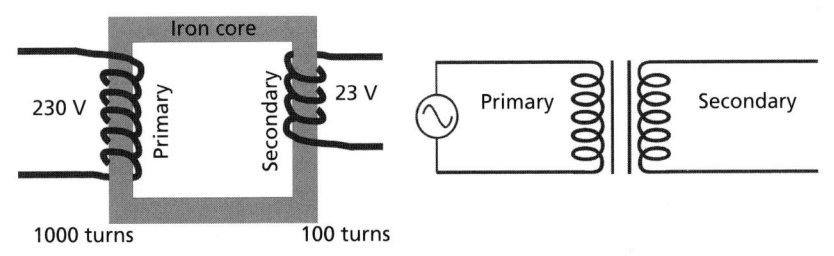

Fig. 4.10 A diagram of a transformer, showing turns around an iron core. Also shown is the circuit diagram of a voltage supplied to the transformer of 230 V RMS being stepped down to 23 V RMS.

Transformers are used for *stepping* down the domestic mains supply to safe voltages, isolating patient circuits and also in electronic filters. The power available on the primary side of the transformer will be similar to the secondary, so if the voltage is stepped down by a certain ratio, then the current must be stepped up by a similar amount. The amount of stepping down (or up) is dependent on the ratio of the turns. In Figure 4.10, if the primary has 1000 turns, and the secondary has 100 turns, then the voltage will be reduced 10 times. If the secondary had 10000 turns, then the voltage would be increased 10 times. By adjusting the number of coils on the primary and the secondary winding it is possible to step up or step down the voltage between the input and the output. The equations are simply $V_s/V_p = N_s/N_p$ and $N_p I_p = N_s I_s$, where V is the voltage, I is the current, s is the secondary and p is the primary.

Other electrical analogies

The water analogy has been already discussed, but there are many well established techniques available for solving electrical circuits by using other analogies. Computer simulations are available that emulate the workings of resistors, capacitors and inductors and so can predict with ease the voltage/current relationship in different parts of the circuit. Many mechanical problems, e.g. in the body or in transducer design, can be made far more understandable if the situation can be made into an electrical analogue and then analysed by a computer An example of this is blood flow.

In a hydraulic system, blood flow is analogous to electrical current, pulsatile blood pressure to (AC) voltage, and vascular resistance to electrical impedance. Parts of the circulation have storage properties and are therefore like capacitors and inductors. The elasticity of the arterial wall acts as *capacitance* and the mass of blood passing through it will behave as an *inductor*. The thin walled non-elastic venous system will tend to be like *resistors*. As there are reactance elements, flow and pressure will not necessary be in phase. If the circulation is modelled out in terms of electrical equivalents, then the system can be modelled on a computer to make good estimates of what is happening at various points around the body. The situation is complex; the driving force (pressure or voltage) is different as the distance from the heart is increased. It starts in the left ventricle, with an almost square wave as the aortic valve opens, with pressure leading the flow. But once the blood starts moving its momentum will tend to keep it moving and the flow will start to lead the voltage. But this can reverse again due to interactions between the *capacitance* and *inductance*. Various electrical analogue models for arterial blood-flow situations have been investigated.

Further reading

Brown, B. H., Smallwood, R. H., Barber, D. C., Lawford, P. V. and Hose, D. R. (1999) *Medical Physics and Biomedical Engineering*, Taylor and Francis.

Horowitz, P. and Hill, W. (1982) *The Art of Electronics*, Cambridge University Press, Cambridge.

McDonald, D. A. (1974) *Blood Flow in Arteries*, 2nd edn, Camelot Press, Southampton.

Strackee, J. and Westerhof, N. (eds) (1993) *The Physics of Heart and Circulation*, IOP Publishing, Bristol.

Chapter 5

Electronics and biological signal processing

> This chapter contains: filters; amplifiers, including patient connected differential amplifiers; signal processing; digital processing; and electrodes and transducers.
> The chapter links with: Chapters 1, 2, 3, 4, 10, 12, 15, 18, 19, 20, 21 and 24.

Introduction

This chapter continues the discussion of electricity but looks at the effect of connecting components together and briefly looks at the *operational amplifier* and *active* circuits (circuits discussed in the previous chapter have been *passive* ones, which involve no electronic circuits). It will then describe how the circuits can be used to process biological signals.

Filters

Simple band-pass filters

If a resistor, an inductor and a capacitor are joined as in Figure 5.1, the magnitude of the voltage across the resistor (V_{out}) will vary as the input frequency, V_{in}, changes, because of the properties of the capacitor and inductor. The *resistive* action of inductor and capacitor oppose each other, and at a certain *resonant* frequency (r in the figure), the total AC resistance (impedance) will be at a minimum. The graph of reactance against frequency demonstrates this, and shows the minimum reactance at the resonant frequency. As the LC and R are forming a voltage divider, the voltage across R will be maximum at this frequency. At other input frequencies, the output voltage will be low. This simple circuit forms the basics of the passive *band-pass* filter, where the filter passes, or lets through, a certain band of frequencies (around the resonant frequency in this circuit). Frequencies lower or higher than the band-pass will be attenuated. Normally operational amplifiers (discussed in the next section) or digital filters are more effective and are used to achieve the same effect. Resonance is important for the understanding of the behaviour of transducers and this is discussed in Chapter 12.

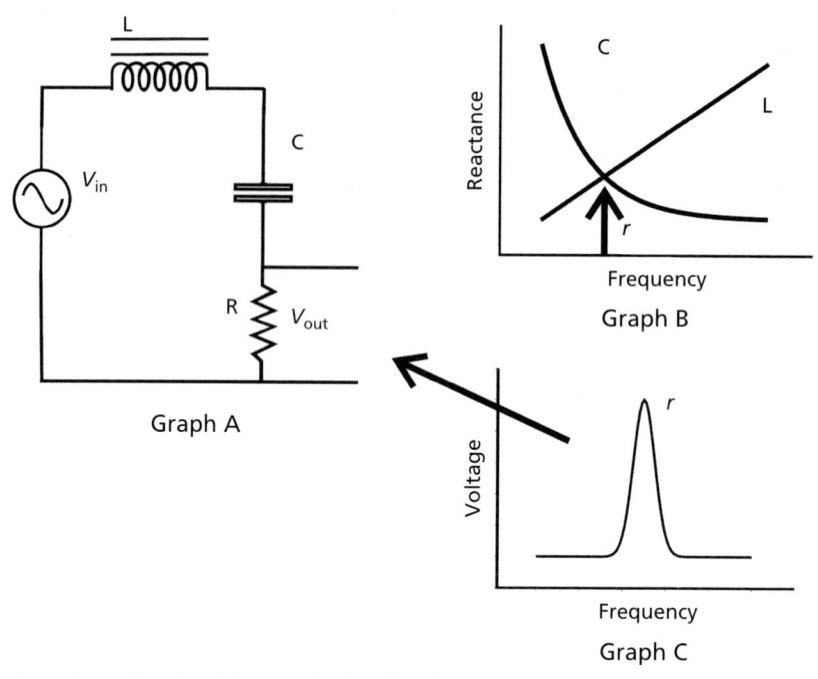

Fig. 5.1 An RLC circuit (shown in A) being driven by a variable frequency oscillator, V_{in}. The frequency response of voltage (V_{out}) against frequency across the resistor R is shown. The reactances against frequency responses are shown for capacitance (C), and inductance (L), showing the point (arrowed) in part B where the combined reactance is a minimum; r is the resonant frequency in both graphs.

Simple low pass filters

If a square wave is applied to the resistor capacitor network as shown in Figure 5.2, the capacitor charges up on the rising edge of the input with a time constant equal to the product of the resistance (Ω) and the capacitance (F), RC, and the output voltage will be $V_{ouput} = V_{input}\left(1 - e^{-t/RC}\right)$. When the

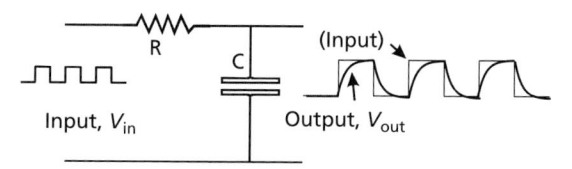

Fig. 5.2 An R–C circuit, showing a square wave input and resultant output across the capacitor. This circuit also acts as a simple integrator.

square wave is in the *off* state, or zero volts, the voltage will fall exponentially with the same time constant. It can be seen from the diagram that, with the appropriate values of R and C, the output from the network is a filtered version of the input.

This RC network is a simple *first order* low pass filter removing the high frequency components of the square wave and letting the low frequency parts through. The *cut-off* frequency of this filter is the frequency value where *attenuation* (i.e. the loss in intensity of the signal) characteristics change. Below the cut-off value, the filter lets all frequency values through with no attenuation and above that value the frequencies are attenuated. If the filter had its cut-off frequency low enough, the waveform would appear as a sine wave, as only the fundamental harmonic of the square wave would be let through, which is a sine wave! The cut-off frequency of the filter is $f = 1/2\pi RC$. Above the cut-off frequency, as the input frequency is increased, the output voltage due to the filter action is reduced by half every time the input frequency is doubled. This type of filter could be useful in filtering EEG signals, so that the higher frequency noise, such as muscle signals is prevented from passing through the filter.

This simple RC circuit also acts as a mathematical integrator with certain restrictions. Integration, as discussed in Chapter 1, is the process by which small elements are summed over a period of time.

Now Q (charge) $= CV$. If this is differentiated, the left side of this equation will be rate of flow of charge, which is current (I), and the right side will become $C\,dV/dt$. Therefore $I = C\,dV/dt$ through the capacitor.

The voltage across R is

$$V_{in} - V_{out}, \text{ so } I = C\,dV_{out}/dt = (V_{in} - V_{out})/R.$$

If V_{out} is much less than V_{in} over most of the input range, or if the time constant RC is large, then $C\,dV_{out}/dt \approx (V_{in})/R$.

This can be written

$$dV_{out}/dt = (V_{in})/CR.$$

If each side is integrated,

$$V_{out}(t) = 1/RC \int^t V_{in}(t)dt + \text{constant}.$$

What this equation indicates is that the RC circuit performs an integral over time for a given input. The restriction is that both the input voltage and the time period must be small compared with the time constant of the circuit (so that V_{out} is much less than V_{in}). This restriction can be overcome by using operational amplifiers, as explained later. The integrator is used in

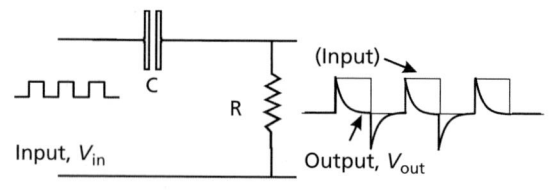

Fig. 5.3 A C–R circuit, showing a square wave input and resultant output across the resistor. The circuit also acts as a simple differentiator.

analogue computation, control systems, feedback, analogue/digital conversion and waveform generation.

Simple high pass filters

If the resistor and capacitor are used the other way round as in Figure 5.3, then this circuit will behave very differently. It will *block* a DC input but will *pass* high frequencies with little attenuation. Again the input frequency at which this situation changes is given by the time constant. This CR combination is called a high pass filter. The combination has another important property. The voltage across C is $V_{in} - V_{out}$, so in a similar manner to the RC circuit discussion,

$$I = C\,d(V_{in} - V_{out})/dt = V_{out}/R.$$

If R and C are very small (so that V_{out} is much less than V_{in}), $V_{out}(t) = RC\,d/dt\,V_{in}(t)$. This means that the RC combination is acting as a differentiator or producing an output proportional to the rate of change of the input waveform.

Differentiation is used extensively in medicine. For example, left ventricular pressure waveforms can be differentiated and used as a means of estimating myocardial contractility. In practice the maximum value of the differential (max dP/dt) divided by the left pressure occurring at the same time as the maximum dP/dt provides a useful index of myocardial contractility.

In order for all these circuits to function, the output V_{out} must not have any additional load or resistance across it. This is impossible in real situations – the circuit must be connected to something! In order to provide *buffering* operational amplifiers are used.

Amplifiers

Operational amplifiers

Operational amplifiers have largely replaced the transistor as building blocks in electronics circuits. They can be made to provide amplification (output/input),

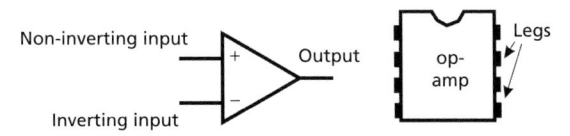

Fig. 5.4 Symbol of an operational amplifier and a plan of mini dual in-line 8 pin package. The package typically measures 10 mm × 5 mm.

filtering, oscillators, and many other functions. The symbol for an operational amplifier is shown in Figure 5.4, and these devices are contained typically in an 8 pin (legs), mini dual-in-line package containing hundreds of components. It provides very high amplification on its own ($>10^6$) and has an extremely high input resistance ($>10^6 \Omega$), which means it does not load or draw any significant current from the preceding circuit. Its inputs are also differential, i.e. it will amplify only a voltage difference and not an absolute voltage. The performance of an operational amplifier is controlled by *feedback*, which is achieved by connecting the output to the input, usually by a resistor or capacitor, which enables the designer to determine the gain and stability of the circuit.

The four basic circuits in which operational amplifiers are used are shown in Figure 5.5(a)–(d). The inverting amplifier circuit is shown in (a), and this has a gain of $-2k/1k = 2$. The non-inverting amplifier (c) has a gain of $1 + R_2/R_1$. The differentiator in (d), and integrator (b) give similar results to the RC and CR circuits described in Figures 5.4 and 5.3, except that the results are much more accurate, and fewer approximations are needed.

The amplifier circuits can be added to the simple RC circuits described earlier to provide practical filtering circuits. However such filters are inefficient because they filter typically only a quarter of the signal amplitude after the signal frequency has been doubled. The ideal filters would totally attenuate the signal amplitude directly after the cut-off frequency. Complicated circuits containing many operational amplifiers are required to produce efficient filters that are near the ideal case.

Single ended amplifiers

Since biological electrical signals, or signals that originate from transducers (see later), are very small in magnitude, they need to be amplified by an amplifier, which provides voltage *gain* (output voltage/input voltage). The simplest amplifier is *single ended* where there is one input terminal and one output terminal with a common to both input and output. A diagram is shown in Figure 5.6.

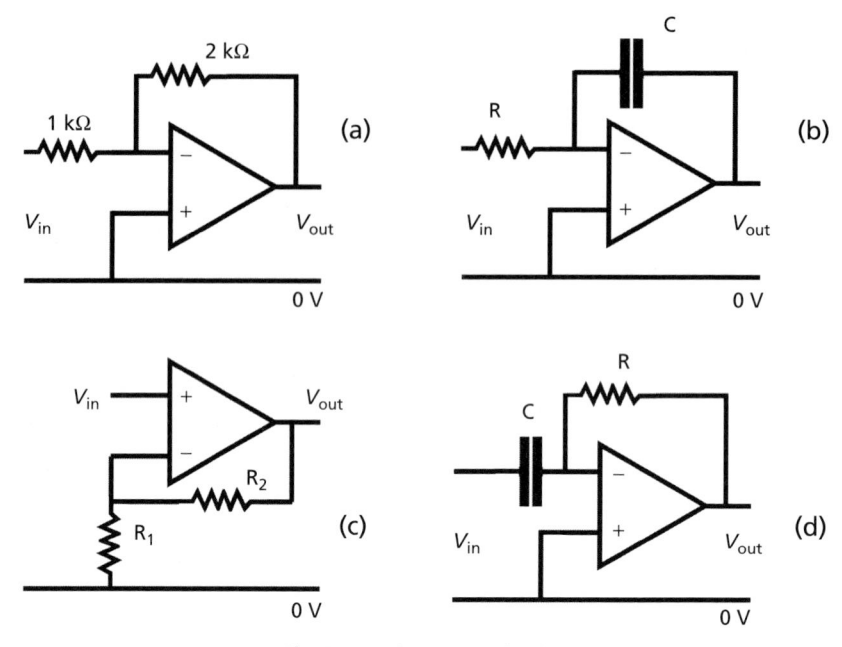

Fig. 5.5 Basic op amp circuits.

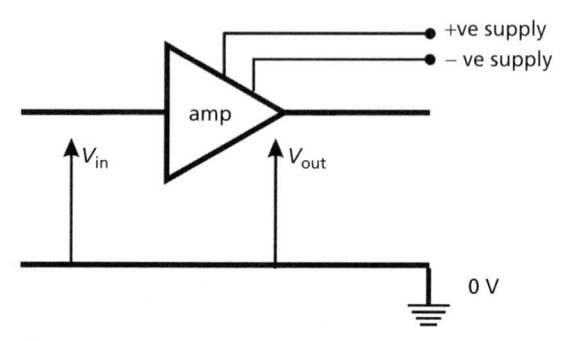

Fig. 5.6 A single ended amplifier in which the common ground is both one side of the input and one side of the output.

This form of amplifier is suitable for audio amplification, for example from a microphone, and the output could be connected to a loudspeaker. The gain in this case can be around 1000 times, similar to that required for an *electrocardiogram* (ECG) amplifier. If this amplifier were connected to a patient via two electrodes, then the loudspeaker would only give out mains

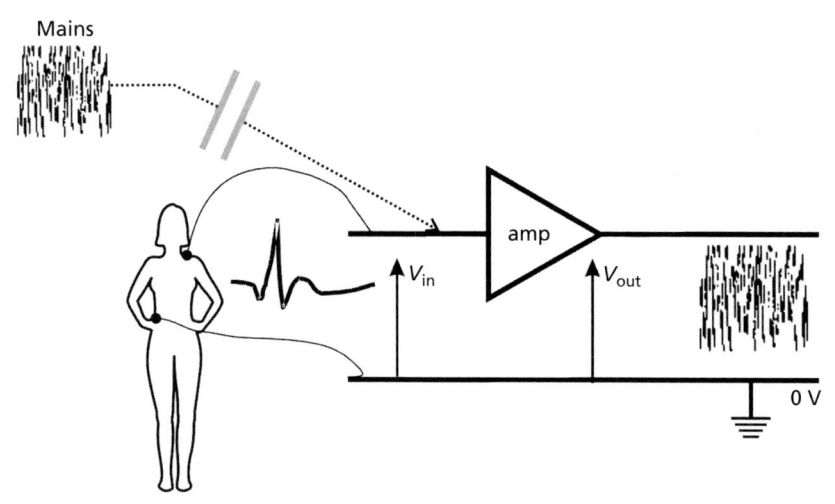

Fig. 5.7 A single ended electrocardiogram amplifier, showing the system amplifying the domestic mains voltage at the input and output.

hum and the ECG signal would be submerged in the noise. The mains voltage from lights, etc. (which is much larger than the ECG voltage) appears at the input via a small *natural* capacitance as shown in Figure 5.7. The amplifier will amplify this as well. Also in this situation the patient is earthed, which is not desirable in terms of patient safety, which is described in the next chapter. One solution would be to make the amplifier battery powered and put the patient in a completely screened room, with no mains apparatus near by. Another solution would be to use a patient isolated differential amplifier.

Patient isolated differential amplifier

The problem in obtaining an ECG signal is that the amplifier must detect the *difference* in potential between the two electrodes applied to the patient whilst ignoring the induced electrical interference signals that appear equally at both electrodes (by capacitance) and are therefore common to both. A differential amplifier does this and has two inputs in addition to the common point. Three electrodes are connected to the patient. For best results the common electrode is ideally placed at a neutral point on the body, and commonly the right leg is used for this in the ECG amplifier. The differential amplifier is shown in Figure 5.8.

If the ECG signal, referenced to the neutral point, is applied between the positive and negative inputs of the amplifier, as shown, then the output of the amplifier will be a magnified version of the ECG. If the ECG signal, or any other

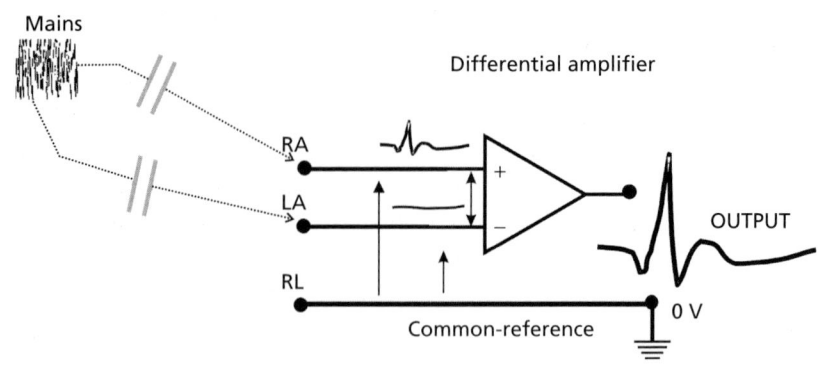

Fig. 5.8 Patient connected differential amplifier measuring the ECG, showing connection to a patient, RA is right arm, LA is left arm, and RL is right leg.

signal is connected to both the active inputs at the same time, then no output will result. The mains signal will thus be appearing at both of the active inputs. At all times throughout the mains voltage cycle, the voltage, with respect to the common input, will be the same value, and so will not be amplified. The facility to ignore signals that are *common* to both input terminals is defined as the *common mode rejection* and the manner in which the amplifier can amplify only the ECG signal and reject the common mains signals is called the *common mode rejection ratio* (CMRR) and is the differential signal gain/common mode signal gain. The common electrode is needed so that the ECG voltages are referred to it and are in the range of the supply of the amplifier. It could be connected to earth, which is undesirable, or it can be *floating* (i.e. not connected to earth). A floating or patient isolated amplifier is shown in Figure 5.9.

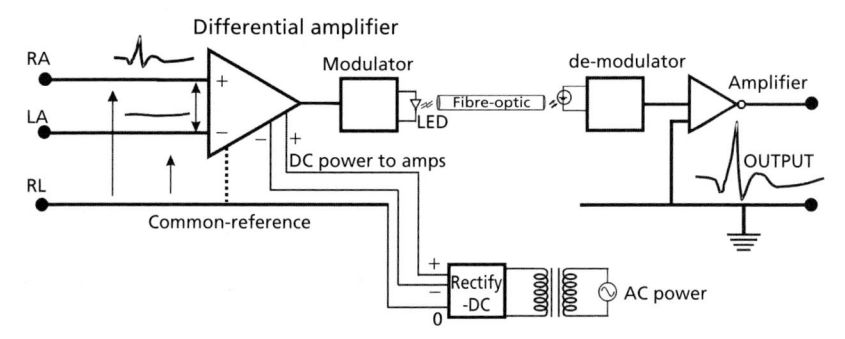

Fig. 5.9 Patient isolated differential amplifier showing isolation provided by fibre-optics, and the power supply arrangements for the patient amplifier. LED is light-emitting diode.

The supply for the patient amplifier (and modulator and associated circuits) is obtained from an isolating transformer, which is *rectified* (and smoothed) to produce DC. This provides isolation from the earth and mains supply as is discussed in the mains safety (Chapter 6). The signal output from the patient amplifier is connected to the main equipment by either an opto-isolator (as in the figure) or another signal isolating transformer. In this way the patient side of the equipment is totally separate from the main equipment and not earthed, but floating. The signal output from the patient connected amplifier first *modulates* a light beam (i.e. the signal fluctuations are impressed on the beam by using a light-emitting diode), and this beam is passed through a fibre-optic cable (this is discussed in Chapter 24). At the end of the cable, the light is converted back to an electrical signal by a photodiode (or similar), and then is *de-modulated* to bring the signal back to its original form. It is then further amplified as necessary. The output now is completely isolated from the patient.

Signal processing

Signal processing is a broad definition, but it includes the processes involved in taking an analogue signal, processing it in the analogue domain, converting it into a *digital* value (see next section), and then either displaying the results or using the results to perform some other function.

Biological signals

The basic signal processing set up is shown in Figure 5.10, and all of the functions have been mentioned, or will be discussed in this chapter. The biological signals can either come directly from the body via electrodes, (e.g. the electrocardiogram) or can be electrical signals produced from a transducer (e.g. a blood pressure transducer). Electrodes and transducers will be discussed in the next section. The electrical signals are amplified by the patient connected differential amplifier, whilst the transducer output (normally via a bridge) is amplified via a single ended amplifier.

The amplifiers used to process these raw signals must have sufficient bandwidth; for example, the ECG has harmonics up to 100 Hz, and so the amplifier must amplify each harmonic equally to maintain the signal accuracy. Also each frequency must be amplified with no time delay or at worst each must have the same time delay. If any of these criteria are not met there will be amplitude and phase distortion.

The signals are then manipulated and processed by the electronic circuits (e.g. filtered or differentiated), quantified by the *analogue-to-digital*

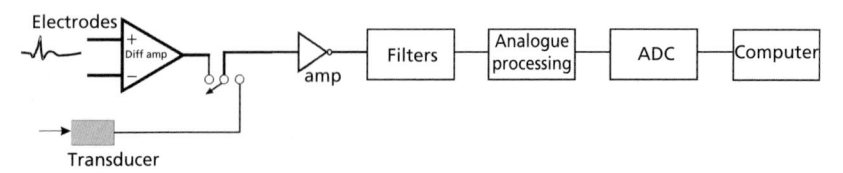

Fig. 5.10 Signal processing block diagram showing the path from electrodes or transducer, to amplify, filter, process and then convert the signal to digital form. ADC = analogue to digital converter, see text.

converter (ADC), and analysed by the computer. If the signal is hidden in the noise, or very noisy, signal processing can often help to improve matters.

Digital processing

The advantages of digital signal processing is that once a signal, such as a voltage, is converted to a series of numbers or *digitised*, the values cannot change, drift or become noisy as in the analogue processes described previously.

The computer can carry out numerous mathematical tasks on the digital data, which are now just a series of numbers – it can digitally filter them, carry out Fourier transforms on them to obtain the power spectrum (one method used is the *Fast Fourier Transform* [FFT]), which is a very efficient algorithm to calculate Fourier transforms, or just measure simple parameters such as the maxima of the signal.

The analogue signal is first converted to a digital value by an ADC. This device takes a *snap-shot* or a *sample* (not the same as a sample of data discussed in Chapter 2) of the waveform at exact time intervals, converts this analogue value to a number and sends it to the computer for storage and processing. The effect of sampling a sine wave is shown in Figure 5.11, and it appears that a staircase waveform is the result. In the reconstruction, this staircase is simply low pass filtered to give the same sine wave again. For an eight-bit system, as shown in the figure, (8 units of memory, each either 0 or 1, giving values of 00000000 to 11111111) the sampled number is between 0 and 255. If the signal is ranging from 0 to 5 V then the digital number produced is the signal multiplied by 255 and divided by 5. For example 2.5 V would be 128 or 10000000 binary. Systems can have many more bits, which can give higher resolutions. For example 16 bits will give 0–65536.

There are mathematical rules about how frequently samples need to be taken from a waveform so that the information of the original analogue waveform is preserved. The minimum sample rate is defined as the *Nyquist* rate and this rate is greater than twice the bandwidth (the highest frequency

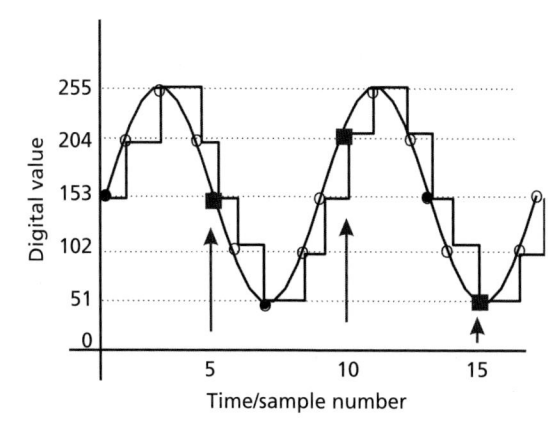

Fig. 5.11 Conversion of analogue signal to digital values. The diagram shows a sine wave sampled at ten equally spaced in time points. The open circles represent a sample, and these samples will have a corresponding digital value. For example, the 5th sample in the diagram has a value of 153, the 10th 204 and the 15th 51.

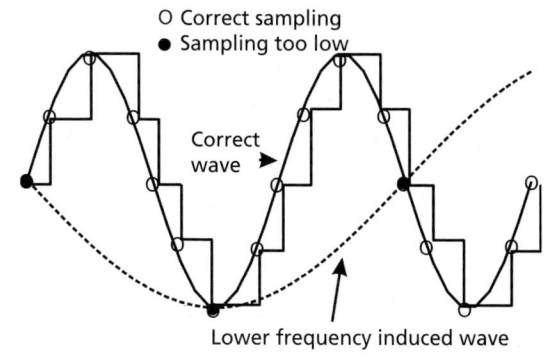

Fig. 5.12 Diagram showing a sine wave sample at above the Nyquist rate (open circles) and below the rate (closed circles). Aliasing occurs when the sampling is too low and results in an aliased wave – a much lower frequency wave than the sampled sine wave.

component) of the signal being processed. At this rate all the frequency information in the original signal will be kept. If it is sampled less, frequency aliasing occurs where low frequency components suddenly appear that are not really there. This is illustrated in Figure 5.12. If a certain sampling rate is fixed, the signal must be filtered to achieve this criterion. In practice, the signal tends to be sampled at between 3 and 8 times the Nyquist rate. No advantage is gained by sampling the waveform at 10 times or more the Nyquist minimum frequency required.

Electrodes and transducers

Electrodes

Electrodes, which are applied directly to the human body, are both used for the measurement of bioelectric events and to deliver current to living tissue in the form of stimulation. This section will consider just the measurement aspect. An electrode may have direct contact (via an electrolyte) to the tissue or the mode may be capacitive (the contact is via an air gap or insulator). The most frequent are the direct contact ones.

When a metallic electrode comes into contact with an electrolyte (a conductive gel), an electro-chemical reaction occurs, which in a conceptual sense resembles a voltage source, a capacitor and a resistor. The voltage source developed by this reaction at the electrolyte interface is called the half-cell potential. Different metals have different half-cell potentials and there is a range of values from near zero to ± 1.7 V. Table 5.1 gives some half-cell values.

If different metals are used for each of the electrodes then large voltages can be produced (as in batteries). For a pair of electrodes used, which is the minimum number needed to obtain a physiological event, the voltage difference should be the difference between the half-cell voltages. This voltage – the voltage needed to *polarise* – should be zero if the metals are well matched. It has been found [Geddes 1967] that silver (Ag) coated with a chloride layer (AgCl) has a low half-cell potential and also has good stability and low drift voltages between the two electrodes connected to the skin via the electrolyte. The electrical *noise* from the Ag electrode dramatically reduces when the

Table 5.1 Some electrode potentials for material used in electrodes (source de Bethune 1964)

Metal	Potential (V)
Aluminium	−1.7
Zinc	−0.8
Iron	−0.4
Nickel	−0.3
Lead	−0.1
Silver	+0.8
Silver/Silver chloride	+0.2
copper	+0.3
platinum	+1.2

chloride layer is applied. AgCl electrodes are therefore fragile because of this property – the noise increases greatly if the Cl layer is chipped or broken, and are also photosensitive. Light change can affect the half-cell potentials.

Electrode impedances

Because of the electro-chemical properties of the electrode, each electrode exhibits impedance. This is dependent on the nature of the *electrical double layer*, which is how the layers of charge are distributed on the metal and electrolyte surfaces and how much separation there is between the two layers. This impedance is often called the *polarisation impedance*. The impedance is normally very low and much less than the skin-electrode impedance. Putting the chloride layer on the electrode alters and increases the impedance but not significantly. Typical Ag–AgCl values are less than 200 Ω for low frequencies and 100 Ω for high frequencies greater than 100 Hz. This is contrasted with values of greater than 1000 Ω at low frequencies for electrodes before chloriding [Geddes 1969].

Electrodes on humans

The electrodes used on humans must be connected to a suitable biological amplifier as mentioned previously. The total impedance of both the electrode impedance and the electrode-skin impedance must be much lower than the amplifier input impedance in order to maximise the transfer of physiological signal from the body and to minimise signal distortion. Also low skin-electrode impedance will minimise the pick up of interference, especially domestic mains supply interference. The total electrode impedance (skin plus electrode) can be thought as a series resistor ($R_{electrode}$) with the physiological signal generator as in Figure 5.13. The combination of the electrode related impedance and the amplifier input resistance ($R_{amplifier}$) form a voltage divider. As the electrode resistance becomes lower than the amplifier input resistance, then the voltage across the amplifier input will be similar to the signal source voltage. It is also important to make the electrode impedance as low as possible so that the induced interference voltage (normally domestic mains) is as low as possible. As $V = IR$, where I is the induced interference current (transmitted from the mains equipment by capacitance – the differential amplifier section) and R is the electrode impedance, then to get the lowest V, R should be low. Obviously reducing the mains interference (lowering I) by switching off the equipment, increasing the distance from mains powered equipment, and using battery powered equipment will help as well. The method of decreasing the skin-electrode impedance is to abrade the skin with a coarse cleaning paste and to

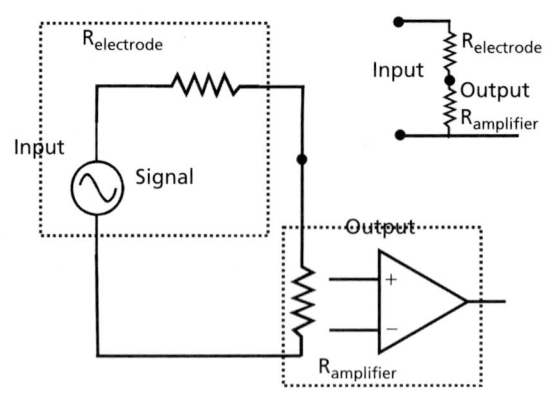

Fig. 5.13 Diagram of a physiological signal generator plus series electrode resistance ($R_{electrode}$) plus amplifier input resistance ($R_{amplifier}$). The voltage divider is formed by the combination of the two resistances with the signal being the input voltage, and the output voltage being the voltage across the amplifier. This equivalent circuit is also shown in the diagram in the top right.

clean the skin with alcohol before applying the electrodes. This reduction in impedance also gives a reduction in motion artefacts.

Transducers

A transducer converts one form of energy into another. In biomedical applications transducers are used to convert physiological signals to analogue electrical signals that can then be processed as described previously.

Transducers come in many forms. The majority fall into the following classes: resistive, inductive, capacitive, photoelectric, piezoelectric, thermoelectric and chemical. Examples of each will be discussed briefly here, and then in more detail in the appropriate chapters.

The variation of resistance has been used extensively to convert temperature and mechanical displacement (pressure) to electrical signals. The resistance of a conductor is dependent on the material, the geometric configuration and the temperature. The choice of material is dependent on its linearity or sensitivity for that purpose. The resistance transducers have a wide range of uses in medicine from strain gauges, thermometers, thermistors to pressure (resistive) transducers.

Inductive transducers are also used in medicine. The inductance of a coil depends on its geometry, the magnetic permeability of the medium in which it is located, and the number of turns of the coil. Changing any of these will give a change in reactance and this can be measured. Only certain changes, however, will have the sensitivity required. An example of its use is the pressure transducer discussed in Chapter 12.

Capacitive transducers are another alternative. These can be very accurate and linear but require complex driving electronics. A capacitor consists of two conducting surfaces separated by an insulator (dielectric), which can be a solid, gas, liquid or a vacuum. They are used in blood pressure transducers (discussed in Chapter 12) and differential capacitive pneumotachographs. Also living tissue can form part of the capacitive transducer.

Photoelectric transducers consist of photovoltaic cells, photoconductive cells, and the phototransistor and they are also used in medicine. The photovoltaic cell typically consists of a thin coating of selenium on an iron backing. Above the selenium is a thin transparent metal film. When this is illuminated a potential difference appears across the barrier. Photoconductive cells consist of a thin film of typically silicon and, when exposed to certain types of radiant energy, its resistance decreases. The phototransistor can have a very fast response time and the collector (the output) amplifies the small current change caused by the light by the same process as transistors do. The most well known use of the phototransistor has been in pulse oximetry, as discussed in Chapter 15. This uses two detectors and each receives different wavelengths of light, obtained by filtering, typically 640 nm and 805 nm.

Piezoelectric transducers rely on the effect of certain materials that develop a voltage difference across them when a pressure is applied or the material is distorted in some way. These transducers are used to detect blood pressure and body sounds, and are used as accelerometers. The piezoelectric transducer can also work in reverse, producing movement when a potential difference is applied. Ultrasound transducers, as discussed in Chapter 10, use both these principles.

Chemical transducers play an important role in the assessment of metabolism. There are two main types: those that measure chemical composition of the blood, tissue and organ fluids, and those that measure the composition of the respiratory gases. Many chemical transducers are electrochemical cells in which the quantity to be measured causes a change in cell potential or a change in current through the cell.

References

De Bethune, A. J. and Loud, S. N. S. (1965) Standard aqueous electrode potentials and temperature coefficients at 25°C, *J. Electrochem. Soc.*, **112**(4), 107C–108C.

Geddes L. A. and Baker, L. E. (1967) Chlorided silver electrodes, *Med. Res. Engng.*, **6**, pp. 33–34.

Geddes L. A. and Baker, L. E. (1969) Impedance-frequency curves for bare and chlorided silver electrodes, *Med. Biol. Engng.*, **7**, pp. 49–56.

Further reading

Brown, B. H., Smallwood, R. H., Barber, D. C., Lawford, P. V. and Hose, D. R. (1999) *Medical Physics and Biomedical Engineering*, IOP Publishing, Bristol.

Horowitz, P. and Hill, W. (1982) *The Art of Electronics*, Cambridge University Press, Cambridge.

McDonald, D. A. (1974) *Blood Flow in Arteries*, 2nd edn, Camelot Press, Southampton.

Strackee, J. and Westerhof, N. (1993) *The Physics and Heart and Circulation*, IOP Publishing, Bristol.

Togawa, T., Tamura, T. and Oberg, P. A. (1997) *Biomedical Transducers and Instruments*, CRC Press, New York.

Chapter 6

Electrical safety

This chapter contains: electricity supply; safety features; leakage currents; earth faults; earth-free supplies; microshock; circuit breakers; heating effects of electrical current; electrolysis; static electricity.
The chapter links with: Chapters 4 and 5.

The electricity supply

Domestic 'mains' systems in the UK use AC at 50 Hz. In the USA the systems use 60 Hz. These frequencies are used as they are efficient frequencies for transmission from power generation to the users and minimise the effect of leakage currents due to capacitance, which is discussed later in this chapter. The mains is initially generated in a power station and the power generated there (volts × amps) is enough to supply a number of hospitals and consumers. The voltage and the potential to do work must be transmitted to the user and this is usually achieved with overhead pylons, or sometimes by underground cables. Both types of cable will be designed to carry current, but the higher the current, the greater power lost to heat (I^2R). It is desirable to keep the current as low as possible to reduce this transmission heat *loss*, and this can be done by making the voltage as high as possible. For a given power, ($V \times I$) there can be a high V and low I or vice versa. The transmission voltage is normally greater than 11 kV. It arrives at a domestic substation and is transformed down to (in the UK) 230 V RMS by a transformer (see Chapter 4). At the substation, as illustrated in Figure 6.1, one connection of the transformer is firmly bound to earth at what is called the *star point*, and this forms the start of what is called the *neutral* lead. The earth connection forms a vital part of electrical safety. The connection on the other side of the transformer is called the *live* lead and this is at 230 V RMS. These two leads are taken to the individual outlets or mains sockets, the live carrying the voltage to the load, and the neutral lead carrying the return current back to the source of the supply. The earth connection of the mains socket is connected back to the star point separately, although sometimes in older installations this can be earthed locally. In this way only one of the socket points is live, and the other is at near zero potential. This is all demonstrated

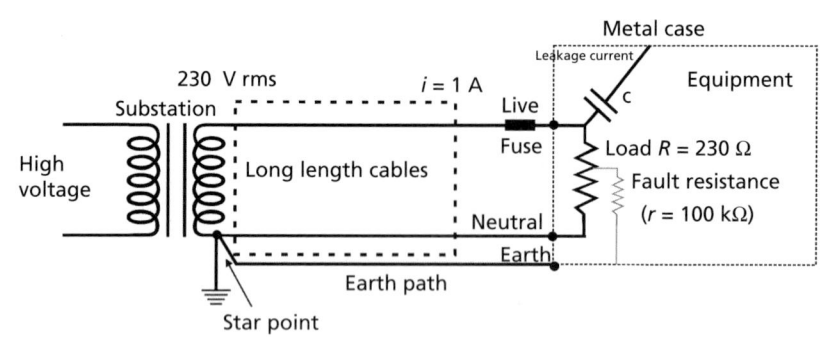

Fig. 6.1 Diagram of the domestic mains supply, supplying class one equipment. The equipment has a load of 230 Ω, and so the normal current would be 1 A. The case of the equipment is earthed. The leakage current by capacitance is shown by C. See text for details.

in Figure 6.1. The neutral connection at the equipment is not exactly zero, due to the resistance of the long length cables (see figure) supplying the mains equipment. Although the resistance of this lead is very low, if it is 0.5Ω and the current flow is (for example) 1 A, then the voltage drop between the transformer and the domestic socket will be 0.5 V. Therefore the neutral point will be 0.5 V.

Electrical safety features

Class 1 equipment

The first safety or basic principle ensures that the electrical device has a metal case which is earthed, as shown in the Figure 6.2. This is class 1 equipment, and can be monitors used in theatre, or simple domestic kettles. The live lead is protected by fuses (or a circuit breaker, which is described later). Any fault that causes a low resistance to appear between live components in the circuit and the metal case, is dealt with by the protective earth pathway which takes the *fault* current to earth. In the Figure 6.1 the fault is shown by a fault 'resistance' to the case of 100 kΩ. If the leakage voltage source was assumed to be 100 V (it can be anything from low voltage to 230 V), then the fault current would be 1 mA. Normally this low resistance path to earth causes an increased live current through the fuse but the fuse would only melt and *blow*, breaking the circuit, if the fault current was in the order of amps, but in this example nothing would happen to the fuse as the fault current is not high enough.

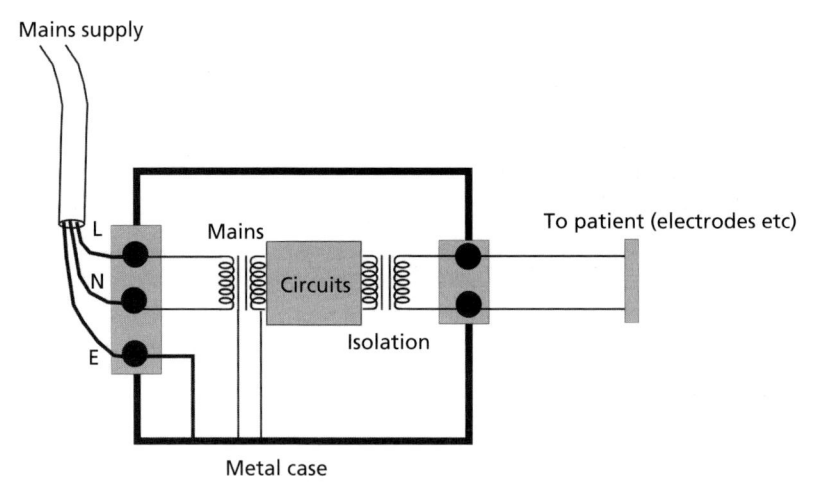

Fig. 6.2 Diagram of a class I equipment that supplies a patient connected part via an isolation transformer. The mains transformer core and one side of the secondary are earthed for safety.

However, class 1 equipment can only be connected to the patient if there is a separate isolating transformer, as shown in Figure 6.2. The patient must not have a direct path to earth and this is discussed later.

Class 2 equipment

Class 2 equipment has no protective earth. Instead the basic insulation is supplemented by a secondary layer of insulation. This equipment is classified as doubly insulated. The power cable has only live and neutral, and the mains connected circuitry is then enclosed in a second insulated layer, so that no conductive exposed parts can be brought into contact with the patient. The case is made of non-conductive plastic.

Class 3 equipment

Class 3 equipment is powered where the mains is stepped down by an external safety extremely low voltage transformer to less than 25 V RMS or less than 60 V DC. Normally the DC situation is supplied by batteries.

Leakage currents

Apart from inappropriate currents caused by faults, there are normally other 'extra' currents in the system caused by *stray capacitance*. Capacitance exists between all conductors, even if the insulation is intact and adequate, and this

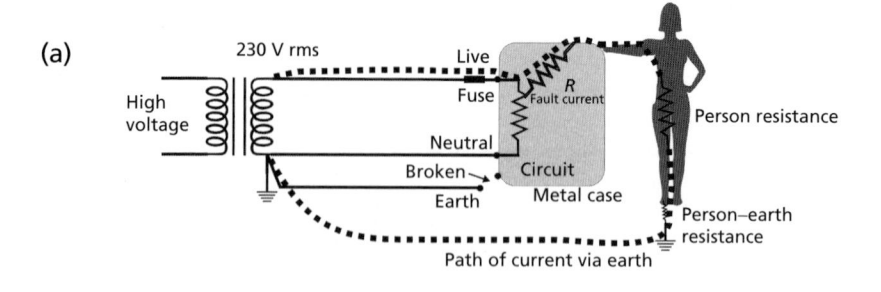

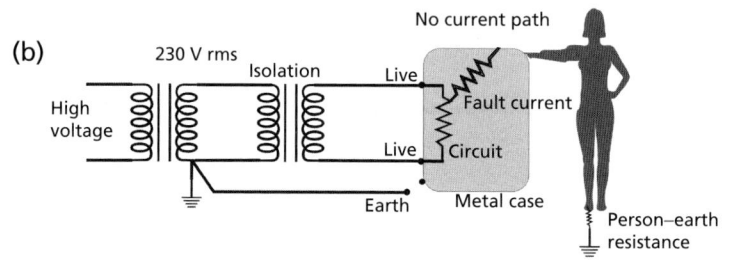

Fig. 6.3 (a) describes a situation where there is a dual fault. The earth connection to the metal case is broken and there is a fault resistance to the case caused by an internal fault. There will be a flow of current through the person, via earth. The magnitude of the current will be dependent on the person resistance and the person–earth resistance. R is the fault resistance. (b) shows a situation where there is an isolation transformer present. In this case, the dual fault situation described in (a) does not give a complete current path, and no current flows through the person.

is the stray capacitance. The closer the conductors are together, the greater the stray capacitance. As the mains voltage is AC, the stray capacitance between the mains wiring and earth results in the flow of 'capacitive' current to earth, because the capacitor acts like an impedance. This occurs both in the supply system and in any equipment connected to it. This current is called leakage current and is illustrated in Figure 6.1. In designing equipment, manufacturers can reduce but not eliminate leakage current. They can keep leakage current within acceptable limits by the careful choice of mains components.

The earthing is vital as it provides a safe low resistance path for the leakage current back to the (earthed) neutral point. Every class 1 device has some leakage current passing down the earth wire: this current is called *Earth Leakage Current*. However, an excessive Earth Leakage Current could pose a possible problem for the earthing system and any circuit breakers as described later. There is a maximum allowed limit for Earth Leakage Current of 5 mA in normal conditions and 10 mA in single fault conditions.

Touch Current, (previously called Enclosure Leakage Current), does present a hazard to users and patients as shown in Figure 6.3(a). Touch Current is leakage current that can find a path to earth through a user who touches the metal case of the equipment. In normal conditions, if the enclosure is conductive and earthed, then the Touch Current will be, for all practical purposes, zero. If the case of the equipment is not earthed, due to a problem or a fault, (as in the diagram) then normal Touch Current will flow through the person. In this condition, where the earth has been disconnected, the upper safe limit is 500 μA.

Apart from the classes 1–3 safety features, if equipment is to be connected to patients (e.g. ECG, etc.), it must be designated type B, BF or CF. Type B may be class 1, 2 and 3 mains powered equipment. This equipment is not suitable for direct connection to the heart, but must have adequate protection against electric shock with regard to allowable leakage currents. The allowable patient leakage currents are shown in Table 6.1. Single fault conditions are when a single condition such as an earth disconnection occurs There can be multiple faults, such as earth disconnection and excessive leakage occurring.

Type BF equipment where that part of the equipment is applied to the patient is isolated from all other parts of the equipment to an extent that allowable patient leakage current (under single fault conditions, i.e. only one of the possible faults being present) is not exceeded even when 110% of the mains voltage is applied between the applied part and earth. This degree of isolation is called an F-type isolated floating applied part. This applied to CF as well.

Table 6.1 Allowable values of continuous leakage and patient auxiliary currents in μA (international regulations defined by IEC 60601–1)

Current	Type B Normal	Type B Single fault	Type BF Normal	Type BF Single fault	Type CF Normal	Type CF Single fault
Patient leakage current and patient auxiliary current DC	10	50	10	50	10	50
Patient leakage current and patient auxiliary current AC	100	500	100	500	10	50
Patient leakage current with reference supply voltage on applied part	–	–	5000	5000	50	50

Type CF is class 1 or 2 mains powered with an internal electrical power source. This equipment provides the most stringent protection against shocks and has an F-type isolated floating applied part. This equipment can be connected to a patient's heart.

Some commonly used symbols are shown in Figure 6.4.

Earth faults

The earth connection can become disconnected due to trauma to the mains cable, for example if it has been continuously stepped on. The plug can be badly wired and also the connection can be from the socket to an old corroded water pipe. Hospital maintenance or the equipment management service should always check the integrity of the earth connection. If an earth fault develops, normally no one would notice any difference. If a small leakage current develops, then the user would feel that the equipment has a *fuzzy* feel to it. If two or more faults develop, i.e. earth lead breakage and severe leakage fault currents (and the order this happens in is not important), then this situation

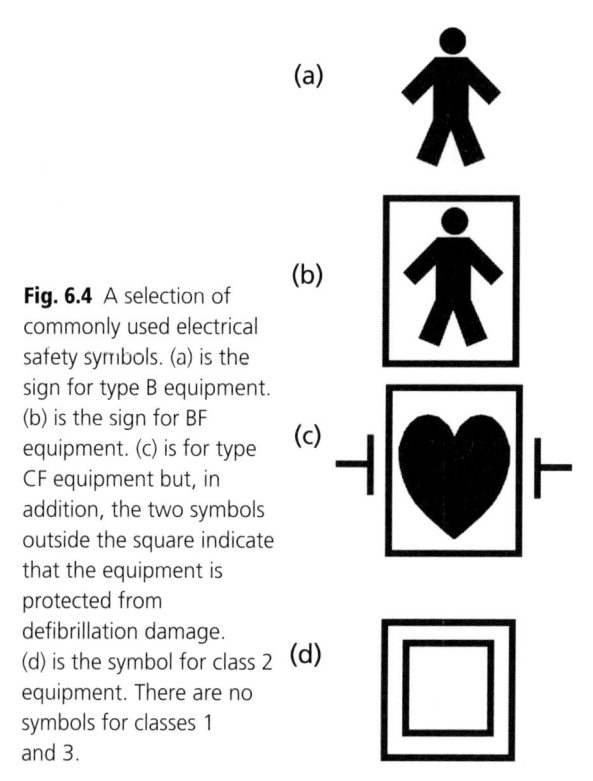

Fig. 6.4 A selection of commonly used electrical safety symbols. (a) is the sign for type B equipment. (b) is the sign for BF equipment. (c) is for type CF equipment but, in addition, the two symbols outside the square indicate that the equipment is protected from defibrillation damage. (d) is the symbol for class 2 equipment. There are no symbols for classes 1 and 3.

is highly dangerous to both the clinician and the patient. In this dual fault situation, if an anaesthetist touches the instrument case, for example, then a connection to earth can be made via the resistance of the person and the circuit will be completed. This is described in Figure 6.3(a).

There will be current flow through the person and the magnitude of it will be dependent on many factors: humidity, sweating, shoe insulation, etc. These factors will change the resistance to earth (person–earth resistance in the figure) and therefore the amount of current through the person. For example, if the anaesthetist is standing on an antistatic floor with one hand touching the faulty equipment and touching nothing else, then the resistance to earth is in excess of 20 kΩ, so that by Ohm's law the maximum current that can flow is $230/20 = 11.5$ mA. Whilst this shock would be painful, it would not kill.

However, if the subject touches the metal equipment with one wet hand while touching an earth conductor such as a cold water tap with the other (or standing in a puddle of water with normal footwear), the resistance to earth would be much lower and possibly less than 2000 Ω. In this case the subject would sustain a lethal electric current of 115 mA and go into ventricular fibrillation. However, this is still a small current for the circuit, so it is likely that the circuit fuse will remain intact. Table 6.2 gives the physiological effects of different currents through the body.

Microshock

The Table 6.2 shows currents through the external regions of the body. However the currents needed to do damage within the body are much less, and very small currents around 50 µA are needed through the ventricular endocardium to produce ventricular fibrillation. The effect of these currents is called *microshock*. The frequency is important as well and low frequencies such as 50 Hz are the most dangerous frequencies (high voltages at high frequencies

Table 6.2 External AC currents through the body, and the physiological effect

Current	Effect
0.5 mA	Threshold of perception
1 mA	Tingling sensation
5 mA	Pain and threshold of cannot let go current
15 mA	Severe pain with local muscle spasm
50 mA	Tonic muscle contraction of respiratory muscles and respiratory arrest
80–100 mA	Ventricular fibrillation

in the MHz range are used in diathermy, and these do not harm the patient, but they are used for controlled cutting and coagulation, and this is discussed in Chapter 21).

Microshock can occur in a few situations: pacemaker wires, cardiac catheters with incorporated electrodes, a catheter filled with conductive medium, or a saline filled cardiac catheter. A fault condition could cause the potential of the catheter or associated devices to rise above zero volts and then any earth path available to the patient would be disastrous. One example of this [Bousvaros et al. 1962] was when a patient developed *ventricular fibrillation* (VF) when the motorised syringe pump connected to the cardiac catheter was switched on. There was a double fault on the syringe driver: the motor was poorly insulated (and therefore had high leakage current) and the earth connection to the device was broken. The patient was earthed by a right-leg ECG electrode that was connected to the earthed chassis of a (now not legal) non-isolated ECG machine. Leakage current flowed along the cardiac catheter, through the patient's heart and right-leg electrode to earth. The equivalent circuit of this situation is shown in Figure 6.5. It can be seen that over 400 µA

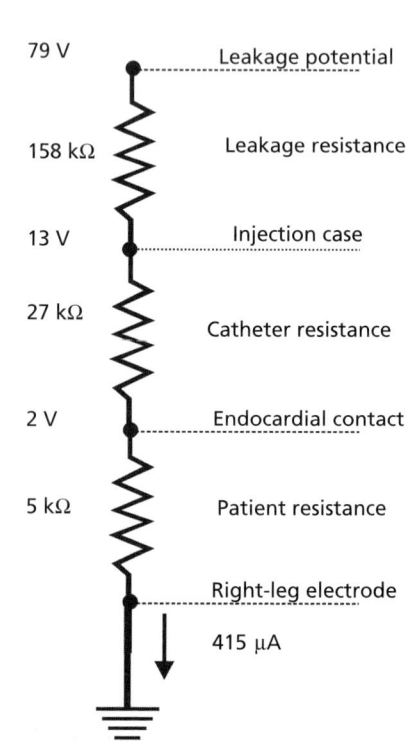

Fig. 6.5 Equivalent circuit of the multiple fault condition described in the text.

flows, which is well above the microshock level. To prevent microshock, stringent leakage levels must be applied to all equipment making contact with the patient either directly or via an endocardiac lead. A stringent code must be observed: all electrical equipment in the vicinity of the patient must conform to a leakage specification of less than 10 µA under normal working conditions; wherever possible such equipment should be battery operated and fully isolated from earth; all electrical circuits making contact with the patient must be electrically isolated from earth; all such equipment in the susceptible zone should whenever possible be coated with insulating material; pacemaker leads or catheters must only be handled with gloved hands and a user should never hold the lead in one hand while using an oscilloscope, monitor or other mains powered equipment in the other.

Circuit breakers

For general protection of equipment, there is a fuse situated in the live line. This will break when a fault situation occurs where the addition of the leakage current to the normal current is high enough to melt the fuse. Automatic electronic circuit breakers that can be reset also exist, and these operate in the same manner as the fuse. These devices (fuse and circuit breaker), however, are not sensitive enough to prevent any shock situation occurring, due to the relatively small currents involved. A *residual current device* (RCD) is a much more sensitive circuit breaker and can break a circuit before a person perceives any pain. This relies on the principle that normally in a correctly working system there is a balance of current in the live and neutral leads. In a fault situation, there is an imbalance of current and this causes an electrically operated switch to break the circuit. This can be in the form of a transformer solenoid type switch or a semiconductor type as shown in Figure 6.6. The sensitively for the RCD can be an imbalance of 100 mA for the least sensitive and up to 10 mA for the most sensitive. In order for the user not to notice any physiological effects, the RCD must operate within 30 ms. Normally a fuse or automatic circuit breaker would be used in addition to break the circuit if very high currents exist.

One problem with using RCDs is that they rely on disconnecting the supply to provide safety. In some medically used rooms this is unacceptable. Consider a situation where a mains lead that has become wet is plugged in at an intensive care unit (ICU) bed. This results in excessive leakage current and the supply is disconnected to protect the users. In doing so it might also shut off the supply to the monitors and ventilators. A solution to this, but this solution can be only

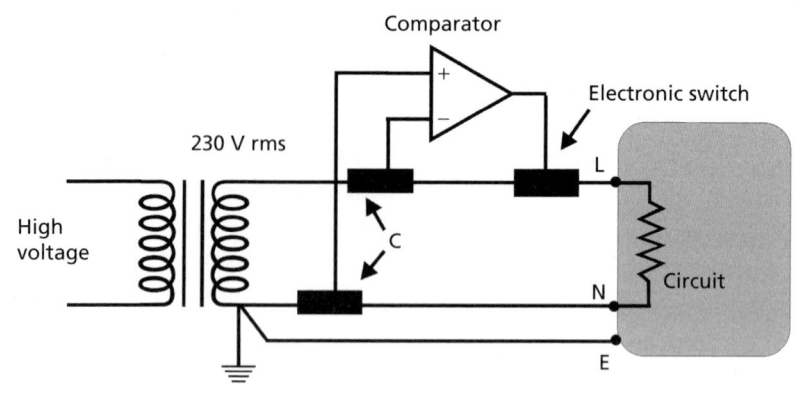

Fig. 6.6 An example of a sensitive circuit breaker. Cs in the diagram are circuit components that give a voltage output proportional to the current flowing through them. These outputs, one from the live line, the other from the neutral line, are supplied to a comparator circuit. This circuit compares the two inputs and gives an output if they are different by a desired amount. This output will operate an electronic switch and break the circuit.

used in strictly controlled situations (such as ICU and theatres), is earth free supplies.

Earth free supplies

The conventional live–neutral–earth system has some advantages especially for domestic use (only one lead is at mains potential; this has advantages, e.g. only one fuse needed) but, as discussed, an earth free mains supply for theatres is possible. A 1:1 isolating transformer is added to the normal system so that no voltage change occurs. The primary winding is connected to the live and neutral and the secondary is not connected to earth and is therefore earth free. This feeds one or more of the power outlets. The earth can be connected to the case as before. In the situation when either of the two live leads (2 now instead of 1) become connected to the case (and the earth connection to the case is not present or high resistance), then the current cannot pass through the patient to the earth situation, as shown in Figure 6.3(b).

Other safety issues of electricity

Heating effects of electric currents

When a current passes through a conductor, work is produced and the energy is converted into heat, in proportion to the resistance. As heat is produced, the temperature of the conductor rises. The rate of heat production is heat per unit time, or J s^{-1} = watts. The effect of a given amount of heat on the temperature of the conductor depends upon the *specific heat capacity* (SHC) of the material (kJ kg^{-1} K^{-1}). Water has a SHC of 4.128 kJ kg^{-1} K^{-1}. Body tissues have a SHC of 3.5 kJ kg^{-1} K^{-1} and for blood it is 3.6 kJ kg^{-1} K^{-1}. If a body is heated by a certain amount, then the temperature reached depends on its mass and SHC. The rate of cooling depends on the rate of heat lost by conduction or dissipation, convection and radiation. The body can be thought of a complex structure composed of multiple elements each with a unique combination of resistance, heat capacity and dissipation. When current passes through these elements, each one produces heat proportional to the current (I^2R) and the temperature reached depends on the dissipation. Surface elements will dissipate heat by radiation and convection. But deeper elements depend on their neighbours' heat production and nearness to the surface. The heating effect of the deep elements depends on the heat dissipating capacity of its neighbours near the surface.

The effect of current density

The current density is the current per unit area, perpendicular to the flow. It is measured in A m^{-2} and sometimes as mA cm^{-2}. The pattern of heating in the body depends on current density and dissipation from different zones to deeper zones with high current density becoming hottest, while those nearer the surface with low density remain coldest. In general, though, high current densities generate the greatest heating, all other things being equal. A large mass of tissue (e.g. the forearm) will heat up less than, say, a small finger.

Current densities of less than 1 mA mm^{-2} are unable to cause damage but burns have been produced by currents of 5 mA mm^{-2} for 10 s or more [Hull 1994]. In diathermy, where radio-frequency currents are used, there is no stimulation, but heating is limited to the current densities –the high density at the positive tip will cause massive heating, but at the negative indifferent plate, the density is lower, so no heating is produced. Burns in any tissue (current) situation are produced when the heat produced is more that is dissipated. This is discussed much more in Chapter 21.

DC and electrolysis

Electrolysis will take place when a direct current is passed through any medium that contains free ions. The positive ions will migrate to the negative electrode and negative ions to the positive one. If two electrodes are placed on the skin and 100 μA DC is passed between them for a few minutes, small ulcers will be found beneath the electrodes. IEC601–01 defines 'DC' as being less than 0.1 Hz. Above this the movement of ions is balanced because the positive and negative electrodes are reversed. The IEC recommendation limits the current to 10 µA.

Static electricity

Static electricity, as discussed in Chapter 4, that builds up on the body can give rise to short-duration shocks. This happens when the body is insulated from the local earth and accumulates a small electric charge. When a person touches something earthed there is a discharge between the person and earth and this flow of current can be enough to cause an unpleasant jolt. However, static discharges have durations measured in microseconds and, as these are of short duration, there is no risk of inducing a harmful stimulation to the heart. For this reason, the antistatic floor resistance cannot be infinite as there needs to be a resistance path for the static to discharge to earth.

References

Bousvaros, G. A., Don, C. and Hopps, J. A. (1962) An electrical hazard of selective angio-cardiography, *Can. Med. Association J.*, **87**, pp. 286–88.

Hull, C. J. (1994) The electrical hazards of patient monitoring. In Hutton, P. and Prys-Roberts, C. (eds), *Monitoring in Anaesthesia and Intensive Care*, WB Saunders Co. Ltd, London, pp. 56–77.

Further Reading

Al-Shaikh, B. and Stacey, S. (2002) Chapter 3 in *Essentials of Anaesthetic Equipment*, 2nd edn, Churchill Livingstone.

Wentworth, Stephanie (ed.) (2009) authors: Graham Aucott, Malcolm Brown, Lindsay Grant, Peter Lee, Fran Hegarty, Justin McCarthy, Roger Jones, Peter Smithson, Azzam Taktak, Colin Walsh and Stephanie Wentworth. Guide to electrical safety testing of medical equipment: the why and the how. IPEM Report Number 97 © Institute of Physics and Engineering in Medicine, ISBN 978 1 903613 36 8.

Chapter 7

Behaviour of fluids (liquids and gases, flow and pressure)

This chapter contains: fluid flow behaviour (laminar and turbulent flow); theory of flow measurement devices; atmospheric pressure (including hypobaric and hyperbaric); partial pressure; diffusion; osmosis.
The chapter has links with: Chapters 1, 3, 8, 9, 13, 14 and 22.

Fluid flow behaviour

A *fluid* can be either a liquid or a gas. Fluids exhibit different flow behaviours depending on their physical properties, in particular viscosity and density. Flow characteristics also depend on the geometry of the pipes or channels through which they flow, and on the driving pressure regimes. These principles can be applied to any fluid, and the complexity of the analysis depends on the flow regimes described in this section [Massey 1970].

Fluid flow is generally described as laminar or turbulent. Laminar flow, demonstrated by Osborne Reynolds in 1867, is flow in which laminae or layers of fluid run parallel to each other. In a circular pipe, such as a blood vessel or a bronchus, velocity within the layers nearest the wall of the pipe is least; in the layer immediately adjacent to the wall it is probably actually zero. In fully developed laminar flow, the velocity profile across the pipe is parabolic, as shown in Figure 7.1, and as discussed in Chapter 1 (see Figure 1.12). Peak velocity of the fluid occurs in the mid line of the pipe, and is twice the average velocity across the pipe at equilibrium, and layers equidistant from the wall have equal velocity. The importance of laminar flow is that there is minimum energy loss in the flow, i.e. it is an efficient transport mode. This is in contrast to turbulent flow, where eddies and vortices (flow in directions other than the predominant one) mean that energy in fluid transport is wasted in production of heat, additional friction and noise. The result is that the pressure drop required to drive a given flow from one end of the pipe to the other is greater in turbulent than in laminar flow.

The shear stress τ, which is the mechanical stress between layers of fluid and between the fluid and the tube wall, is proportional to the velocity gradient across the tube (dv/dr) of the fluid layers. The constant of proportionality between these two variables is the *dynamic viscosity, η*.

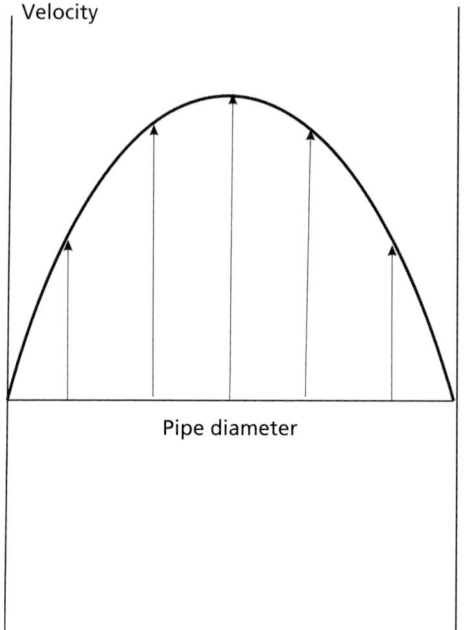

Fig. 7.1 Fluid velocity profile across a tube in laminar flow.

Expressed mathematically,

$$\tau = \eta \frac{dv}{dr}.$$

Where dynamic viscosity is constant, the fluid is a *Newtonian fluid* such as air or water. Blood is a suspension of particles in a liquid and is therefore a non-Newtonian fluid, which means that η is not constant and depends on the time dependent rate of shear between layers. Dynamic viscosity has units of N s m^{-2} or kg s^{-1} m^{-1}. There also exists the property *kinematic viscosity*, ν, where $\nu = \eta/\rho$, where ρ is the fluid density. The flow will decrease as the viscosity (η) increases, which is intuitive. As temperature increases, the viscosity of liquids decreases, while that of gases increases. The reason for this can be thought of as a decrease in the tightness of the intermolecular bonds of liquid decreasing viscosity, and an increase in the random kinetic energy of gas molecules increasing viscosity.

Fluid flow occurs only when there is an appropriate pressure gradient between two ends of a pipe. At a low pressure gradient (*low* is a context sensitive term here, depending on the fluid and the pipes being studied), the flow rate is directly proportional to it and the flow is laminar. At a critical point (*critical flow rate, critical velocity*), the flow becomes turbulent, and the flow rate is subsequently approximately proportional to the square root of the pressure gradient, and to the reciprocal of the square root of the fluid density and to the reciprocal of the square root of tube length. One of the tools used by hydraulic engineers for modelling fluid flow is a series of dimensionless ratios, one of which is *Reynolds' number, Re*, where

$$Re = \frac{V D \rho}{\eta},$$

and V is a characteristic velocity, and D is a characteristic dimension, usually tube diameter.

Whatever the size of the pipe and whichever fluid is used, in a smooth circular pipe, such as a blood vessel or a bronchus, fluid flow changes from laminar to turbulent when Re exceeds approximately 2000. Turbulence may occur intermittently, locally and prematurely at lower values of Re if the tube is not smooth or straight locally, for example at bends, orifices, and changes of diameter. The resistance to fluid flow down a tube can be thought of as

$$\frac{pressure\ gradient(\Delta P)}{flow(Q)},$$

an hydraulic analogy to Ohm's law in electrical terms, $R = V/I$ (see Chapter 4). In laminar flow, this ratio is constant. In thinking about the variables that might govern flow it is intuitive that increasing ΔP or pipe diameter, or decreasing η or pipe length will increase Q. The *Hagen–Poiseuille* equation for laminar flow, which can be deduced from first principles, embodies these concepts, and states:

$$Q = \frac{\pi r^4 \Delta P}{8 \eta L},$$

where r and L are pipe radius and length respectively. (See Chapter 3, where the Hagen–Poiseuille equation has been used to discuss dimensions.) Note the fourth power of radius in the equation, which means that this influences changes in Q to a much greater extent than the other factors. This has clinical importance in relation to flow of fluids down intravenous cannulae, gas flow down endotracheal tubes, and the significance of bronchospasm. A doubling of radius multiplies flow rate by 16. Resistance to turbulent flow is less easily represented mathematically but, as indicated above,

$$Q \propto \frac{\sqrt{\Delta P}}{\sqrt{\rho L}}.$$

These expressions can be used to predict that the velocity at which transition from laminar to turbulent flow occurs is directly proportional to viscosity and inversely proportional to density and to tube diameter. In other words, flow becomes turbulent at a higher velocity (e.g. remains laminar for longer) with more viscous and less dense fluids, and in smaller diameter tubes. As indicated in the Hagen–Poiseuille equation the more the variables change to increase Q ($\uparrow r$, $\uparrow \Delta P$, $\downarrow \eta$, $\downarrow L$), the more likely the flow is to approach turbulence. Turbulent flow also occurs where fluid flow is accelerated or decelerated at changes in diameter or direction of the pipe, such as at entry and exit orifices or bends, and is associated with kinetic energy loss.

Theory of flow measurement devices

While a pipe can be thought of as having a length many times its diameter, an orifice can be considered as a structure with a length many times shorter than its diameter, and fluid behaviour is correspondingly different. Nevertheless, flow of fluid through an orifice has some uses, not least in the measurement of flow. If a U-tube manometer is placed with the outlets each side of the orifice, as shown in Figure 7.2, the pressure downstream is found to be lower than that upstream by an amount proportional to the flow squared; this confirms incidentally that the flow is turbulent through such an orifice.

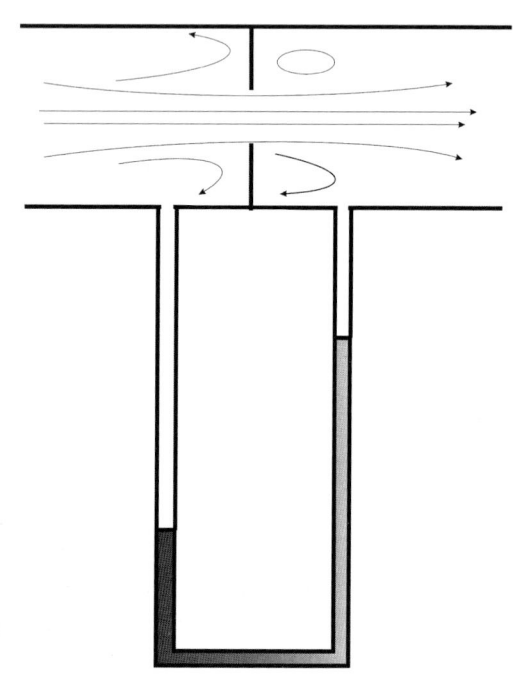

Fig. 7.2 Orifice meter flow measurement device.

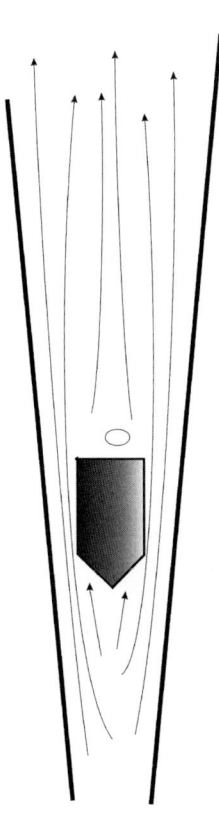

Fig. 7.3 The Rotameter flow measurement device.

If the orifice is turned on its end and the geometry altered slightly, it begins to resemble a flow meter familiar to anaesthetists, the *Rotameter*, shown in Figure 7.3, a device that has its origins in the brewing industry. By altering the tube to a tapering section and inserting a float in the tube, the transformation to a *constant pressure, variable orifice flow meter* is complete. In this application constant pressure caused by the weight of the bobbin is supported by the flowing gas that escapes through the orifice created by the annular gap between the bobbin and the tube wall. An algebraic expression for the upthrust on the bobbin by the gas flow is a function of annular cross-sectional area, gas density, gas viscosity, velocity gradient across the annulus and velocity squared [Hutton et al. 1994]. The higher the flow, the higher up the tapered tube the bobbin is supported to allow the gas to escape through the widening annulus. As in an orifice, the flow here is turbulent and fluids of different densities, such as oxygen and helium, flow at different rates. When gas flows are lower and the bobbin sits lower in the tapered tube, the annular gap is more like a pipe than

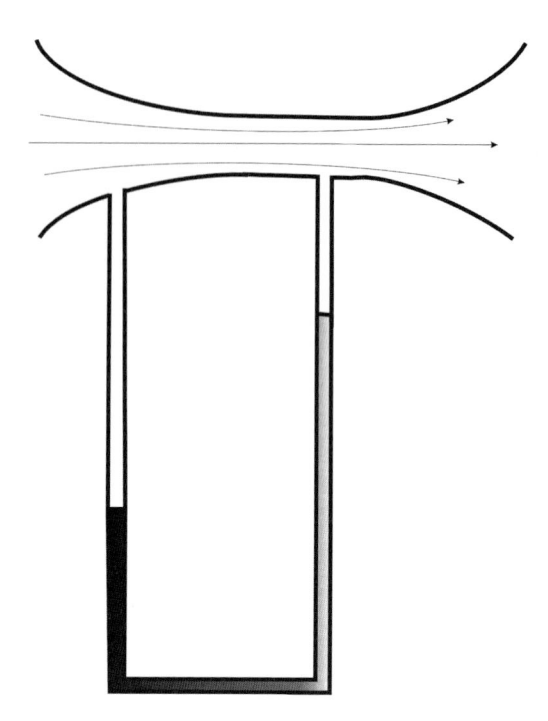

Fig. 7.4 The Venturi flowmeter.

an orifice, and the flow is likely to be laminar. Under these conditions, the fluids with similar viscosities, such as oxygen and helium, will flow at similar rates. Therefore Rotameter tubes are calibrated for specific gases at specified temperatures over a range of gas flow rates.

If the orifice flowmeter is evolved into a slightly different shape, it becomes a venturi, with a smooth transition in pipe diameter at the throat, as shown in Figure 7.4. The pressure at the throat is lower than the upstream pressure for the following reason: a fixed volume flow rate of fluid has a certain mean velocity upstream of the throat. For continuity, the mean velocity has to increase through the reduced cross-sectional area of the throat in order to maintain the same flow rate. The energy increase associated with that velocity change is kinetic energy, which has a velocity squared term in it. For conservation of energy, an increase in kinetic energy is accompanied by a loss of potential energy (see Figure 3.2).

It was *Bernoulli* who explained that, for fluids, the hydraulic potential energy term is expressed as a pressure term, and because kinetic energy increases, the pressure decreases at the throat of the venturi. Bernoulli's equation, which should be thought of as a way of describing conservation

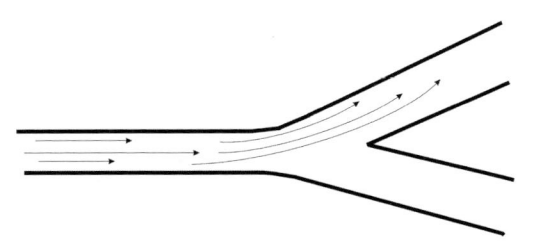

Fig. 7.5 The Coandă effect.

of energy is:

$$\frac{P_1}{\rho} + \frac{v_1^2}{2} = \frac{P_2}{\rho} + \frac{v_2^2}{2}.$$

A venturi can be used as a flow meter like an orifice, but with greater accuracy, because the flow through it is more likely to be laminar with a linear relationship between pressure difference across it and the flow through it as predicted by Hagen–Poiseuille. It can be seen that, if a measurement of $(P_1 - P_2)$ is obtained from a manometer, this gives an equation for the two unknowns v_1 and v_2; the other equation is the continuity equation for flow rate $Q = A_1 v_1 = A_2 v_2$, where v_1 and v_2 are the velocities at cross-sectional areas A_1 and A_2 respectively of the venturi at 1 (entrance) and 2 (throat) of the venturi, giving Q. A venturi has other uses too; for example the pressure drop at the throat of a venturi can be used to entrain another fluid. This is the principle behind medical suction devices, or to provide patients with oxygen enriched air to breathe [Macintosh 1987], or as a means of ventilating a patient through an open bronchoscope. Another related application of the lowering of pressure in a narrowed section of a tube is the Coanda effect, shown in Figure 7.5. where there is a branching of the tube downstream of the low pressure, the flow will tend to adhere to one wall rather than the other; this might be the reason for some causes of hypoxia, where some downstream alveoli are not supplied with gas; it can also be used as a fluidic valve mechanism to switch flows between inspiratory and expiratory ports in a ventilator (see Chapter 26).

When ultrasound is used to measure cardiac output (see Chapters 10 and 13), the device usually also produces a result for aortic gradient. It does this by measuring v_2 (the aortic velocity), assuming v_1 the velocity of blood in the ventricle is zero, and deduces $(P_1 - P_2)$ from Bernoulli's equation. In addition the principle of a venturi is identical to the one describing lift over the curved surface of an aerofoil.

Pressure

Atmospheric pressure is the pressure exerted on us by the air around us; the weight of the atmosphere above us, acted on by gravitational acceleration, means that it behaves like a pressurised container. The pressure in a pressurised container can also be stated in numbers of atmospheres. If the pressure in a cylinder is quoted as 5 ATA, this means it is 5 ATmospheres Absolute, which means that the pressure inside the cylinder will be 4 atmospheres above atmospheric or 4 atmospheres gauge pressure. If the cylinder pressure is quoted as gauge pressure, then its absolute pressure is gauge pressure plus atmospheric pressure.

The numerical value of atmospheric pressure at mean sea level (m.s.l.) is 1.0 bar, or 1.013×10^5 Pa, or 101.3 kPa, or 760 mmHg on a mercury column, or about 10 m on a water column. A reasonable approximation of atmospheric pressure at m.s.l. is 100 kPa, which allows convenient calculation of partial pressures of components in gas mixtures from volume percentage concentrations using Dalton's law of partial pressures (see below).

Columns of liquid in closed tubes, with a vacuum above the fluid level (sometimes referred to as a Torricellian vacuum), are sometimes used to measure pressure. To deduce the expression for pressure in such a column, consider a column of fluid of height h, cross sectional area A and density ρ as shown in Figure 7.6.

The volume of this fluid column is $A\,h$, and its mass is $\rho\,A\,h$. The force exerted by the fluid column at its base, is thus $\rho\,g\,A\,h$. As the figure shows, the base of the fluid column coincides with the surface of the surrounding reservoir, which is acted on by ambient pressure. This pressure can be calculated as the force exerted at the base of the column divided by the area of the column:

$$P = \frac{\rho\,g\,A\,h}{A}$$

or $P = \rho\,g\,h$ N m^{-2} (Pa), which is how we rather loosely refer to pressure as mm Hg or cm H$_2$O. If a dimensional analysis is done on this expression (see Chapter 3 for principles), it will be seen that the units are those of pressure. Pressure difference is sometimes as useful a measurement as absolute values are, and this can be measured using a U-tube manometer, as for example in the venturi or orifice flow meters described above. In a manometer mercury is often the fluid used, and $\rho = 13.6 \times 10^3$ kg m^{-3}, $g = 9.81$ m s^{-2}. At m.s.l., $h = 0.760$ m (760 mm).

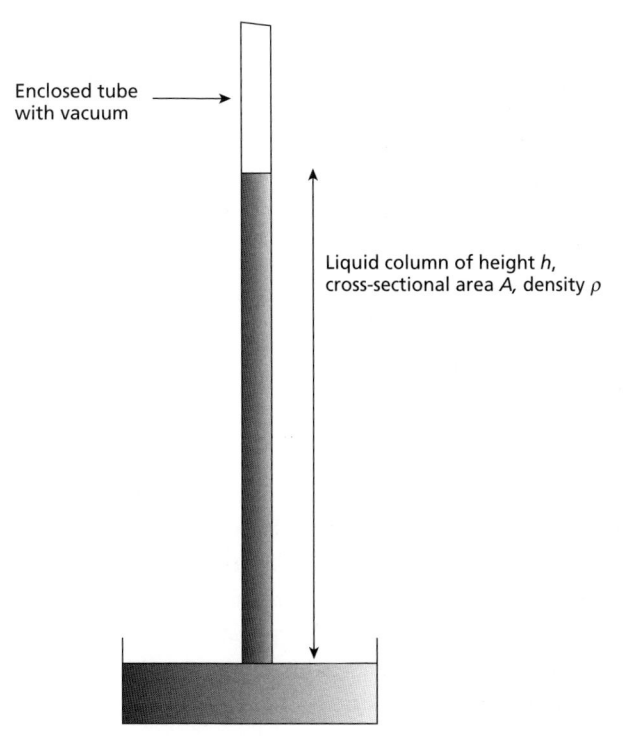

Fig. 7.6 A column of fluid used to measure pressure.

Therefore,

$$\text{pressure at m.s.l.} = 13.6 \times 10^3 \times 9.81 \times 0.760$$
$$= 1.013 \times 10^5 \text{ N m}^{-2}$$
$$= 101.3 \text{ kPa}.$$

Hypobaric and hyperbaric atmospheres

To further illustrate concepts of pressure, consider changes in ambient pressure. Air is compressible and its density diminishes with altitude, making the variation in pressure with altitude a non-linear relationship. The situation is further complicated by the fact that in the troposphere at least (the lowest part of the atmosphere with which we are most familiar), there is a linear temperature change of $-6°\text{C km}^{-1}$ increase in altitude (in aviation this is often quoted as $-2°\text{C}$ per 1000 ft of altitude). At the cruising height of commercial aircraft of 10 000 m, the ambient pressure is approximately one quarter that at m.s.l. (25 kPa).

The alveolar gas equation [West 1979] states that, to an approximation

$$P_{alveolarO_2} = P_{inspiredO_2} - \frac{P_{CO_2}}{R},$$

where R is the respiratory quotient,

$$P_{inspiredO_2} = F_{iO_2} \cdot (P_{ambientO_2} - S.V.P._{H_2O, 37C}),$$

where $S.V.P._{H_2O, 37C}$ is the saturated vapour pressure of water at body temperature, 6.3 kPa, and does not vary with altitude, only temperature (see Chapters 8 and 9). At 10 000 m altitude,

$$P_{inspiredO_2} = 0.21 \times (25 - 6.3) \text{ kPa}$$

$$= 3.7 \text{ kPa}.$$

Clearly the P_{CO_2} will be driven down by the resulting hypoxaemia, but the physiological situation is one of extreme hypoxia in the alveoli, likely to lead rapidly to loss of consciousness and even death. Furthermore the ambient temperature at this altitude is $-50°C$, and is itself a threat to life.

To counteract the possibility of hypoxia, aviators breathe oxygen in increasing fractional concentration with increasing altitude. This continues to be beneficial up to 11 000 m altitude, at which point breathing 100% oxygen at that ambient pressure gives the same alveolar oxygen partial pressure as breathing air at m.s.l. Above this height $P_{alveolarO_2}$ falls further unless the oxygen is breathed through a pressurised mask, a sort of aviators' C.P.A.P. (Continuous Positive Airway Pressure).

Atmospheric pressure continues to increase below ground level, such that it is 270 mbar greater than that at m.s.l. at the bottom of a 2 km mineshaft. The physiological danger here is one of nitrogen loading, and dissolving of nitrogen in body tissues rather than hypoxia, and this is discussed in detail in Chapter 9 on solubility.

Partial pressure

Assume that a container of volume V contains compressed air, which is a mixture by volume of one part oxygen and four parts nitrogen (20% and 80% respectively), and has a pressure of 5 bar (shown in Figure 7.7). Let us then assume that the O_2 and N_2 constituents are separated (for example by a molecular sieve, a physico-chemical device for separating components of a gas mixture), stored separately, and then put back into the volume V sequentially and separately. If the pressure of each of these components were measured in turn, they would be respectively $P_{O_2} = 1$ bar and $P_{N_2} = 4$ bar, and these

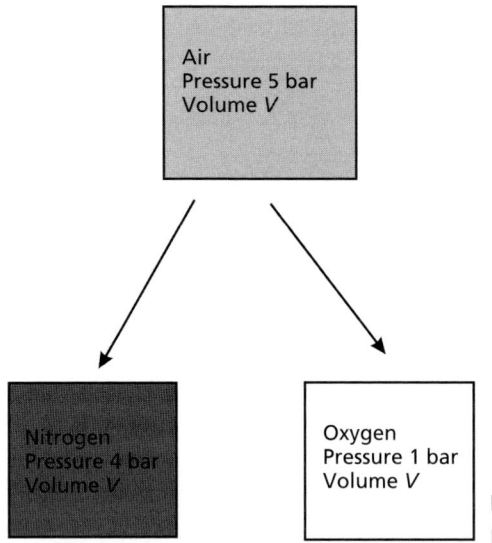

Fig. 7.7 Dalton's law of partial pressures.

are the partial pressures of the constituents of a mixture of gases. In this case $P_{O_2} + P_{N_2} = P_T$, the total pressure.

Dalton's law of partial pressures states that in such a mixture of gases each gas exerts the same partial pressure within the mixture as if it alone occupied the whole volume of the mixture, and the sum of partial pressures of the constituents equals the total pressure of the mixture. Moreover, the ratio of partial pressures of the constituents to the total pressure in the container equals the ratio of partial volume to total volume. Put algebraically, if a gas mixture consists of gas components a, b, c, ... n, then,

$$P_a + P_b + P_c + \cdots + P_n = P_{Total}$$

and, applying Boyle's law (see Chapter 8),

$$\frac{V_i}{V_T} = \frac{P_i}{P_T},$$

where $i =$ a, b, c, ... n.

Thus by Dalton's law of partial pressures, the atmosphere, which is composed of 21% oxygen and 78% nitrogen by volume, therefore contains an oxygen partial pressure of about 21 kPa and a nitrogen partial pressure of about 78 kPa, the difference between the sum of these two partial pressures and atmospheric pressure (101 kPa) being due to the presence of argon.

The concept of partial pressure is important for a number of reasons, not least because the anaesthetic or metabolic effect of a gas depends on its partial pressure rather than its concentration (see Chapter 9, concerning vaporiser output at altitude).

Diffusion

Molecules in liquids, and to a greater extent those in gases, are sufficiently far apart with enough kinetic energy that they move randomly and widely in the space available. Such motility allows the mixing of two families of molecules (say oxygen and nitrous oxide, or two different concentrations of saline), initially separated by a barrier that may be simply an intervening medium or it may be a semipermeable membrane. The mixing, or diffusion, can occur up to the point where the species are evenly mixed. A molecular transfer process takes place between two gas mixtures or species across an intervening medium or a membrane, such as the alveolar-capillary membrane, or within the alveolus itself. The rate of diffusion depends on the concentration gradient across the medium or the membrane and, as indicated in the previous section, this really means partial pressure difference when a gas is being considered, or concentration gradient in the case of liquid diffusion (see Figure 7.8, see below).

The greater the concentration difference of a substance between two parts of a container, whether or not separated by a membrane, the greater the rate of diffusion per unit area from a higher to lower concentration regions, and this applies even if the membrane separates two different media, e.g. a gas

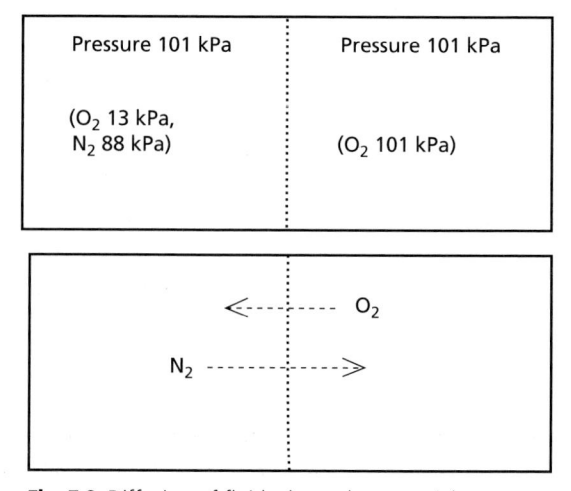

Fig. 7.8 Diffusion of fluids depends on partial pressure.

mixture and blood in the case of the alveolar-capillary membrane. This is *Fick's law of diffusion*, which also says that the rate of diffusion is directly proportional to the membrane area and inversely proportional to its thickness. Note that as the molecular species diffuses, the concentration or partial pressure gradient diminishes and so does the rate of diffusion. Clinical examples include oxygen and carbon dioxide diffusion across the alveolar membrane in opposite directions. As discussed in an earlier section, bulk transport of a fluid takes place along a pressure gradient. Diffusion, or molecular transport, can occur without the presence of a global pressure gradient if the concentration gradient or partial pressure gradient of the individual species exists as shown in Figure 7.8. If a semi-permeable membrane separates two different groups of molecules at equal pressures, and the membrane is permeable to both groups, diffusion can occur across the membrane. Such semi-permeable membranes vary in both their absolute permeability to and their rates of diffusion of different substances.

It will be recalled that the diffusion capacity of the lung is tested using carbon monoxide, as the diffusion of this gas across the alveolar-capillary membrane is the rate limiting step for the uptake of this particular gas. Another clinical example of diffusion is the risk of diffusion hypoxia following an anaesthetic, which includes a significant concentration of nitrous oxide; following a nitrous oxide anaesthetic there is a diffusion gradient for N_2O from the tissues to the lungs; if a patient is breathing air, 78% of which is nitrogen, there is a diffusion gradient for N_2 from the lungs into the tissues; however, N_2 is less soluble than N_2O in tissues, which means the predominant diffusion is by N_2O from the tissues into the lungs, which displaces N_2 from the lungs and its accompanying oxygen. For the same reason we avoid the use of N_2O in patients with body spaces containing nitrogen (air), such as a pneumothorax. Solubility of gases in liquids is another important topic in relation to the behaviour of gases and liquids, particularly in relation to altering diffusion gradients, and will be discussed in another section.

There is another aspect that governs the rate of diffusion and that is the size of the molecules. Bigger molecules diffuse more slowly than smaller ones. *Graham's law of diffusion* states that the rate of diffusion of a gas is inversely proportional to the square root of either its density or its molecular weight.

Osmosis

Imagine a semi-permeable membrane separating two families of molecules (perhaps a glucose solution and water), which is permeable to one (say water but not glucose). Water diffuses across the membrane and dilutes the glucose

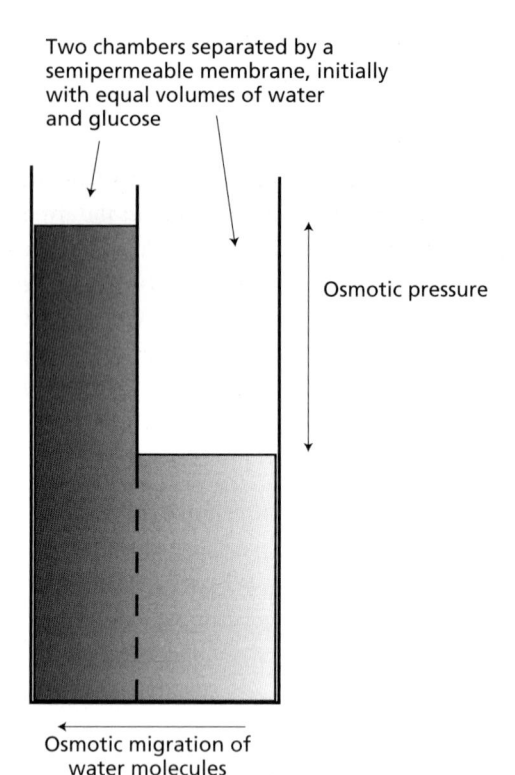

Fig. 7.9 Osmosis and osmotic pressure.

Two chambers separated by a semipermeable membrane, initially with equal volumes of water and glucose

Osmotic pressure

Osmotic migration of water molecules

solution until the hydrostatic pressure on that side of the container has risen to a level higher than that on the water side of the membrane, as shown in Figure 7.9. This hydrostatic pressure difference prevents further diffusion of water across the membrane, and is called the osmotic pressure. The process of water diffusion is called osmosis in this context.

The generation of an osmotic pressure depends on the number of particles in a solution, and their size does not matter, assuming the membrane is not permeable to them. An NaCl solution has Na^+ and Cl^- ions, two ions in solution per molecule of solute, and therefore has twice the osmotic power of, say, a glucose molecule, which does not dissociate, but is a much larger molecule. A 0.5 molar solution of NaCl has the same osmotic power as a 1 molar solution of glucose. One mole of a substance which does not dissociate, dissolved in 22.4 L of solvent, exerts an osmotic pressure of 1 bar gauge pressure (this is a corollary of Avogadro's hypothesis) [Davis et al. 2003]. The term 'osmolarity' is used to describe the potential for exerting osmotic pressure because of the different types of molecules that exist in the body fluid compartments, each with a different degree of osmotic potential. The osmolarity of the plasma, the interstitium and most other body fluids is around 300 milliosmol L^{-1}, and we give intravenous fluids that are isotonic (with equal osmolarity) to avoid fluid shifts by diffusion between, say, the circulation and the extracellular or intracellular space. The plasma also contains large molecule proteins; these contribute only a small proportion of the osmotic pressure of the plasma, the majority being due to electrolytes; the capillary membrane is impermeable to protein but not to electrolytes, so protein molecules form the osmotic pressure to drive water molecules into the plasma, sometimes referred to as 'oncotic pressure'. At the arteriolar end of the tissue microcirculation the intravascular hydrostatic pressure usually exceeds oncotic pressure so that fluid loss to the tissues occurs; at the venular end of the tissue microcirculation, the intravascular hydrostatic pressure is exceeded by the oncotic pressure, and tissue fluid returns to the circulation. Osmotic pressure within the renal tubules governs the way the kidney handles water and electrolytes.

The term *osmolality* is often used rather than osmolarity, and is defined as millimol solute per kg of solvent (rather than millimol per litre of solution); this is preferred because it avoids the need to consider volume changes in solutions due to temperature change. When icy roads are gritted, the presence of particles in the water lowers the freezing point of water by 1.86°C for each milliosmol per kg of water. Osmolality, usually of urine or plasma, is measured using the same principle of freezing point depression in an osmometer. This device allows a few ml of solution to be collected in a container while being cooled and stirred; initially the temperature falls below the freezing point (supercooling) until the freezing process starts in the water of one part of the solution; stirring allows the freezing process to spread throughout the solution, during which the temperature rises to the value that is the freezing point for the solution.

Dissolving such a solute in a solvent and raising its osmotic pressure also reduces the vapour pressure of the solvent, and this is *Raoult's law*. This is thought to be because the presence of the solute molecules reduces the surface area of the solvent at its surface to allow them to vaporise, thereby reducing the vapour pressure [Davis 2002] (see Chapter 9); the corollary of this is that the boiling point of the solution is raised.

References

Davis P. D. and Kenny, G. N. C. (2003) *Basic Physics and Measurement in Anaesthesia*, 5th edn, Butterworth Heinemann, Edinburgh, pp. 75–85.

Dorrington, K. L. (1989) *Anaesthetic and Extracorporeal Gas Transfer*, Clarendon Press, Oxford, pp. 4–7.

Hutton, P. and Prys-Roberts, C. (eds) (1994) *Monitoring in Anaesthesia and Intensive Care*, WB Saunders, London, p. 186.

Macintosh, Mushin & Epstein (1987) *Physics for the Anaesthetist*, 4th edn, Revised Mushin, M. W. and Jones, P. L., Blackwell Scientific Publications, Oxford, pp. 73–76.

Massey, B. S. (1970) *Mechanics of Fluids*, 2nd edn, Van Nostrand Reinhold, London.

West, J. B. (1979) *Respiratory Physiology, the Essentials*, 2nd edn, Williams and Wilkins, Baltimore, p. 67.

Chapter 8

Thermodynamics: heat, temperature and humidification

> This chapter contains: the gas laws; specific heat capacity; latent heat of vaporisation and of crystallisation; temperature; thermometry; heat transfer; keeping patients warm; humidity and humidification.
> The chapter links with: Chapters 1, 7, 9 and 14.

Introduction

In continuing the concepts of energy and work, heat is a form of energy, and heat and work are related by the laws of thermodynamics. The first law states that the amount of mechanical work achievable out of a thermodynamic system going through a cyclic process is proportional to the amount of heat put into it, which is intuitive. The second law states that the amount of work gained from the system is always less than the amount of heat put into it; in other words there is some energy wasted, which is also believable. Heating a substance gives its constituent molecules increased kinetic energy, which can be put to mechanical use, for example in driving a turbine. Such an increase in energy either raises its temperature or changes its state. The increasing energy causes increasing velocity of movement of molecules and eventually breaks intermolecular bonds, giving the molecules more freedom of movement. As characteristic temperatures are reached, e.g. melting point or boiling point, the energy increase goes into changing the substance's state into a more energetic one, e.g. solid to liquid, or liquid to gas, without any associated temperature change. Loss of heat from a substance involves the reverse processes. The gas laws have their origin in this concept of the kinetic theory of gases, and are an inherent part of consideration of the effects of heating and cooling, compression and expansion of gases.

The gas laws

Pressure has been defined in Chapter 7, but to understand the concept, one that is taken for granted, it is necessary to start with the idea of molecules of matter undergoing a change in momentum (the physical definition of momentum being mass × velocity, see Chapter 3). It is unnecessary here to deduce the following equation, but the proof can be found in any physics or engineering text [Macintosh et al. 1987; Dorrington 1989; Hill 1976]).

$$PV = \frac{Nmv^2}{3}$$

is the equation based on the molecular theory of gases from which a concept of pressure P (force/area) is derived. V is the gas volume, N is a constant referring to the number of molecules of gas, and all other terms are as used previously; a constant mass of gas is assumed. Notice that both sides of this equation contain energy terms: pressure × volume on the left hand side, and a kinetic energy term on the right hand side. This expression suggests that pressure is inversely proportional to gas volume, directly proportional to total mass and to the square of molecular velocity. Conceptually, doubling the velocity doubles the momentum changes, but it also doubles the number of intermolecular collisions, hence the squared relationship to pressure. Molecular kinetic energy is a function of temperature (absolute temperature, measured in kelvin). Thus at constant temperature: $PV =$ constant, and this is Boyle's law.

Each P–V curve corresponds to a constant value of T (an isotherm) for a fixed temperature and two points on the same curve can be thought of as being governed by the same rectangular hyperbola, i.e. $P_1 V_1 = P_2 V_2$; these changes are isothermal (constant temperature) by definition. The graphical relationship between P and V is shown in Figure 8.1(a). Under these conditions the equation may be used to work out the volume of gas released (V_2) at atmospheric pressure ($P_2 = 1$ bar) from a cylinder of known size (size E, $V_1 = 5$ L), whose contents are stored at known pressure ($P_1 = 138$ bar absolute). A family of P–V curves for different values of T may be drawn and usefully used to illustrate the gas's behaviour for a range of temperatures and this is shown in Figure 8.1(b). It can be seen that, for increasing temperature, the curves become shifted upwards to the right, suggesting increased thermal energy as indicated by increased area under the P–V curves. Under these conditions the combined gas law is sometimes written as $PV = nRT$, with each curve corresponding to a different value of T, where n is the number of moles of gas, R is the universal gas constant and T is the temperature in kelvin.

Just as Boyle's law assumes constant temperature, there are other conditions that can be preset, and the gas law can be expressed differently. For example:

at constant pressure, $V =$ constant$_1$ × T and this is Charles' law;

at constant volume, $P =$ constant$_2$ × T and this is Gay-Lussac's law.

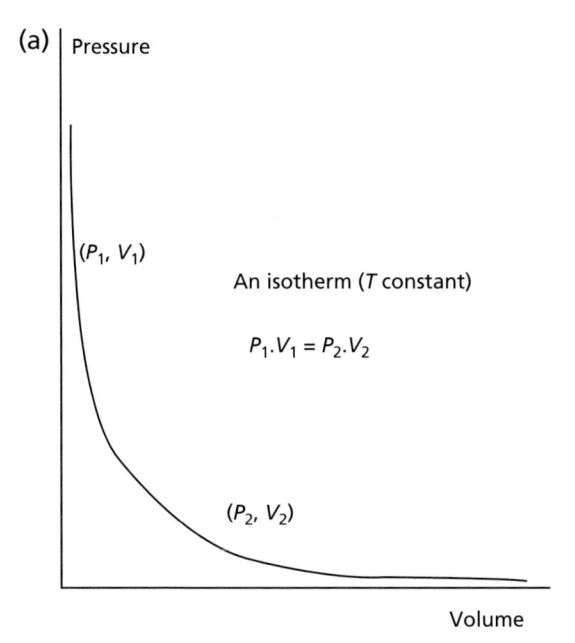

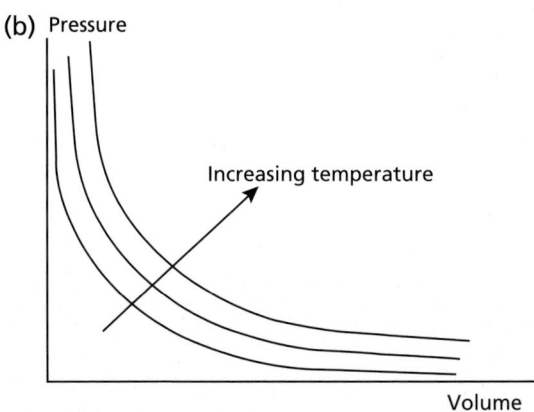

Fig. 8.1 $PV = nRT$: (a) showing a single isotherm (T constant). (b) showing several curves, each corresponding to a value of T.

Each of these represents a straight line relationship between temperature and either volume or pressure, and these are shown in Figures 8.2(a) and 8.2(b). In the days when these laws were deduced, having done some measurements and obtained a straight line, consideration was given to the practical significance of the points at which volume or pressure became zero. This was when the

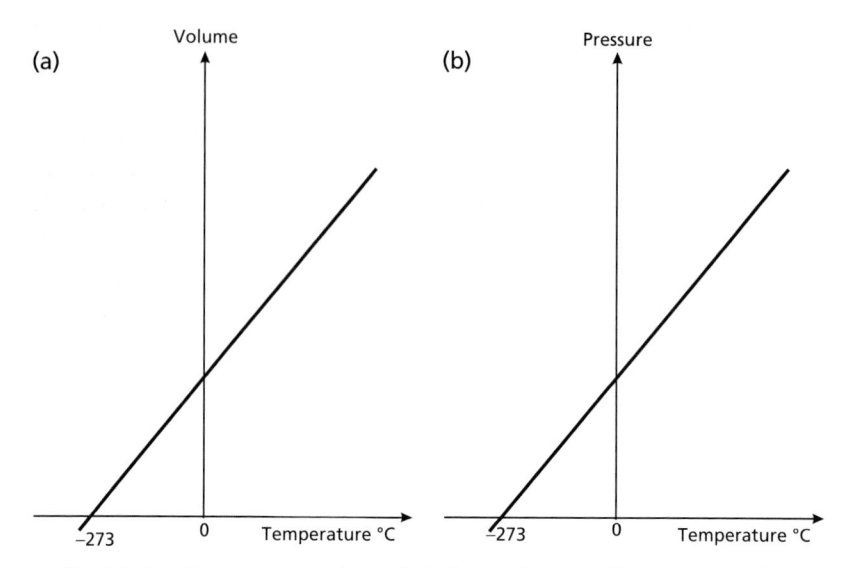

Fig. 8.2 Graphical representations of (a) Charles' law and (b) Gay-Lussac's law.

concept of *absolute zero* of temperature was recognised and from which the Kelvin temperature scale evolved.

As indicated above, the combined gas law is often expressed as $PV = nRT$, and another way of expressing this is:

$$\frac{P_1 V_1}{T_1} = \frac{P_2 V_2}{T_2}.$$

This equation is used to calculate changes in conditions of a perfect gas when heated or cooled, or expanded or compressed. We usually think of such changes as being isothermal, when $T_1 = T_2$. It must be remembered that temperature is calculated in K (K = °C + 273) when the gas laws are being utilised. Gases can have energy applied to them in different ways. Rapid compression of a gas results in molecules being forced together without adequate time for heat exchange to occur, and they are unable to move adequately fast for their particular level of energy, and so the energy is converted into thermal energy as a temperature rise; such compression is termed *adiabatic*. If the compression is applied slowly enough, the excess thermal energy is dissipated to the environment outside the container, the temperature rise is minimal or zero, and the compression is isothermal.

If, instead of compressing the gas, heat is applied, then if the volume of the container is kept constant the pressure will rise. Alternatively the volume of the container may be allowed to increase by way of a piston, such that the pressure remains constant. Under these different scenarios, the pathway of pressure and volume that a gas follows will differ depending on the circumstances indicated.

A perfect gas is one that is assumed to behave in a way described above, with perfectly elastic collisions between molecules and container walls, with no intermolecular forces of attraction, and the size of molecules being negligible. However a real gas is not a perfect gas under all circumstances and there are, for example, times when a vapour behaves like a gas. Deviations start to occur from the rectangular hyperbolic relationship between P and V of a perfect gas towards the bottom left hand corner of the curve, close to where the axes meet as shown in Figure 8.3; in other words, at low volumes, low pressures and low temperatures (these are context sensitive terms depending on the substance being considered), where cool, slow moving gas molecules are compressed initially into a vapour, then into a liquid, when the volume shrinks dramatically

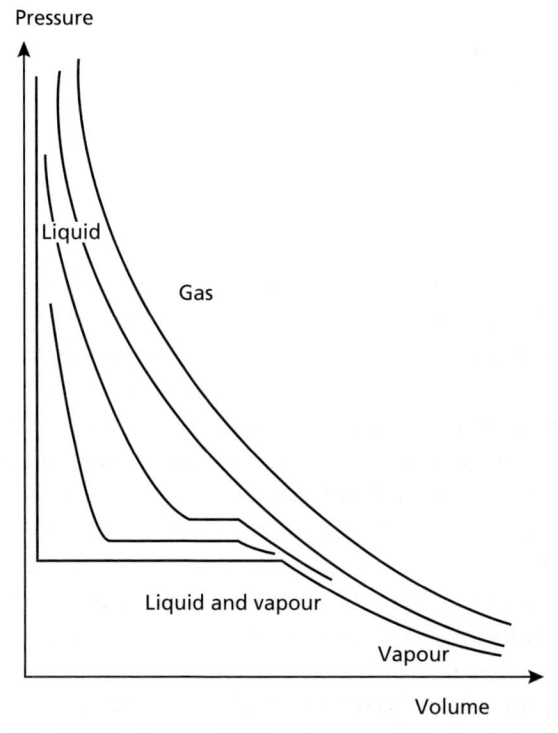

Fig. 8.3 The bottom left hand corner of the pressure–volume curve.

as the molecules coalesce, then becomes incompressible as a liquid, when the pressure rises rapidly.

From this position if the substance is heated to a higher temperature and the volume is allowed to expand, it is still possible to subsequently re-compress the molecules into a liquid. However if the temperature of the substance is raised beyond a particular point, called the *critical temperature* for that substance, i.e. to a P–V curve displaced upwards and to the right, then no matter how much compression occurs it cannot be re-liquefied. The substance has too much thermal energy for compression to result in coalescence of fast moving molecules into liquid. The substance can be considered to be a gas rather than a vapour above this critical temperature. Oxygen is thought of as a gas under most circumstances because its critical temperature is low, −116°C. Water, on the other hand, has a critical temperature of 374.1°C, so it is difficult to experience water in its high energy state as anything other than steam (a vapour, not a gas). Nitrous oxide has a critical temperature of 36.5°C, which may be close to room temperature in some climates. The *critical pressure* is the pressure needed to liquefy a substance at just below its critical temperature.

Sublimation occurs when a substance goes straight from the solid state to the vapour state without going through the liquid state. This can occur, for example with water under very low pressures, and is used to carry metabolic heat from an astronaut's life support system to the ambient vacuum. The *triple point* of a substance occurs when solid, liquid and vapour states of a substance are in equilibrium, as discussed below in relation to water.

Specific heat capacity

The specific heat of a substance in J kg^{-1}°C^{-1}, is the quantity of heat required to raise 1 kg of the substance through a temperature rise of 1°C. Different substances have different specific heats and in general solids and liquids have higher specific heats than gases, which means they need more heat energy to raise their temperature by a given amount.

Another way of thinking about specific heat is the amount of work done on 1 kg of a substance to raise its temperature by 1°C. Since work can be done on a gas, either by keeping pressure constant or volume constant, specific heats for gases are quoted under these two conditions respectively and are known as C_p and C_v. While the gas law $P.V = \text{constant}$ for isothermal processes is usually used, the law $P.V^\gamma = \text{constant}$, where $\gamma = C_p/C_v$ is used under less well defined thermodynamic processes. For air $C_p = 1.00$ J kg^{-1}°C^{-1}, $C_v = 0.714$ J kg^{-1}°C^{-1}.

Latent heats of vaporisation and of crystallization

The amount heat required to increase the energy of a fluid to change its state from liquid to vapour without any temperature change is called the *latent heat of vaporisation*. At the other end of the spectrum involving change of state, *heat of crystallization* is liberated with no temperature change when a substance crystallizes or freezes from a liquid to a solid state. Likewise the same heat has to be supplied to ensure melting from solid to liquid state with no temperature change.

Temperature

Heating up a solid, liquid or gas gives their molecules increased kinetic energy, and this results in a temperature rise. Thus a temperature change is the effect of heating or cooling a body. The section on the gas laws above has already described how temperature change can occur due to pressure and volume changes under *adiabatic* conditions, where no additional heat is exchanged with the gas, and under *isothermal* conditions, where the process occurs slowly enough that temperature remains constant. The same pressure and volume changes can be achieved by heat supply. The S.I. unit of heat is the same as that of energy, namely J, although the kilocal or Cal is still sometimes used, where 1 Calorie = 4.186 kJ.

The temperature scale used now tends to be °C (after Anders Celsius, 1742, or the centigrade scale), chosen because under standard conditions water freezes at 0°C and boils at 100°C. However, gas law calculations require the use of the Kelvin scale of temperature (after Lord Kelvin, 1848), where 0 K = −273°C, *absolute zero*, and one degree change Kelvin equals one degree change Celsius. This can be deduced by plotting volume against temperature (Charles' law), or pressure against temperature (Gay-Lussac's law), and extrapolating to where volume or pressure reach zero. Temperature is that property by which we may quantify the heat energy of a substance. The Kelvin scale was adopted as an international temperature scale in 1954 [Rogers and Mayhew 1970]. To do this a fixed point of physical equilibrium is chosen as the reference point for the temperature scale, the triple point of water. This is the point at which all three phases of water (ice, water, steam) are in equilibrium with each other, and although the pressure at which this occurs is very low (0.006 bar), the temperature at which this occurs is only fractionally greater than that of the ice point at atmospheric pressure (1.013 bar). It is considered to be a reproducible physical state against which to calibrate a temperature scale. The internationally agreed temperature 'number' of the triple point of water is 273.16. Hence the unit of thermodynamic temperature (the Kelvin) is the fraction 1/273.16 of the thermodynamic temperature of the triple point of water. By this definition the difference in temperature between the ice point for water and the steam point is 100 units, which makes the range almost identical with the previous empirical Celsius scale. Almost identical but not precisely, because this scale has its datum at 273.15 K, i.e. 0.01 K below the triple point. Although the unit on the thermodynamic Celsius scale is identical to that on the Kelvin scale, it is usual to denote 273.15 K as 0°C.

The temperature at which human metabolic processes work most efficiently is 37°C. If an anaesthetist allows the patient's temperature to fall intraoperatively, there are physiological consequences.

Thermometry

Traditionally temperature was measured by the mercury in glass thermometer, where mercury expands with heating and rises in a glass column on a linear scale, with a maximum operational range of −39°C (the freezing point of mercury) to 250°C. An alternative liquid is alcohol, with a maximum operational range between −117°C and 78°C. Mercury is very dense and covers a wide range of temperatures, while alcohol is less dense and expands the scale of measurement over a small range of temperatures.

The thermistor is a semiconductor device, whose electrical resistance decreases with temperature rise, more or less linearly (when it forms one arm of a Wheatstone bridge) over a reasonable temperature range. It is certainly adequate for measuring body temperature, and forms the functional basis of the tympanic membrane thermometer and can be used for measuring naso-pharyngeal temperature [Edge et al. 1993, Roth et al. 2008].

The thermocouple is a device consisting of two strips of dissimilar metals, connected together at both ends, each end being immersed in a fluid at different temperature; one is likely to be a reference temperature of 0°C. This arrangement forms an electrolytic cell, and the *Seebeck effect*, important in the thermocouple function, is the induction of a voltage (and subsequently a current), proportional to the difference in temperature between the two junctions; the relationship between temperature difference and induced voltage is a nonlinear one.

Liquid crystal displays, with temperature sensitive colour changes in the display, have also been used for skin temperature measurement. They have been found to demonstrate hysteresis and are draught sensitive [MacKenzie et al. 1994].

Other eponymous effects in relation to heat include:

Joule effect: the heating effect of an electrical conductor when current flows through it.

Peltier effect: when electrical current is passed through a single junction of two dissimilar metals, heat is released or absorbed, depending on the direction of current flow.

Thompson effect: the liberation or absorption of heat caused by the passage of electrical current along a conductor with a temperature gradient.

Heat transfer
Allan 1995, Sullivan *et al.* 2008

Patients gain and lose heat by convection, conduction, radiation, vaporisation in the respiratory tract and in sweat.

Convection

Convection is heat transfer through a fluid medium such as air or water. This occurs because the warmer molecules change their position within the fluid, for example by floating to the top while cooler ones sink.

Conduction

Conduction is heat transfer through a solid medium. If a metal object is heated at one end, this heat is transferred towards the other end as a result of the temperature gradient between the ends. This occurs because thermal energy is transferred between neighbouring atoms.

Radiative heat loss

Radiation from bodies, to different extents emit, absorb, reflect or refract electromagnetic energy, depending on temperature and the nature of their surface. Such energy includes light waves and also heat as infrared waves. The human loses 70% of body heat by radiation.

Vaporisation

In the respiratory tract body heat is given up to body water as latent heat of vaporisation, which results in vaporisation of respiratory tract water, which carries heat away. The same applies when sweat is produced and vaporised by surface body heat.

Keeping patients warm

Operative surgery and anaesthesia usually involves heat loss from the patient, through exposed tissue, vasodilatation, cold intravenous fluid administration, and cold anaesthetic gas ventilation. Anaesthesia also attenuates the efficiency of the patient's thermoregulatory system mediated through the hypothalamus. Therefore it is important that conductive heat loss is not made worse by laying the patient on a highly conducting surface, or convective heat loss is not worsened by placing the patient in a circulation of cold air. Since the patient's physiological well-being depends on maintaining a core temperature of 37°C, active steps have to be taken to keep the patient warm.

Forced warm air blankets are now used to cover the patient as far as surgical access will allow. The air temperature is maintained at between 32 and 38°C and it is pumped into channels in a thin walled plastic blanket. This reduces the convective temperature gradient and accompanying heat loss, and also acts as a radiative reflective surface. The blankets have proved to be a very efficient means of warming the patient [Müller 1995, Stanger et al. 2009]. A conductive heat pad has also been re-introduced as an intraoperative heating device, but has been found to be less effective than forced air warming [Leung et al. 2007].

Intravenous fluids are warmed by passing the fluid giving set through some form of heat exchanger [Johnson et al. 1994]. This can be achieved by wrapping the coils of an extension of the giving set around a heater or putting them inside a heating device. Alternatively, the giving set extension is formed as a coaxial pair of tubes, through the inner one of which passes the fluid intended for intravenous administration, and through the outer one of which circulates warmed water. More recently a conductive heat exchanger has been introduced that applies a heating element directly to the fluid channels [Turner et al. 2006].

As indicated in the section on low flow breathing systems (Chapter 25), breathing gas can be warmed and humidified by the use of a heat and moisture exchanger, described below. Furthermore, the use of low anaesthetic gas flows reduces the heat loss through that route and the chemical process of CO_2 absorption by 'Soda Lime' is exothermic.

Humidity and humidification

Air that is totally saturated with water vapour at a given temperature is said to be fully saturated, or to have relative humidity of 100%. The absolute humidity is the number of gram of water per cubic metre of air, or mg l^{-1}. At 37°C a fully saturated sample of air (100% humidity) has an absolute humidity of 44 mg l^{-1}. This mass of saturated water vapour can be shown by the gas laws to have an associated vapour pressure (its SVP) of 6.3 kPa (47 mm Hg). A relative humidity (r.h.) of 50% implies half of this mass of water vapour and also a partial pressure of water vapour at this temperature of half the SVP, so relative humidity can also be described as the ratio of the vapour pressure of the sample to the SVP at that temperature. As air cools, the SVP of the water vapour falls. This means the air is able to contain less water vapour, so the relative humidity increases for a given water content. It also means that water condenses out of the air as its temperature falls. As the air is reheated, the SVP of water rises and the water evaporates again.

The humidity of air can be measured by either the *Regnault hygrometer* or the *wet and dry bulb hygrometer* [Mushin et al. 1987]. This consists of an enclosed thin silver tube containing ether and a thermometer to measure the temperature of the ether. There is also a tube through which air can be pumped into the ether. As this happens, the ether evaporates and its temperature and that of the silver tube surface falls. Cooling continues until the air adjacent to the outside surface of the tube becomes saturated with water vapour, which condenses on its outside surface (the *dew point*). This is the temperature at which the air, containing the amount of water vapour it does at room temperature, becomes fully saturated with water vapour. Reference to a table shows the partial pressure of water vapour for full saturation at the lower temperature, which is the same as the unsaturated vapour pressure at room temperature. The room temperature SVP can also be referenced, and the ratio of the two vapour pressures is the relative humidity.

The wet and dry bulb hygrometer consists of a pair of thermometers, the bulb of one of which is covered in water soaked cloth. Evaporation of the water results in a fall in temperature of the said thermometer, at a rate that depends on the humidity of the surrounding air, being greater at lower humidity. The difference between the temperature readings of the two thermometers is related to the relative humidity and is read off a table.

Humidification

Natural humidification occurs when inspired air or gas is heated by the respiratory tract itself. The heating of the air gives greater water vapour carrying capacity. At the same time evaporation of water from the mucosal surface of the respiratory tract increasingly humidifies the air, increasing the water vapour pressure until the SVP for water is reached at body temperature P_s, namely 6.3 kPa (47 mm Hg) at 37°C. The air is now fully saturated (100% humidity) at that temperature.

Where the ambient pressure P_{amb} is 101 kPa, the fractional concentration of water vapour is:

$$\frac{P_s}{P_{amb}} = \frac{6.3}{101} = 6.2\%, \quad \text{equivalent to 44 mg l}^{-1}.$$

At 21°C, fully saturated air contains water vapour at 2.5 kPa (19 mmHg), equivalent to 2.4% or 18 mg l^{-1}.

Humidification is important in the operating room; a minimum of 50% r.h. is required to ensure electrical antistatic precautions. Within the respiratory tract, it is important to have adequate humidification to avoid drying of secretions, particularly in small airways and small patients. Secretions can also block the endotracheal tube. There is also considerable heat loss from the patient in breathing in cold, dry gas. Body heat can be lost at the rate of 18 W if high flow breathing systems are used, delivering cold, dry respiratory gas to the patient. It is therefore highly desirable to heat the gas to body temperature or slightly above, and to humidify it to 100%.

Humidification systems

Older systems were designed to pass a proportion of the breathing gas over the surface of warm water, rather like the crudely designed anaesthetic vaporizer referred to in Figure 9.2a, before delivering the gas to the patient down a delivery tube. In the tube the humidified gas cools, and some of the water vapour condenses, releasing the heat of condensation; this minimises the heat loss from the gas. The system was relatively complex and needed different design constraints for small bore or longer tubing, or different gas flow rates if inefficiency or scalding were to be avoided [Oh *et al.* 1993; Harrison *et al.* 1995; Ognino 1985].

Modern devices are simple heat and moisture exchangers (HME), in which porous material absorbs heat from the exhaled gas, allowing water to condense out and be stored in or adjacent to the porous material. When the patient inhales, the heat is returned to the incoming gas, and the condensate evaporates. The work of breathing in some of them can be increased when they are wet [Quinn *et al.* 1994; Hedley *et al.* 1994], the need for an HME is less in low flow anaesthesia than in high flow [Henriksson *et al.* 1997] and the presence of an HME can hinder anaesthetic gas flow [Da Fonseca *et al.* 2000].

Nebulisers are used to produce small liquid droplets, either to humidify or to deliver medication to the respiratory tract. They can be of three types;

- gas driven nebulisers consist of a gas jet delivered through a venturi, which draws water from a capillary in the low pressure region just beyond the venturi. The droplets then hit an anvil, which breaks them up into microdroplets of 2–4 μm diameter;
- spinning disc nebulisers allow microdroplets to be formed from larger droplets being broken up by the centrifugal force of a spinning disc;
- an ultrasonic nebuliser allows water to be dropped on to an ultrasonic transducer head, vibrating at, say, 3 MHz, producing 1–2 μm droplets [Al-Shaikh *et al.* 2002].

References

Allan J. R. (1995) The thermal environment and human heat exchange, Chapter 16 in *Aviation Medicine*, 2nd edn, Butterworth Heinemann, Oxford, pp. 219–34.

Al-Shaikh, B. and Stacey, S. (2002) *Essentials of Anaesthetic Equipment*, 2nd edn, Churchill Livingstone, London, pp. 106–107.

Da Fonseca, J. M. G., Wheeler, D. W. and Pook, J. A. R. (2000) The effect of a heat and moisture exchanger on gas flow in a Mapleson F breathing system during inhalational induction, *Anaesthesia*, 55, pp. 571–73.

Dorrington, K. L. (1989) *Anaesthetic and Extracorporeal Gas Transfer*, Medical Engineering Series, Oxford Science Publications, Oxford, pp. 4–7.

Edge, G. and Morgan, M. (1993) The Genius infrared tympanic thermometer, *Anaesthesia*, 48, pp. 604–607.

Harrison, D. A., Breen, D. P., Harris, N. D. and Gerrish, S. P. (1993) Performance of two IC humidifiers at high gas flows, *Anaesthesia*, 48, pp. 902–905.

Hedley, R. M. and Allt-Graham, J. (1994) Heat and moisture exchangers and breathing filters, *Br. J. Anaesth.*, 73, pp. 227–36.

Henriksson, B-Å., Sundling, J. and Hellman, A. (1997) The effect of a heat and moisture exchanger on humidity in low flow anaesthesia, *Anaesthesia*, 52, pp. 144–49.

Hill, D. W. (1976) *Physics Applied to Anaesthesia*, 3rd edn, Butterworths, London, pp. 226–28.

Johnson, A. L. M. and Morgan, M. (1994) An assessment of a blood warmer for high and low flows. *Anaesthesia*, 49, pp. 707–709.

Leung, K. K., Lai, A. and Wu, A. (2007) A randomized controlled trial of the electric heating pad vs. forced air warming for preventing hypothermia during laparotomy, *Anaesthesia*, 62, pp. 605–608.

Macintosh, Mushin and Epstein (1987) *Physics for the Anaesthetist*, 4th edn, Revised Mushin, M.W. and Jones, P.L., Blackwell Scientific Publications, Oxford, pp. 73–76.

MacKenzie, R. and Asbury, A. J. (1994) Clinical evaluation of liquid crystal skin thermometers, *Br. J. Anaesth.*, 72, pp. 246–49.

Mûller, C. M. *et al.* Forced air warming maintains normothermia during orthoptic liver transplantation, *Anaesthesia*, 50, 1995, pp. 229–32.

Mushin, W. W. and Jones, P. L. (1987) *Physics for the Anaesthetist*, 4th edn, Blackwell Scientific Publications, Oxford, pp. 125–27.

Ognino, M, Kopotic, R. and Mannino, F. C. (1985) Moisture conserving efficiency of condenser-humidifiers, *Anaesthesia*, 40, pp. 990–95.

Oh, T. E., Lim, E. S. and Bhatt, S. (1991) Resistance of humidifiers and inspiratory work imposed by a ventilator–humidifier circuit, *Br. J. Anaesth.*, 66, pp. 258–63.

Quinn, A. C., Newman, P. J., Allt-Graham, J. and Fawcett, W. J. (1994) In vivo assessment of the work of breathing through dry and wet heat and moisture exchangers, *Br. J. Anaesth.*, 72, suppl, European Society of Anaesthesiology Congress Abstracts, p. 9.

Rogers, G. F. C. and Mayhew, Y. R. (1970) *Engineering Thermodynamics Work and Heat Transfer*, Longman, London, pp. 20–31.

Roth, J. V. and Braitman, L. E. (2008) Nasal temperature can be used as a reliable surrogate measure of core temperature, *J. Clin. Monit. Comput.*, 22, pp. 309–314.

Stanger, R. *et al.* (2009) Predicting the efficacy of convection warming in anaesthetised children, *Br. J. Anaes.*, **103**, pp. 275–82.

Sullivan, G. and Edmondson, C. (2008) Heat and temperature, BJA Continuing Education in Anaesthesia, *Critical Care & Pain*, **8**, pp. 104–107.

Turner, M., Hodzovic, I. and Mapleson, W. W. (2006) Simulated clinical evaluation of four fluid warming devices, *Anaesthesia*, **61**, pp. 571–75.

Chapter 9

Solubility, vaporisation and vaporisers

This chapter contains: solubility of molecules in liquids; solubility of gases in gases; vaporisation; the basis of vapour production by a vaporiser; vaporisers. The chapter links with: Chapters 7 and 8.

Introduction

In discussing humidity in the preceding chapter, the concept of equilibrium between water and its vapour has been introduced as a thermodynamic concept. The concept of vaporisation of other liquids such as volatile anaesthetic agents follows on naturally from that, but first of all it will be worth taking a detour through a discussion on solubility of gases and vapours in their own and other liquids [Davis 2003].

Solubility of molecules in liquids

To maintain simplicity in the discussion on humidity, no mention was made of the presence of air or other gas above the surface of the water, only the water vapour. Depending on the solubility of the gas in the liquid, a variable amount of the gas dissolves in the liquid, whether that be air in water or carbon dioxide in blood. As will be discussed in the section on vaporisation, some molecules of gas enter the liquid and some leave it, depending on their individual kinetic energies, until equilibrium is reached. If the pressure inside the container with the gas or vapour and liquid is increased, then the partial pressure of the gas above the liquid surface increases; this increases the population density of gas molecules, resulting in more of the gas molecules dissolving in the liquid. *Henry's Law* states that for a fixed temperature the solubility of a gas in a liquid is proportional to its partial pressure in equilibrium with the liquid. Note the condition of constant temperature because, in addition, solubility decreases with increased temperature. This occurs because an increase in the thermal energy of the dissolved gas molecules increases the partial pressure of the gas and encourages it to come out of solution (see below on vaporisation). Thus gas bubbles are more apparent in liquids that are heated.

A historical clinical example of the relevance of ambient pressure and nitrogen solubility in body tissues is in decompression sickness associated with tunnel workers. Modern examples include underwater diving and, to a lesser extent, aviators and space walking astronauts. Nitrogen is a compressible gas and goes into solution in body tissue spaces under compression if the miner, tunnel worker, or diver is breathing ambient air. The longer they are working under increased ambient pressure, the more nitrogen becomes dissolved in body tissues, according to Henry's law. The danger lies in coming to the surface, when the nitrogen is decompressed and forms bubbles in those tissues, exactly analogous to opening a bottle of carbonated drink. The collated pathological effects of such bubble formation in tissues are known as decompression sickness, known colloquially as 'the bends'. The same problem occurs in aviators whose cockpits explosively decompress above 7500 m or in a space walking astronaut who has been breathing air at 101 kPa in the space station and has to step into a 30 kPa space suit ventilated with oxygen at $F_{iO_2} = 1.0$; without an appropriate prebreathing protocol to denitrogenate the body, decompression sickness could result. The problem of decompression sickness is more highly accentuated in diving, where for every 10 m increase in depth in water, the diver's ambient pressure increases by 1 atm. Therefore, for a depth of 30 m, the diver is required to breathe gas at 4 ATA. If the diver is breathing air, and stays at depth beyond the time recommended in diving tables (guidance tables to help divers avoid decompression sickness), significantly hazardous nitrogen loading occurs. Failure to unload the nitrogen bubbles through the lungs adequately slowly by a slow or staged ascent to the surface will result in decompression sickness. This is usually treated by repressurising the victim in a hyperbaric chamber, thus redissolving the nitrogen in the tissues and having the diver breathe oxygen for a number of hours, then slowly decompressing until the nitrogen bubbles have been safely excreted from the lungs. Furthermore, pressurised nitrogen has anaesthetic properties, pressurised oxygen has toxic effects on the central nervous system and the lungs, and an increase in the density of the breathing air leads to an increase in respiratory work. All these factors make air less than ideal as a breathing gas for diving at depths greater than 50 m. Instead a helium/oxygen mix is preferred, as it is clinically where upper airway obstruction exists.

As already mentioned, different gases dissolve in liquids to different extents. Nitrous oxide is much more soluble in all liquids than nitrogen, which is why its use may cause diffusion hypoxia post operatively and why its use is avoided in patients with pneumothoraces. For example at ambient pressure and body temperature, one litre of water dissolves 0.14 l of nitrogen, but 0.39 l of nitrous oxide. By the same token, one litre of blood dissolves even more nitrous oxide under the same conditions, 0.47 l.

There are two measures of solubility. One is the *Bunsen* solubility coefficient, which is the volume of gas dissolved per litre of liquid at the relevant temperature, and the partial pressure of the gas is corrected to standard temperature and pressure (STP). A more useful measure is the *Ostwald* solubility coefficient, which is the volume of gas dissolved per litre of liquid at the relevant temperature, uncorrected. This is therefore independent of pressure, because although Henry's law states that the amount of gas dissolved in a liquid is proportional to the ambient pressure, if this is measured as a volume at that ambient pressure, it is found by Boyle's law to be constant, which is the relevance of the Ostwald solubility coefficient.

This brings the discussion to the concept of the *partition coefficient*, which is the ratio of the amount of a substance in one phase (e.g. a vapour), dissolved in another substance, usually in another phase (e.g. a liquid, although it can also describe the solution of one liquid in another), when the two phase samples are of equal volume and in equilibrium with each other. For example the blood–gas partition coefficient for nitrous oxide at body temperature is 0.47, because, given one litre of nitrous oxide in equilibrium with one litre of blood, 0.47 litre of N_2O dissolves in the blood. The concept of equilibrium is important, because this means that the partial pressure of the vapour or gas above the liquid is the same as its partial pressure in the dissolved state in the liquid. The word *tension* is sometimes used synonymously with partial pressure when a gas or vapour is in solution in a liquid. If the partial pressure of the vapour or gas above the liquid is raised, Henry's law dictates that the amount dissolved in the liquid will increase until the tensions are in equilibrium. For a relatively insoluble substance this tension equilibrium will be reached more quickly. Hence for an anaesthetic agent that is relatively insoluble in blood, such as N_2O, tension equilibrium is reached quickly, and this is reflected in the fact that the alveolar partial pressure rapidly reaches the inspired partial pressure. For anaesthetic agents that are highly soluble in blood, such as halothane, the high blood solubility (2.5) means that more of it is transported by the circulation away from the alveolus to the effector site and the partial pressure or tension of the agent in the alveolus only slowly approaches the inspired partial pressure.

Another important partition coefficient for anaesthetic agents is the oil–gas partition coefficient. Because the solubility in oil is similar to the solubility in nervous tissue, and the anaesthetic action appears to depend on this, a high oil–gas partition coefficient is an indicator of anaesthetic potency. For example N_2O is not a potent agent and has an oil–gas partition coefficient of 1.4, whereas sevoflurane is and has a coefficient of 53. The *Overton–Meyer hypothesis* suggests a near linear relationship between the logarithm of lipid solubility and anaesthetic potency.

Solubility of gases in gases

As already indicated in Chapter 7, gases and liquids are simply variants of fluids, so it is possible therefore for gases to dissolve in gases if pressure and temperature are appropriate. A clinical example of this is *Entonox*, a 50:50% mixture of oxygen and nitrous oxide. It is normally a gas mixture unless its temperature falls below −6°C, when the mixture separates (see Chapter 22).

Vaporisation

A liquid can be thought of as incompressible; at a molecular level it is because the molecules are so close together that very little further approximation of them is possible. If these molecules are given sufficient energy in the form of heat, they acquire more kinetic energy as shown by an increase in temperature. The molecules near the surface of the liquid are thus able to loosen themselves from the attractive forces of neighbouring molecules and leave the liquid surface with their excess kinetic energy. These molecules are considered to form a vapour, with an increased distance between them, compared with those in the liquid state. This allows the vapour to be thought of as compressible. If the liquid is in a closed container, a state of saturated equilibrium exists, with as many molecules leaving the liquid state as re-entering it. Within this closed container, the kinetic energy of the vapour molecules creates a pressure within it. This is called the *Saturated Vapour Pressure* (*SVP*), which increases with increasing kinetic energy (temperature), and represents the pressure at which a vapour and its associated liquid can coexist in equilibrium in a closed container at that temperature. Applying external pressure to a vapour will compress those molecules together and reform them into a liquid (as well as increase the amount of vapour dissolved in the liquid as described above). As the temperature of a vapour (which is not necessarily in association with a liquid) is raised, there comes a point where, no matter how much pressure is applied to it, it is no longer possible to liquefy it; the molecules have too much kinetic energy to allow this. The temperature at which this occurs is known as the *Critical Temperature*. A vapour at a temperature above its critical temperature is considered to be a gas in terms of its physical behaviour at a molecular level. For oxygen the critical temperature is −116°C (157 K), and hence it is a gas at all imaginable ambient temperatures, even at pressures of gases stored in cylinders. In contrast, the critical temperature of nitrous oxide is 36.5°C (309.5 K), and the pressure required to liquefy it is 74 bar just below this critical temperature. Thus in temperate or cold climates, N_2O exists in liquid form in cylinders. Water has a critical temperature of 374.1°C (647 K), which means that it is a vapour under conditions met by an anaesthetist, when most vapours behave very much like gases.

Just as pressurising a vapour below its critical temperature results in liquefying of the vapour, sufficient depressurisation of a liquid results in vaporisation. The energy required for this process, the *Latent Heat of Vaporisation*, in the absence of another energy source, is taken from the heat contained in the liquid itself, its container and surroundings. Hence a cylinder containing nitrous oxide that is suddenly opened to allow rapid depressurisation and consequent outflow of N_2O will become cold; water vapour on the cylinder surface will turn to ice. Note that the latent heat of vaporisation results only in a change of state, not in a temperature change. The heat of vaporisation is non-linearly related to the temperature of the liquid, being greater at lower temperatures and falling as temperature rises. Once the critical temperature is reached, the heat of vaporisation is zero since, by definition, the substance can exist as a vapour without the addition of any further heat. Figure 9.1 shows this relationship for nitrous oxide. Hence a cylinder containing nitrous oxide liquid at 20°C, which is slowly emptied, will have less heat removed initially for the vaporisation process than later on when the liquid is cooler. The astronaut's life support system uses the process of depressurisation of ice and water to a vacuum to remove metabolic heat from a spacesuit.

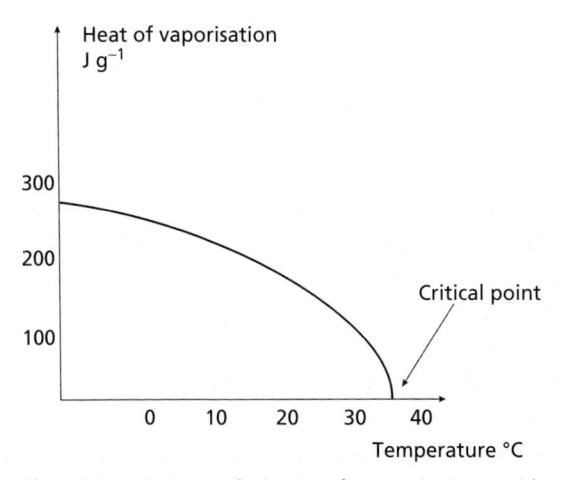

Fig. 9.1 Variation of heat of vaporisation with temperature.

Figure 9.2 shows (a) a simple vaporiser without any device to regulate its output and (b) a graphical representation of its output of volatile agent with time. A stream of gas is made to pass over the surface of the volatile liquid. As the output switch is turned on, the initial output is due to the saturated vapour pressure of the vapour in the container (1) [MacIntosh *et al.* 1987]. As the gas flow continues, only a small proportion of it comes into saturation contact with the liquid, and the average output falls rapidly (2) [Palayiwa *et al.* 1991]. The subsequent gradual fall in output is because of the fall in temperature of the liquid due to the heat of vaporisation and results in a diminishing amount of vaporisation of the liquid (3) [Nielsen *et al.* 1993]. As the temperature falls further and approaches that of the surroundings from which the heat of vaporisation is drawn, it becomes the constant temperature that determines the vapour pressure, and hence the output from the vaporiser (4) [White *et al.* 1996]. It is clear that the output from the vaporiser also depends on the flowrate of carrier gas over the liquid surface in the chamber; a very low carrier

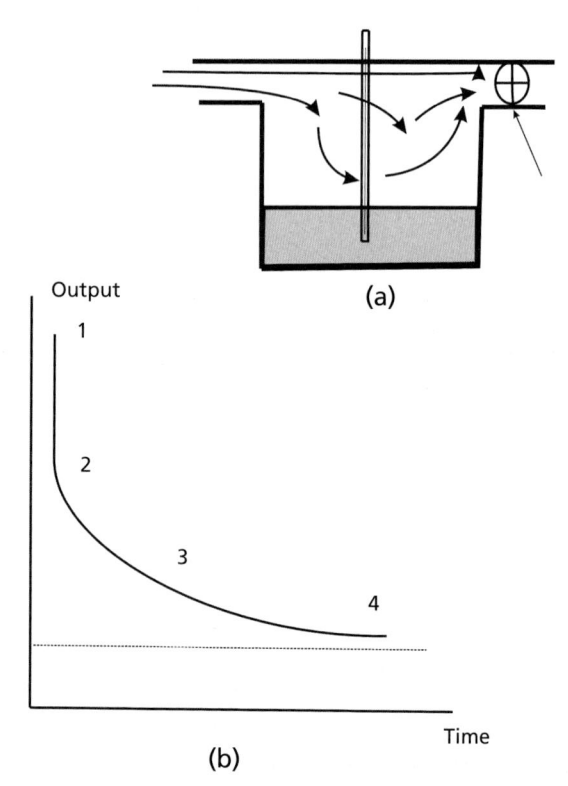

Fig. 9.2 (a) An uncompensated anaesthetic vaporiser. (b) Vaporiser output with time.

gas flowrate might well be expected to give a lower output than a high flowrate, all other things being equal.

The basis of vapour production by a vaporiser

Figure 9.3 shows a vaporiser in an ambient pressure of P_{amb} (101 kPa), containing a volatile liquid whose saturated vapour pressure (SVP) is P_s. If the vaporiser is efficient and there is maximum access to the liquid by the gas, then the maximum fractional output concentration in the vaporising chamber is

$$\frac{P_s}{P_{amb}}.$$

If the vaporiser dial is set to a small fraction f of this output by increasing the bypass flow, then the output from the vaporiser becomes

$$\frac{fP_s}{P_{amb}}.$$

Thus if f is a factor that constitutes, say, a 1% setting on a vaporiser calibrated at sea level, then at sea level

$$\frac{fP_s}{101} = 0.01 \quad (1\%) \text{ by calibration,}$$

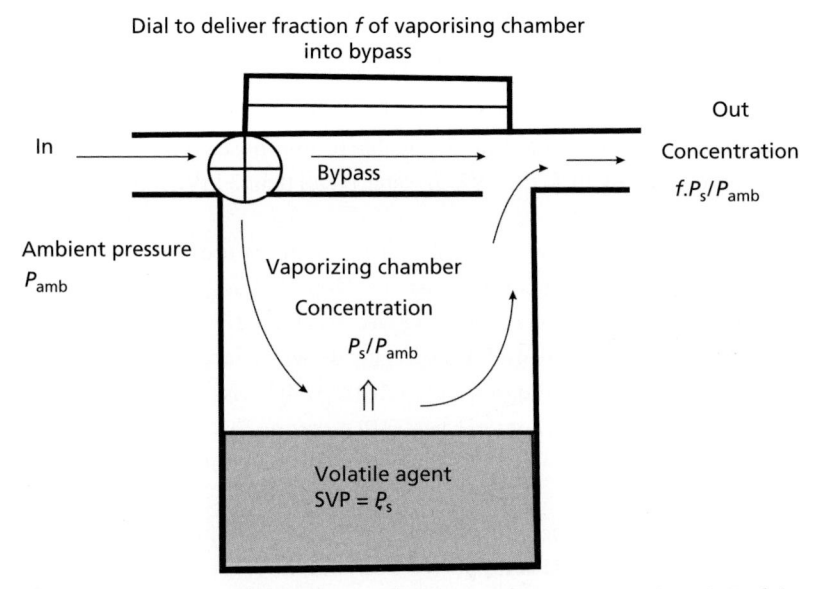

Fig. 9.3 A vaporiser and its output in relation to ambient pressure P_{amb}, SVP of the agent P_s, and dial setting fraction f.

and the partial pressure exerted is 1% of 101 kPa or 0.01 × 101 kPa = 1.01 kPa by Dalton's law of partial pressures.

If the same vaporiser is taken to Salt Lake City at an altitude is 2000 m where the ambient pressure is 80 kPa, with the same setting f on the vaporiser, giving a reading of '1%', the output is in fact

$$\frac{fP_s}{80},$$

somewhat greater than the true output at sea level. The partial pressure of the volatile agent delivered from the vaporiser output, however, is still fP_s, the same as at sea level, assuming no ambient temperature change. Since it is the partial pressure of the volatile agent that is important for pharmacological and anaesthetic purposes, the vaporiser does not need recalibrating at altitude.

It has been stated that the saturated vapour pressure is the pressure at which a vapour is in equilibrium with its associated liquid at a given temperature in a closed container. If the container is opened and the temperature of the liquid/vapour complex is raised, then eventually bubbles of vapour appear in the liquid itself. The liquid body itself can be considered to have started to vaporise. The vapour pressure has risen with temperature until it equals ambient pressure, and the liquid is said to be *boiling*.

Water has an SVP of 101 kPa at 100°C, and oxygen has the same SVP at −183°C (90 K). Therefore, at normal ambient temperatures of between 0 to 30°C (273 to 303 K), the SVP of oxygen greatly exceeds 101 kPa, and the oxygen rapidly boils away. A vacuum insulated flask is used in hospitals and other settings to provide a stable liquid source of oxygen, to keep the temperature and the SVP as low as possible to minimise boiling and wastage (see Chapter 22). At 20°C the SVP of water is less than 3 kPa, and at 37°C it is 6.3 kPa (see Chapter 8).

Figure 9.4 shows how SVP varies with temperature for a number of commonly used volatile anaesthetic agents. Note that desflurane has a higher SVP than most volatile anaesthetic agents, reaching 101 kPa at 23°C. This means that, to be clinically useful, a specially designed vaporiser is required to prevent uncontrolled vapour output [Graham 1994; Weiskopf et al. 1994]. The desflurane vaporiser incorporates a heater that ensures that the liquid vaporises fully and is released to the rest of the vaporiser in a controlled fashion (see below).

As an experiment to demonstrate the nature of vapour pressure, it is known that atmospheric pressure is supported by a column of mercury 760 mm high with a vacuum above it (see Figure 7.6). If some water is introduced into the base of the mercury column, it rises to the top, vaporises in the vacuum at the

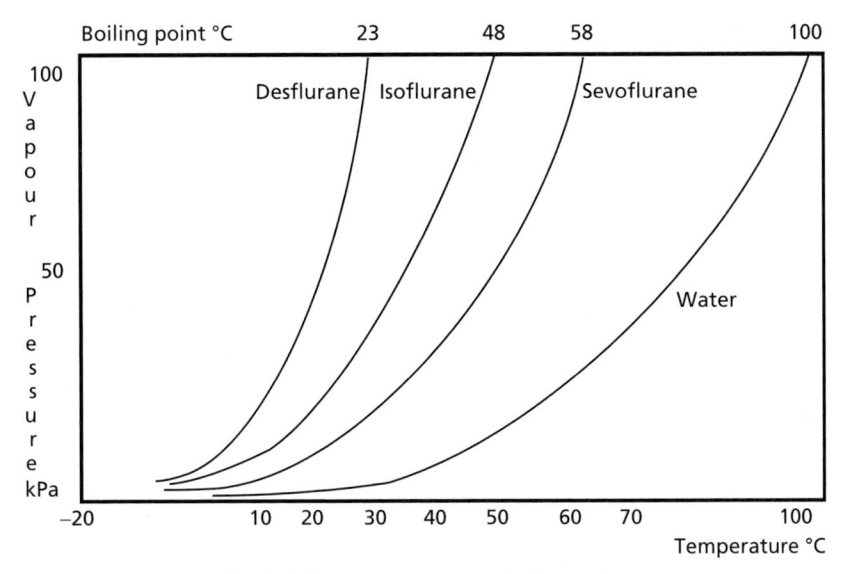

Fig. 9.4 Vapour pressure and boiling point.

top, and depresses the mercury column to, say, 742 mmHg. Thus water vapour at this temperature is 18 mmHg [Macintosh *et al.* 1987].

Vaporisers

The principle of SVP and vaporisation has been discussed and the predicted output from a simple, uncompensated plenum vaporiser was used to demonstrate it. In order to deliver accurately a few per cent of a volatile agent in a gas stream, either a well designed vaporiser or reliable volatile agent monitoring is required, and these days both are often incorporated together. Not all anaesthetic workstations have agent monitoring, but most modern vaporisers are engineered to a high standard to ensure a controlled output, unlike that shown in Figure 9.2.

In order to provide a controlled output, most vaporisers include some specific design features. The first change made to the simple plenum shown above (the plenum is the vaporiser's main chamber) is to provide two routes for the gas supply: one to the vaporisation chamber, the other as a bypass route; the ratio between these two flows – the *splitting ratio* – is what the control knob dictates. The maximum amount of volatile agent vaporised depends on the SVP of the agent, its temperature and that of the gases passing through the chamber, the gas flow rate, the amount of contact between the gases and the liquid, and the geometry of the vaporisation chamber. To a small extent it also depends on the gas components because of differences in density and viscosity, and because, for example, nitrous oxide is more soluble in the liquid agent than air is; there is also a small variation with altitude, again because of density change in the carrier gases. If the vaporiser is switched off, all the gas passes through the bypass; if it is switched on to maximum output, a maximum gas flow passes through the vaporisation chamber. At other dialled concentrations, a variable proportion passes through both pathways.

The second design change is to provide a means of maximising contact between the carrier gas and the liquid. The principles of these are shown in Figure 9.5 and include wicking, multiple baffles and a nebuliser, and elongating the gas pathway to maximise exposure.

The third design change is the provision of temperature compensation to prevent a fall in output that would otherwise occur, such as that shown in Figure 9.2. The earliest method was to attach the vaporiser to a material with a large heat capacity to minimise the temperature fall during the vaporisation process, such as a large mass of metal or water [Zuck 1988]. Figure 9.6 shows two later temperature compensation designs. One is a bimetallic strip, acting as a cap over the gas entry port to either the bypass chamber or the vaporising chamber; as the chamber temperature falls during vaporisation, the bimetallic strip bends and either restricts the flow to the bypass chamber or allows more

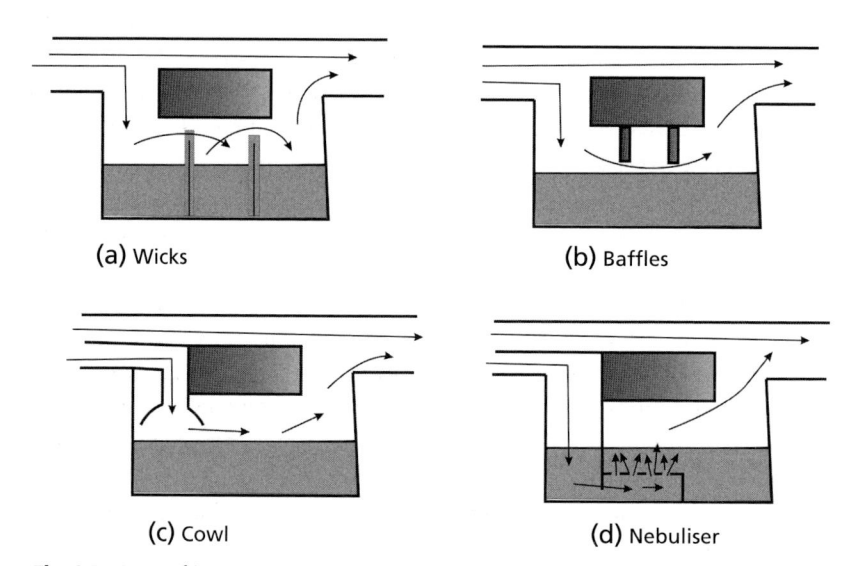

Fig. 9.5 Ways of increasing exposure of gas flow to anaesthetic agent in a vaporiser.

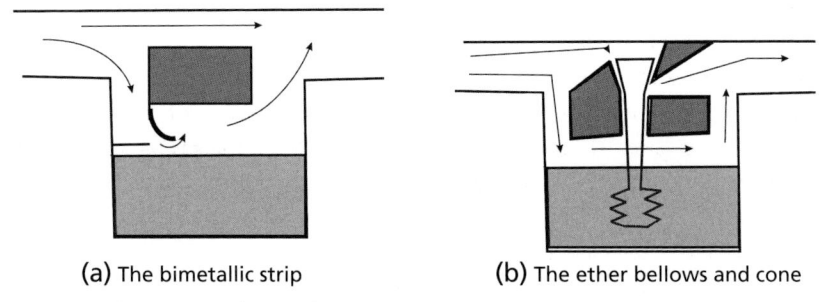

(a) The bimetallic strip   (b) The ether bellows and cone

Fig. 9.6 Two designs for temperature compensation in vaporisers.

gas to enter the vaporisation chamber, thus compensating for reduced temperature. Another design uses an aneroid bellows, which sits in the vaporisation chamber and is connected by a rod to a cone in an orifice in the bypass chamber; as the temperature falls in the vaporisation chamber, the bellows contracts, reducing the bypass flow, thereby increasing the splitting ratio.

If the pressure of the gas downstream of the vaporiser increases intermittently, for example due to the presence of a ventilator, some of this downstream gas, which contains volatile agent, can re-enter the vaporising chamber retrogradely, to be discharged again with an additional, unmetered vapour load. This is a hazardous anomaly, which needs a compensation device. It can take the form of either a one way valve downstream of the vaporiser, or an additional length of tubing between the vaporiser and the ventilator to prevent the *pumping effect* of retrograde movement of gas into the vaporiser. The apparently complex design of modern vaporisers aims to achieve all of these properties and also to maintain the output constant across a wide range of carrier gas flowrates.

The simplest vaporiser in historical terms is the *Boyle's bottle* [White et al. 1996], shown in Figure 9.7. Figure 9.8 shows a modern *Tec5* vaporiser and the following features are noted: the bypass chamber (7) passes underneath the vaporising chamber (4); there is an elongated, coiled inlet passage (2) upstream of the vaporising chamber to prevent the pumping effect described above; there is a helical wick (3) to maximise carrier gas contact with liquid agent; the temperature compensation mechanism is a bimetallic strip (8) that restricts flow through the bypass chamber. Figure 9.9 shows a modern *Dräger Vapor* and it has similar features to the Tec. A series of baffles (1) in the upstream inflow tube counteracts the pumping effect. A spiral wick (2) with a sleeve dips into the liquid agent and the wick is made of a synthetic material that has a high capillarity, but does not absorb the agent. The concentration

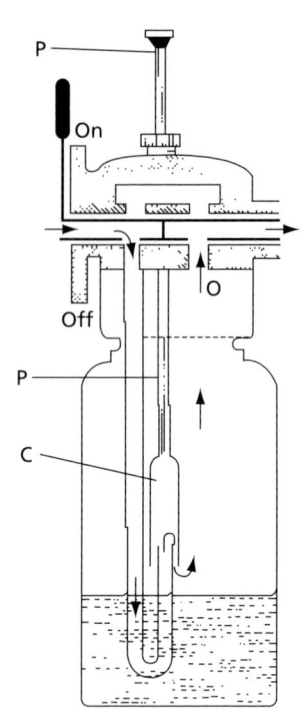

Fig. 9.7 The Boyle's bottle vaporiser.

control valve (5) is directly linked to the control knob, and the mixing chamber is where the vapour is mixed with the bypass gas. The bimetallic temperature compensation device consists of an inner rod (8) made of a relatively non-expansile material, coated in brass, and the combined structure dips into the liquid; as the temperature of the liquid falls, the brass coating contracts, pulling the inner rod up to constrict the bypass flow.

On acquisition of any vaporiser, the information presented to the user includes calibration curves or the equivalent for output against temperature and gas flow rate. Output from vaporisers has been shown to vary by more than ±15% [Nielsen et al. 1993]. Note also that modern vaporisers have a keyed filling port to prevent the chamber being filled with the wrong agent, but should still not be overfilled [Palayiwa et al. 1991]. Halothane requires the chemical Thymol as a preservative and the accuracy of the Fluotec 3 vaporisers is reduced by the presence of the Thymol [Gray 1988], because Thymol blocks the wicks, which also means the service interval for a Halothane vaporiser is shorter than for vaporisers of other agents.

In recent years, with anaesthetic machines incorporating integral agent monitoring including modern electronics, new vaporisers are becoming less

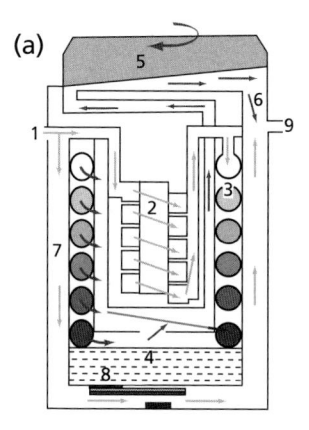

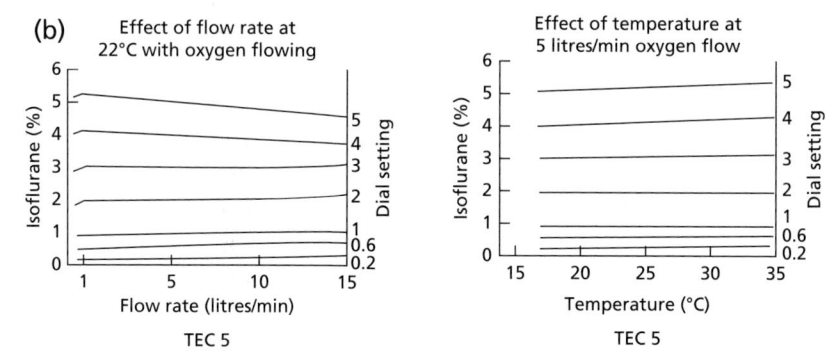

Fig. 9.8 (a) A cross-sectional view of the Tec5 vaporiser. (b) Effect of gas flowrate and temperature on Tec % vaporiser output.

mechanically engineered, the vaporiser instead incorporating solenoid flow control valves, electronic control and a closed loop feed-back mechanism. The vaporiser output is measured by the agent analyser and, if there is any discrepancy with the set value, a servo-control mechanism changes the splitting ratio appropriately. Such a vaporiser, shown in Figure 9.10, is used on the modern *Datex-Ohmeda* anaesthetic machine. The vaporising chamber is now a cassette, which contains lamellae made of a synthetic material that behave like a wick, with metal plates between them to make the gas pathway more convoluted in order to aid vapour contact. At the back, where the cassette interfaces with the anaesthetic machine backbar, there are inflow and outflow valves and a temperature sensor. The filling port is at the front, and there is a valve to the main cassette that closes when the cassette is full. There are magnets on the front that help the control unit identify which agent is in use. The cassette

144 | SOLUBILITY, VAPORISATION AND VAPORISERS

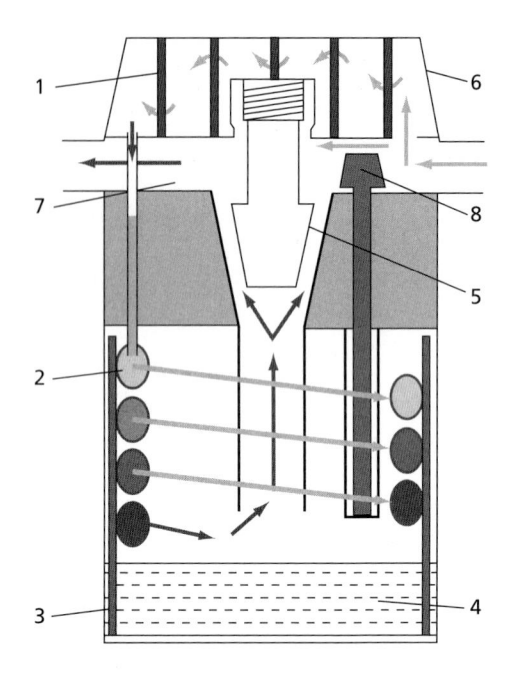

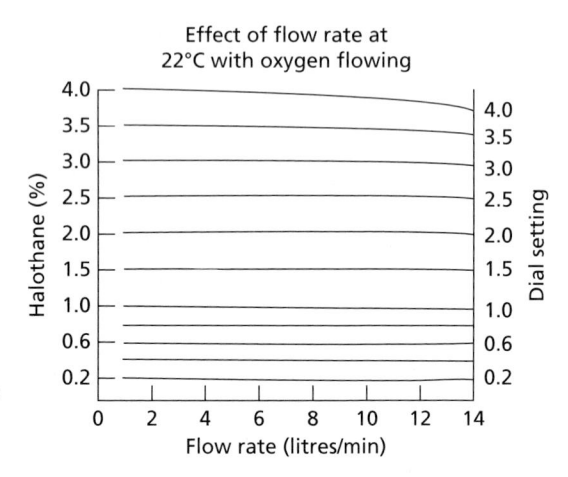

Fig. 9.9 Cross-sectional view and output variation with gas flowrate for the Draeger Vapor vaporiser.

control unit, which is part of the anaesthetic machine, has additional inflow and outflow check valves, a bypass flow measurement device and an agent flow measurement device. When a certain concentration is dialled up by the user, the fresh gas flow is split in a calculated ratio. The small proportion that is diverted to the cassette passes through the one way valves and into the cassette, where it picks up vapour. It then exits the cassette whereupon it passes through

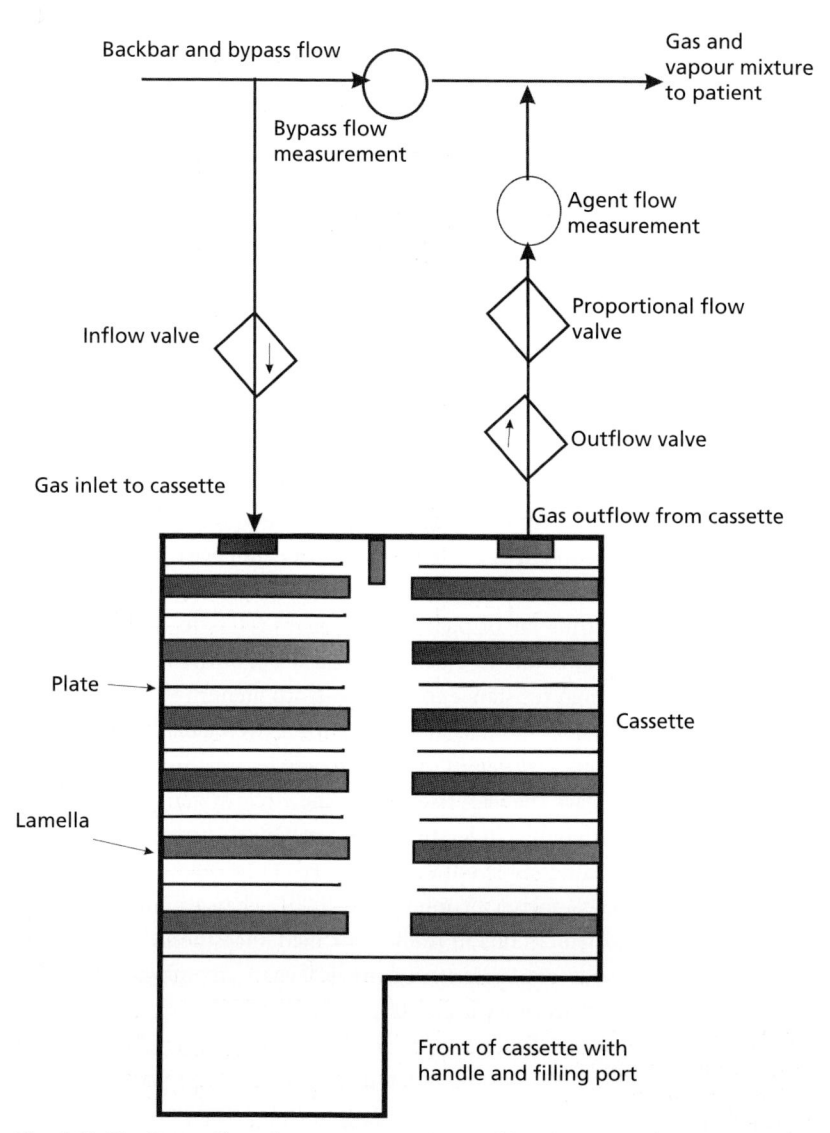

Fig. 9.10 The Datex Ohmeda cassette vaporiser and backbar connections, plan view.

a flow control valve to a flow measurement device and rejoins the bypass flow, which is also measured. A microprocessor uses the information from these flow measurements and the temperature sensor to calculate the amount of agent to be added to the bypass flow to achieve the dialled concentration and alters the flow control valve accordingly.

Because desflurane has such a low boiling point, special measures have to be taken in the design of the vaporiser to achieve a reliable vapour output. The chamber containing the liquid contains a heater which maintains the agent temperature at 39°C, at which temperature the SVP of the agent is 194 kPa. When an output concentration is dialled up, a shut-off valve downstream of the exit from the vaporising chamber opens and admits vapour under pressure into the downstream pipework. An electronically controlled pressure regulator reduces the pressure of the aliquot of vapour to that which normally exists in a plenum vaporiser (1–2 kPa). The vapour then passes to a variable flow restrictor linked to the concentration control knob, whence it joins the bypass gas flow and leaves the vaporiser. The pressure of the vapour is proportional to the required flow of the agent. There is a differential pressure sensor in the pipework that compares the pressure coming out of the vaporising chamber with the dialled concentration and acts on the control valve appropriately.

There are also devices for injecting liquid agent directly into the gas stream. This overcomes the problem of the temperature dependence of the SVP of the vapour.

This description thus far of plenum vaporisers refers to those mounted on continuous flow anaesthetic machines with a pressurised gas supply, in which the relatively high resistance to gas flow from plenum vaporisers is not a limiting design consideration. When used with a circle breathing system, a plenum vaporiser is normally used outside the circle breathing system itself, but it is possible to place the vaporiser inside the circle system instead, when its output is very different. Under these circumstances a vaporiser through which the patient breathes air spontaneously needs to be of low resistance and is called a *draw-over* vaporiser [Young *et al.* 2000]. They are often also used with simpler to-and-fro circuits in military or field anaesthesia situations, or in countries where sources of pressurised medical gases are unavailable. While they are not designed to be as accurate as other devices, they are certainly adequate for clinical use. Several types are in use [English *et al.* 2009] and, given their flexibility, care must be taken in the way they are used [Donovan *et al.* 2007].

It should also be remembered that despite accurate output from a vaporiser, some of the volatile agent is absorbed by anaesthetic tubing, although this is less true of modern agents and modern tubing material than formerly [Smith *et al.* 2002].

References

Davis, P. D. and Kenny, G. N. C. (2003) *Basic Physics and Measurement in Anaesthesia*, 5th edn, Butterworth-Heinemann, Edinburgh, pp. 65–74.

Donovan, A. and Perndt, H. (2007) Oxford miniature vaporiser output with reversed gas flows, *Anaesthesia*, **62**, pp. 609–614.

English, W. A. *et al.* (2009) The Diamedica drawover vaporiser: a comparison of a new vaporiser with the Oxford miniature vaporiser, *Anaesthesia*, **64**, pp. 84–92.

Graham, S. G. (1994) The Tec 6 vaporizer, *Br. J. Anaesth.*, **72**, pp. 470–73.

Gray, W. M. (1988) Dependence of output of halothane vaporisers on thymol concentration, *Anaesthesia*, **43**, pp. 1047–49.

MacIntosh, Mushin & Epstein (1987) *Physics for the Anaesthetist*, 4th edn, Revised Mushin M. W. and Jones P. L., Blackwell Scientific Publications, Oxford, pp. 101–24.

Nielsen, J. *et al.* (1993) Accuracy of 94 anaesthetic agent vaporisers in clinical use, *Br. J. Anaesth.*, **71**, pp. 453–57.

Palayiwa, E. and Hahn, C. E. W. (1991) Overfill testing of anaesthetic vaporizers, *Br. J. Anaesth.*, **74**, pp. 100–103.

Smith, C. *et al.* (2002) Leakage and absorption of isoflurane by different types of anaesthetic circuit and monitoring tubing, *Anaesthesia*, **57**, pp. 686–89.

Weiskopf, R. B., Sampson, D. and Moore, M. A. (1994) The desflurane (Tec6) vaporizer, *Br. J. Anaesth.*, **72**, pp. 474–79.

White, S. A. and Strunin, L. (1996) Anaesthesia with the Boyle's bottle vaporizer, *Anaesthesia*, **51**, pp. 939–42.

Young, D. A., Brosnan, S. G. and White, D. C. (2000) A semiquantitative analysis of the dynamics of a Goldman type vaporizer, *Anaesthesia*, **55**, pp. 557–70.

Zuck, D. (1988) The Alcock chloroform vaporiser. An early calibrated, temperature compensated plenum apparatus in its historical context, *Anaesthesia*, **43**, pp. 972–80.

Further reading

Davey, A. (2005) Vaporizers. Chapter 5 in *Ward's Anaesthetic Equipment*, 5th edn, WB Saunders Co. Ltd, pp. 65–86.

Chapter 10

Ultrasound and Doppler

This chapter contains: properties of ultrasound; visualisation of needle positioning; Doppler (duplex, frequency shift spectrum, blood flow); measurement of absolute blood flow; high power ultrasound.
The chapter links with: Chapters 4, 5 and 13.

Introduction

Ultrasound has many uses in areas of medicine associated with anaesthesia. It is used for imaging, visualisation of needle and catheter positioning, therapy and, together with the Doppler effect, for measurement of flow velocity. Real-time information can be obtained with ease, and with the low energies used, diagnostic equipment exposures are not thought to be a safety issue, either for the operator or the patient.

Properties of ultrasound

Ultrasound is a form of mechanical energy that consists of high frequency vibrations at frequencies above human hearing range (> 20 kHz) and up to frequencies in the tens of MHz range. The frequencies used are dependent on issues such as the penetration and resolution required. It is thought that low intensity ultrasound passes through living tissue without altering tissue function. Higher energy can produce heating and cavitation, both of which can alter cell function.

Ultrasound is generated by electrically inducing a deformation in a piezo-electric crystal, which compresses and decompresses the medium to which it is coupled (see Figure 10.1(a)) at a rate equal to the frequency of the driving voltage. The pressure changes travel through the medium in a longitudinal direction and the distance between the points of maximum pressure, or compression, is known as the wavelength. Figure 10.2 shows the relationship between the period of the wave and the wavelength. The length of this distance is dependent on the *elasticity* (compressibility) and the density of the medium, and the delay between the movement of adjacent particles in the medium. As shown in the figure, the wavelength (λ) and the transmission frequency (f) are

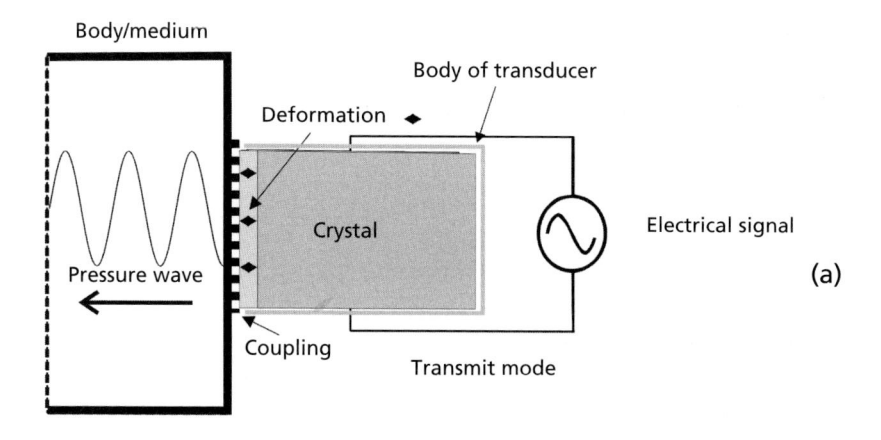

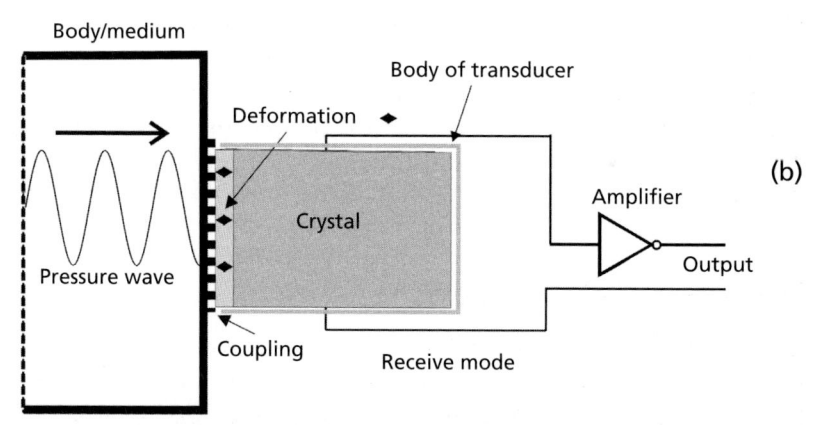

Fig. 10.1 Production of ultrasound from a piezo-electric crystal. The crystal is surrounded typically by a metal shell (body of transducer), except on the left hand side. An electrical signal input causes the crystal to deform (a), and this creates a pressure wave into the body with which it is mechanically coupled. Coupling requires the use of a special gel to transfer the movement of the transducer to the body as the movement is not transmitted by air. In (b) the piezo-electric crystal also allows the reverse process to occur, and a pressure wave on to the crystal causes an electrical signal to be induced, which is further amplified.

related to the propagation velocity c by $c = f\lambda$. The magnitude of the wave is the difference between the maximum and minimum pressure values. The wave propagates by movement of particles: it cannot travel in a vacuum and it does not *ionise* (see Chapter 29) the medium through which it travels.

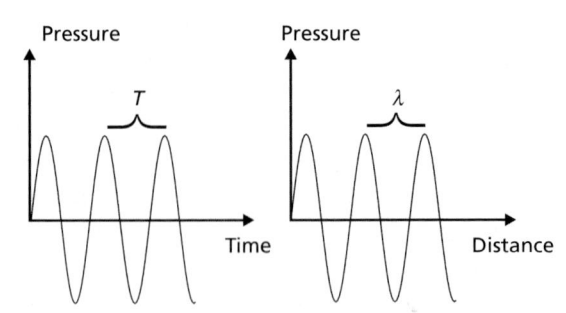

Fig. 10.2 Diagram showing the period and wavelength of ultrasound pressure waves. T is the period of the waveform (the time for one complete oscillation), and is the reciprocal of the frequency, $T = 1/f$. The wavelength λ, is the distance occupied by one oscillation. λ is related to the speed of the wave in the medium c, and the frequency by $c = f\lambda$.

The propagation of an ultrasound wave is not constant throughout the body. Various parts allow the passage of the wave at different velocities. Also the wave is attenuated differently by the various tissue types. For example, in soft tissues the ultrasound wave has a velocity of between 1460 and 1630 m s^{-1} whereas in bone it is 2700–4100 m s^{-1}. Bone attenuates the wave about ten times more than tissue.

The wavelength of the beam determines the limits of resolution. The intensity of the beam determines the sensitivity of the instrument and governs the number and size of echoes recorded. Attenuation increases linearly with frequency in soft tissues but in air and water increases with the square of the frequency. The greatest penetration is achieved with the lowest frequency but because the wavelength is long, the resolution is poor. A compromise is to use the highest frequency that will ensure adequate penetration of the tissues or organs being investigated. For example a transmission frequency of 3 MHz is normally used for the kidneys, 10 MHz is used for ophthalmic investigations, and 20 MHz is used for catheter tip intravascular devices. The absorption of ultrasound by the tissues results in the generation of heat that, in the case of diagnostic work, can cause temperature increases of the order of 1°C. It is the reflection of the ultrasound beam from the junction between two tissues or from tissue–fluid or tissue–air interfaces that forms the principle of the majority of diagnostic techniques. The same piezo-electric crystal is usually used as the receiver for ultrasound, when the transmission is switched off. The pressure changes coupled to the transducer will induce electrical signals proportional to the pressure (see Figure 10.1(b)).

Imaging is carried out using pulsed echo ultrasound to produce a real-time two dimensional B scan. In this scanning method many transducers are used, either in an angular sector scanner way or as a linear array. The sector probe (Figure 10.3(a)) has fewer transducers and the rotation is mechanical whereas the linear array type (shown in Figure 10.3(b)) has typically over 300 elements and the beam *movement* is carried out electronically. Mechanical devices are now out-dated although they are still used. Figure 10.3(a) demonstrates the concepts of how mechanical sector scanner functions, and Figure 10.3(b) shows the linear array transducer. The pulse echo transducer is shown at three

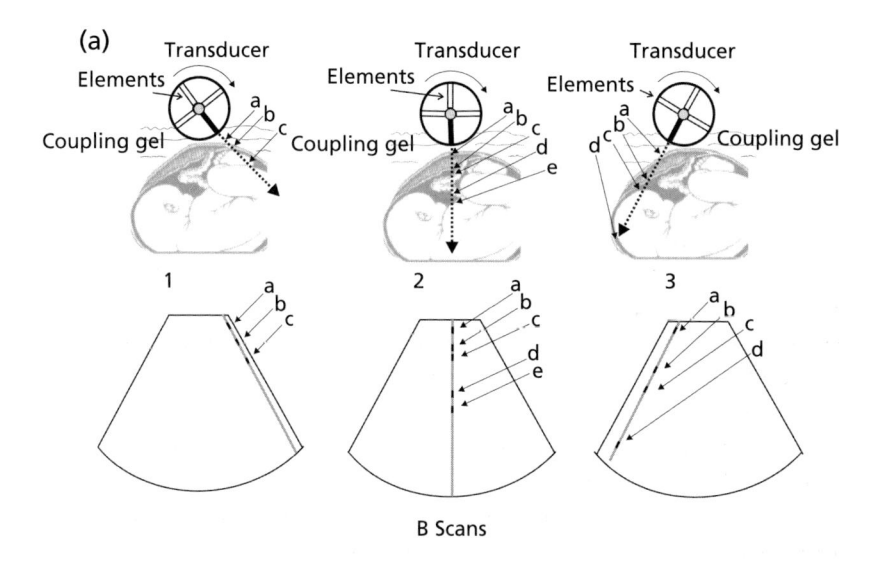

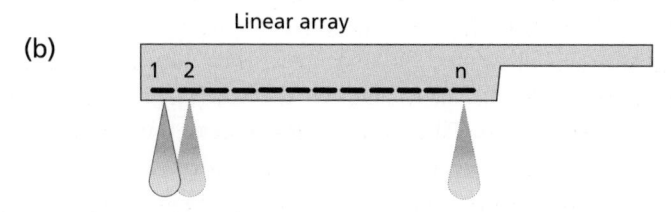

Fig. 10.3 The use of two-dimensional B-scanning transducers, using (a) a four sector mechanical transducer, and (b) a *n* element linear array. In (a) the same transducer scans the same anatomical site, only in each case the elements of the transducer have rotated through 45°. The sectors 1–3 show the build up of the received image showing the interface boundary echoes produced from a, b, c as appropriate. (b) shows a linear array of *n* elements. First element 1 is made to transmit and then receive by an electronic switch, and then element 2, and so on until element *n*.

different paths, although many paths in reality make up the complete sector (the complete sector is shown in the figure as position 1 to position 3). At each position the transducer sends out ultrasound pulses and the return of these echoes from the various boundaries gives depth information. The position of the transducer in the arc is known and linked to the display, so the echoes will have the correct position. The brightness of each echo point will be related to its strength. A schematic version of this two-dimensional scan is shown at the bottom of the diagram. The real-time display often uses at least 100 scan lines (vertically) that are *refreshed* 30 times per second to appear as a flicker-free picture to the observer.

Other scanners can be miniaturised and inserted into the body to provide images. Successful imaging has been carried out of most parts of the body but, for the anaesthetist, the most important areas are now visualisation of needle positioning, echocardiography, carotid artery and internal jugular vein visualisations, and foetal imaging.

Visualisation of needle positioning

Historically ultrasound has been used by radiologists to guide needle, catheters and guide-wire placement, but it is now being increasingly used by anaesthetists for vascular access, nerve blockade, drainage of plural or ascitic fluid collections and percutaneous tracheostomy. Ultrasound allows identification of the target structures and gives real-time guidance to place needles and catheters precisely [Chapman *et al.* 2006]. Probably the largest use of ultrasound with anaesthetists is with regional anaesthesia, and for central venous cannulation. Current National Institute for Health and Clinical Excellence (NICE) guidelines recommend the use of ultrasound for these purposes.

The key to success with using ultrasound guided techniques is acquiring considerable real-time scanning experience. Useful practice can be gained by using water baths and phantoms (inert structure with ultrasound sensitive recognisable structures buried in it). One of the important areas is in the assessment of needle position. There are a number of factors that influence needle visualisation [Schafhalter-Zoppoth *et al.* 2004]. Larger needles are more easily seen than smaller ones, particularly in cross-section, but smaller ones may produce fewer artefacts. Needles placed perpendicular to the ultrasound beam are the easiest to see, and the tip can often be seen due to its irregular surface, when the shaft can not. The introduction of the needle in a short in–out or side to side motion can help with visualisation. Also when the needle shaft crosses the beam, an acoustic shadow will be seen, and this can help. The injection of fluid has been also shown to enhance needle and catheter tip visualisation, which is helpful in nerve blockade to confirm the site and spread of the anaesthetic. When small bubbles are in a solution (as is often the case when using a small syringe) and this solution is injected into a fluid filled vessel, then the visualisation is enhanced. This is one of the principles of the use of ultrasound contrast media, where bubbles can enhance the images. The bubbles can be increased by agitation or repeated injection between two syringes with a nearly closed three-way tap. However, injection of air or microbubbles in solid tissues is not helpful because the acoustic shadow that they produce spoils underlying deeper structures. This may be relevant for regional anaesthesia, so pre-warming local anaesthetic solutions may help reduce bubbles.

Doppler

When an ultrasound beam encounters a stationary target, it is partially reflected back towards the transmitter and the carrier frequency of the wave remains the same. When the reflector, which can be a group of moving blood cells in a vessel, is moved towards the transmitter, however, it encounters more oscillations per unit time than its stationary equivalent, so the frequency is apparently increased. This is demonstrated in Figure 10.4.

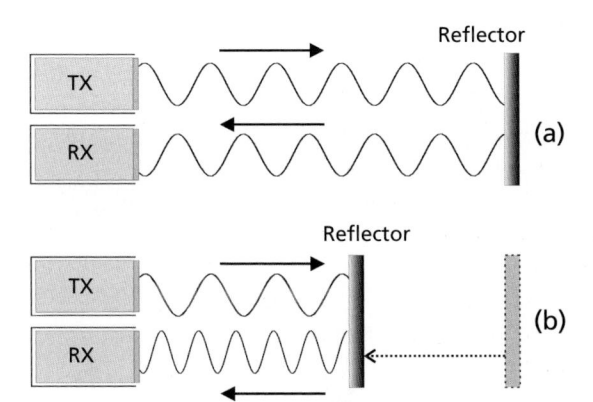

Fig. 10.4 This figure demonstrates the Doppler effect. In (a) the transducer (TX) transmits an ultrasound wave and it is reflected by a reflector to be received by the receiving (RX) transducer. In (b) the reflector has moved from its initial position to a position nearer the transducer and on the return path the waveform is compressed, which means the frequency increases.

Doppler velocimetry is a non-invasive technique for measurement of blood velocity (not flow) within the body. For a transmitted frequency of f_t, a wavelength λ, and the velocity of sound in the medium c, $f_t = c/\lambda$.

If the beam hits an object that is moving directly towards the transmitter at velocity v, the frequency of the waves arriving at the reflector (f_r) will now be:

$$f_r = (c+v)/\lambda.$$

The reflector will now act as a source that is moving towards the transmitter, the actual frequency sensed by the transmitter (in receiver mode) will be $f_r = (c + 2v)/\lambda$.

The apparent increase in frequency is given by:

$$f_r - f_t = (c + 2v)/\lambda - c/\lambda = 2v/\lambda = 2v f_t/c.$$

This equation applies as long as the speed of the reflector is small compared with the speed of sound. Putting some numbers into this equation, in tissue, $c = 1500$ m s^{-1}, $f_t = 2$ MHz, $v = 0.1$ m s^{-1}, the difference frequency will be 267 Hz. This difference signal is audible and therefore the Doppler output could be amplified and delivered to a loudspeaker or earphones. The *sound* pattern of pulsatile blood flow in different disease conditions can be very descriptive and a trained operator can recognise these changes. Quantification of the Doppler waveform will be discussed later.

Normally the Doppler beam is applied at an angle to the blood vessel as shown in Figure 10.5. The resulting frequency shift is now multiplied by $\cos \theta$. For the maximum frequency shift the transmitter/receiver must be aligned with the direction of flow ($\cos \theta = 1$). If the beam is perpendicular to the flow then no signal will be produced ($\cos \theta = 0$). The effect of changing this angle is often erroneously neglected when making Doppler measurements.

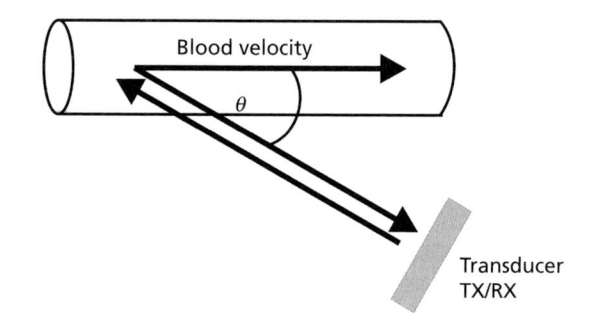

Fig. 10.5 This figure illustrates the angle θ between the ultrasound beam and the moving blood.

Duplex Doppler

Ultrasound scanning and Doppler can be combined enabling the user to visualise the cross-sectional anatomy of interest by using the grey-scale image, and by placing a cursor over the particular vessel to be measured. This cursor identifies the position on the image of a range-gate of Doppler ultrasound (the position at which Doppler signals will be detected). The user switches over to the Doppler mode, which normally freezes the image and then blood flow calculations can be made from the desired site. The angle between the flow lines and the beam can also be measured.

A more advanced technique is colour flow (Doppler) imaging. This technique uses a conventional black and white ultrasound image, and a real-time colour image, which represents motion detected by Doppler in the same scan plane. They are combined to create a composite image. This allows the investigator to visualise flow and anatomy simultaneously with much greater clarity than before. Also advances in digital technology have improved both Doppler and ultrasound images.

The use of the Doppler frequency shift spectrum

The Doppler difference signals are in the audible range and so storing them for later analysis is straightforward on an audio cassette recorder or a chart recorder. A more efficient and quantitative method of analysing the Doppler signal is to produce the Doppler power spectrum or frequency shift spectrum of the audio signal. The Doppler shift frequency is proportional to the velocity, and the power in a particular frequency band is proportional to the number of blood cells moving with velocities that produce frequencies in that band. The Doppler power spectrum should have the same shape as a velocity distribution plot for the flow in the vessel. An example of this is shown in Figure 10.6. The horizontal axis represents time, the vertical axis is the Doppler frequency shifts at that time and the grey scales give the power at that frequency or the number of blood cells moving.

The *shape* of the maximum (peak) frequency shift curve can give information that a simple mean flow measurement, which may also be obtained by other clinical methods, can not. In the assessment of arterial disease the pulsatile components of a flow waveform are much more sensitive to proximal stenosis than mean flow [Lee 1978, Farrar 1979], and allows a quantification of the severity of the disease.

The Figure 10.7 shows two waveforms shapes from a femoral artery, one from a normal artery (a) and the other from a high impedance downstream state (b), as can be found in a stenosed artery. It would be convenient for

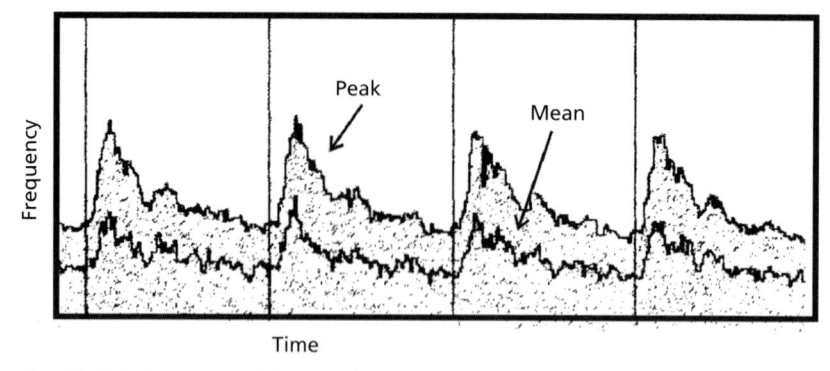

Fig. 10.6 A frequency shift Doppler power spectrum produced from a Doppler spectrum analyser connected to a Doppler probe. The vertical axis is the positive frequency shift spectrum (it can be negative as well for, for example, the opposite direction of flow). The vertical lines are where the analyser calculates that the new cardiac cycle has begun. The peak is the peak frequency shift spectrum and the mean is the mean frequency shift spectrum.

quantification of disease state if each different waveform shape could be represented by a single number, or series of numbers. When the cardiac cycle is pulsatile the maximum frequency shift waveform comprises two parts; the systolic and diastolic parts, having different frequency characteristics. The systolic component has a fast rise of frequency with respect to time, whereas the diastolic part has lower frequency components. A stenosed artery has different frequency dependent characteristics from a normal artery. The two sections will change in differing proportions depending on which type of artery is being examined. In the absence of a pulsatile waveform a downstream stenosis would result in a decrease in both flow and the amplitude of the spectrum throughout the cycle (i.e., the same reduction in both sections).

Two popular indexes for giving a *index* of upstream resistance or impedance are the *pulsatility index* (PI) and the *resistive index* (RI). The PI is calculated as $(S - D)/mean$ as shown in Figure 10.7. The RI is calculated as $(S - D)/S$, where S is the maximum frequency shift during systole and D is the maximum shift during diastole. D can also be negative for reverse flow situations.

There are many applications of Doppler velocimetry of interest to anaesthetists, for example renal blood flow, uteroplacental artery flow, transcranial middle cerebral artery, hepatic blood flow, cardiac output and many others. Doppler velocimetry can be used as a non-invasive measure of renal blood flow and impedance, allowing measurement of the effects of certain drugs on the renal circulation [Stevens 1989, Tooley 1994, Munn 1993].

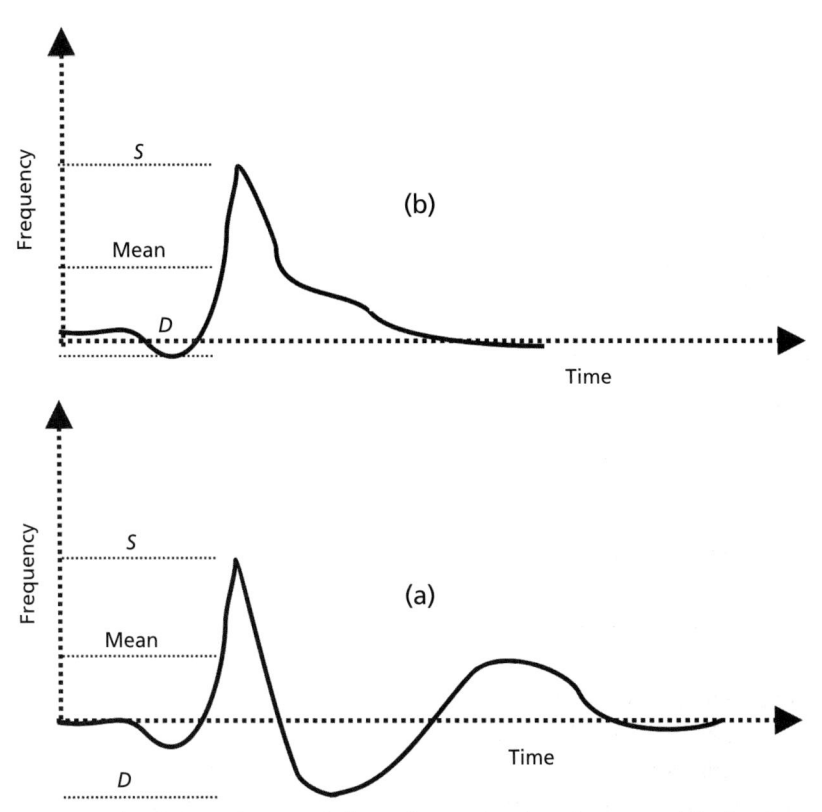

Fig. 10.7 This diagram shows typical waveforms taken from frequency shift plots from a common femoral arteries from a normal subject (a), and (b) a patient with peripheral disease. In (a) the frequency goes negative, indicating reverse flow. The horizontal axis is time and the vertical axis is the positive and negative frequency shift spectrum. S is the maximum frequency shift during systole, D is the shift during diastole, and the mean is the mean shift over the whole cycle.

Measurement of absolute blood flow

The measurement of the blood flow comprises measurement of the angle between the ultrasound beam and the vessel, estimation of the mean diameter of the vessel, and estimation of the mean velocity of the flow in the direction of the ultrasound beam. The simplest formula involves converting the beam parallel to the vessel axis, by knowing the angle, and multiplying this by the vessel cross-sectional area to give mean flow. Deciding which is the mean velocity can create problems as the blood velocity profile in arteries is complex. For example, near the aortic valve it resembles plug flow but down-stream the flow is more parabolic. Even with so many causes of error a reasonable estimation of volumetric blood flow can be made by using the formula:

$$mean\ flow = kA f_d c/(2f_o \cos\theta),$$

where k is a factor to convert the velocity to its mean value and must depend on where the blood flow is in the body. For parabolic flow $k = 0.5$. A is the cross-sectional area of the vessel and is usually calculated using the internal diameter of the vessel (d) and the formula $\pi d^2/4 f_d$ is the Doppler shift, f_o is the transmission frequency, c is the velocity of sound and θ is the angle of the ultrasound beam to the flow line.

There has been some commercial interest in trying to use this simple formula in real-time stand alone Doppler instruments, such as the sternal notch device [Donovan 1987] and trachea device [Hansen 1992]. One of the problems associated with these devices is the assumption on the angle. If this is near zero, as in the case of the sternal notch device, this is not so much a problem, but in the other instrument mentioned the angle is assumed to be near 54° and the errors associated with variations to this angle are much higher and this may be why the device is very inaccurate.

High power ultrasound

Much higher levels of ultrasound, however, can be used in a practical way. Lithotripsy is one of these and it often proves to be the best method for the removal of kidney, uretonic and gall bladder stones. No invasive surgery is required but anaesthesia is generally used to minimise the discomfort to the patient. The technique usually employs shock waves created by the rapid vaporisation of fluid that surrounds a spark gap, the wave being focused by an elliptical reflector such that maximum intensity coincides with the position of the stone. Alignment of the focus can be performed using ultrasound imaging. The stone fragmentation typically requires about 4000 shocks.

Conclusions and future directions

Four-dimensional ultrasound (real-time 3D scans with Doppler flow information) is now commonly used in specialist clinics and could become commonplace in the anaesthetic domains. The equipment for general scanning is becoming much smaller and portable. This has enabled ultrasound technology to be introduced, and it is now well used in operating theatres, intensive care units and pain clinics. The areas where once ultrasound was only the domain of the radiologist and ultrasonographer such as chest ultrasound and echocardiography, are likely to become part of the anaesthetist's domain.

References

Chapman, G. A., Johnson, D. and Bodenham, A. R. (2006) Review article: Visualisation of needle position using ultrasonography, *Anaesthesia*, **61**, pp. 148–58.

Donovan, F. D., Dobb, G. J., Newman, M. A. *et al.* (1987) Comparison of pulsed Doppler and thermodilution methods for measuring cardiac output in critically ill patients, *Critical Care Medicine*, **15**, pp. 853–57.

Farrar, D. J., Green, H. D. and Peterson, D. W. (1979) Noninvasively and invasively measured pulsatile haemodynamics with graded arterial stenosis, *Cardiovascular Research*, **13**, pp. 45–47.

Hausen, B., Schafers, H., Rohde, R. *et al.* (1992) Clinical evaluation of transtracheal Doppler for continuous cardiac output estimation, *Anesthesia and Analgesia*, **74**, pp. 800–804.

Lee, B. Y., Assadi, C. and Madden, J. L. (1978) Haemodynamics of arterial stenosis, *World Journal of Surgery*, **2**, pp. 621–29.

Munn, J., Tooley, M., Bolsin, S., Hronex, I., Lowson, S., Wilcox, J. (1993) The effect of metoclopramide on renal vascular resistance and renal function in patients receiving a low-dose infusion of dopamine, *British Journal of Anaesthesia*, **71**, pp. 379–82.

Schafhalter-Zoppoth, I., Mcculloch, C. E. and Gray, A. T. (2004) Ultrasound visibility of needles for regional nerve block: An in vitro study, *Regional Anesthesia and Pain Medicine*, **29**, pp. 480–9.

Stevens, P. E., Gwyther, S. J., Boultbee, J. E. *et al.* (1989) Practical use of duplex Doppler analysis of the renal vasculature in critical ill patients, *The Lancet*, **1**, pp. 240–42.

Tooley, M., Greenslade, G., Halliwell, M. and Bolsin, S. (1994) A pilot study to assess the validity of Doppler pulsatility and resistance indices as correlates of renal blood flow changes produced by dopamine in normal subjects, *The British Journal of Radiology*, **67**(799), pp. 733–34.

Further reading

Evans, D. H., McDicken, W. N., Skidmore and R. Woodcock, J. P. (1989) Waveform analysis and pattern recognition. In *Doppler Ultrasound, Physics, Instrumentation, and Clinical Applications*, John Wiley & Sons Ltd, Chichester, pp. 163–87.

Wells, P. N. T. (ed.) (1993) *Advances in Ultrasound Techniques and Instrumentation*, Churchill Livingstone, New York.

Chapter 11

Principles and standards of anaesthetic monitoring

Contents: recommendations for standards of monitoring (oxygenation, airway and ventilation, circulation, depth of anaesthesia, temperature, neuromuscular function, post-anaesthesia care).
The chapter links with: Chapters 12, 13, 14, 15, 16, 18 and 19.

Introduction

The World Federation of Societies of Anaesthesiology (WFSA) adopted standards relating to the safe practice of anaesthesia in 1992 and such standards had already been proposed by a number of countries in order to cut the morbidity due to anaesthesia itself. In the modern era it is easy to forget that historically anaesthesia and surgery did indeed have associated morbidity and mortality and there was very little assistance from technology to monitor patients.

The evolution of these standards is based on two main requirements of monitoring. The first is to record anticipated deviations from normal values, which require accurate measurement to ensure patient safety. The second is to warn of unexpected, life-threatening events that, by definition, occur without warning, and could affect the fit, young patient as easily as the old and infirm. All international standards stress the importance of the continual presence of a fully trained and accredited anaesthetic person, and one Australian study demonstrated that many mishaps occur in the absence of such a person [Runciman 1988]. This applies to general and regional anaesthesia, sedation and recovery. Because perceptions of safety and standards vary throughout the world, despite the presence of an International Standards Organisation, debate about the minimum requirements for monitoring continue. Central to the maintenance of these standards is the quality of persons entering the specialty, the quality of training programmes, and the continuing education of specialists throughout a professional lifetime [Sykes 1992].

It is difficult to determine with certainty the effect that additional technological monitoring has on safety. One clear example is the inability of the trained human eye to detect cyanosis, this human failure occurring maximally

Table 11.1 Classification of critical incidents as a percentage of the total

Incident	%
Equipment failure	13.4
Human error	68.2
Disconnection	13
Other	5.4

at 81–85% oxygen saturation. Clearly, the pulse oximeter has improved the quality of cyanosis detection.

Numerous studies all over the world have shown that mortality due to anaesthesia itself fell significantly between the 1950s and the 1980s, by which time extensive technological monitoring was being introduced, and training programmes had been very much improved.

Utting [1987] reviewed 750 cases of death and cerebral damage reported to the British General Medical Council between 1970 and 1982 that were thought to be the result of errors in technique. Of these, 37% were thought to be accidental, therefore not correctable, and 63% due to error, and therefore correctable. Two thirds of these were concerned with problems of the airway, endotracheal tube, ventilation or hypoxia. A number of these incidents, especially those concerned with airway obstruction, occur in the immediate postoperative period. It is also true that by far the greatest proportion of incidents, including death, are due to human error (Table 11.1) [Cooper 1978] and, of the critical incidents reported, problems with airway control and inspired gas composition constituted the majority.

It is probably true that the market availability of monitoring has driven clinical standards, rather than the other way around. Nevertheless, it is hard to prove that monitoring standards improve safety in anaesthesia, but it has long been agreed that the controlled trials required for this purpose would be unethical [Eichorn 1989]. A review was published of 1.3 million anaesthetics taken from 1976 to 1988; it reported that in the three years since monitoring standards had been imposed in the USA, there had been only one accident and no deaths, as compared with 10 accidents and 5 deaths in the preceding nine years. Analyses like these are often difficult to interpret, however. Runciman [1993], in another major study, concluded that the problems that give the most risk to patients have been identified as those relating to hypoxic gas mixtures, gas flows, breathing systems, endotracheal tubes, the airway, and ventilation; that most of these problems can be avoided following the guidelines outlined below; and that, in many countries, the expense of this additional monitoring is far less than the medico-legal consequences of failing to provide them.

It was also concluded [Webb *et al.* 1993, Holland *et al.* 1993, Runciman *et al.* 1993] that more than 70% of anaesthetic incidents occurring during induction, maintenance and recovery from anaesthesia are detectable by monitor and that, of these, over 80% would be detectable by the correct use of a pulse oximeter. When used in conjunction with a capnometer, over 90% of monitor-detectable incidents are detectable, and this argues strongly in favour of the combined use of these two monitors. One recent study showed that increasing the amount of physiological monitoring to a level above that of modern standard critical care monitoring [Watkinson *et al.* 2006] did not improve adverse outcomes or mortality. On the other hand, novel ways of automatically integrating patients' physiological information [Tarassenko *et al.* 2006] can lead to earlier detection of untoward physiological events.

Given the plethora of monitoring data available to the modern anaesthetist, it is possible to be overwhelmed with information to the extent that a suboptimal response to a critical incident is affected. Work has been carried out to try to determine the best way to present data to the anaesthetist, and although reaction times to graphical displays are more rapid, it is appropriate to have a mixed graphical–numerical display [Kennedy *et al.* 2009, Chakrabati *et al.* 2009], and the evolution of optimal graphical displays continues [Görges *et al.* 2008].

With regard to anaesthesia on a world-wide basis however, it is reasonable to say that safe anaesthesia does not require a mass of complex equipment and the greatest degree of safety for the available resources is achievable by a careful, conscientious anaesthetist who balances risk, benefit and cost in the context in which he or she works. At the time of writing the WFSA is undertaking a worldwide project to extend the availability of pulse oximetry [Walker *et al.* 2009].

Before outlining the present list of international monitoring standards, it is worth discussing the development of alarm systems. Because of a massive increase in the amount of information available to anaesthetists on account of monitors, there can be a cacophony of alarm noises that make alarm systems unpopular, thus risking their non-use, or at least the possibility that the anaesthetist's attention will be directed towards the alarm itself and away from the patient's problem [O'Carroll 1986, Tarassenlo *et al.* 2006, Edworthy *et al.* 2006]. The current international standard on alarms is IEC 60601–1, parts 1 to 8, which has been developed by clinicians, engineers and psychologists. It is important that alarm design is appropriate for the intended

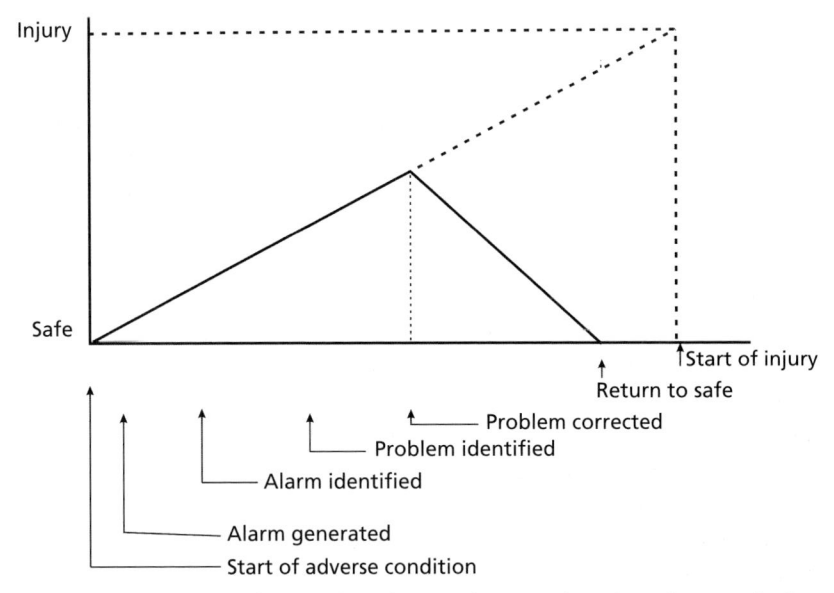

Fig. 11.1 Development of a critical incident in relation to detection, alarms and safety [Schreiber and Schreiber 1987].

users [Nazir *et al.* 2000; Barton 1991] and that the alarm device is appropriately positioned to be safe [Pryn *et al.* 1989].

The anatomy of a critical incident, shown in Figure 11.1, forms a sound basis on which to design alarm systems.

Recommendations for standards of monitoring

The Association of Anaesthetists of Great Britain and Ireland (AAGBI) issues a short summary of its recommendations for standards of monitoring during anaesthesia and recovery, which it regularly reviews and updates [AAGBI 2007]. A more detailed outline of monitoring requirements is given below to enhance the reader's understanding of these requirements.

Oxygenation
Oxygen supply

Supplemental oxygen is usually given to patients. The anaesthetist should ensure the integrity of the oxygen supply and, when nitrous oxide or other supplemental gases are used, the concentration of oxygen in the inspired gas should be monitored throughout each anaesthetic with a monitor fitted with a low oxygen concentration alarm. There should also be an oxygen supply failure alarm and a device to protect the patient against the delivery of a hypoxic gas mixture. In addition a system should be in place to prevent the misconnection of gas sources. Ensuring the integrity of the oxygen supply means ensuring that pipeline connections are intact, that oxygen pipeline pressure is adequate, that there are no cross-connections, and that at least one spare oxygen cylinder with adequate contents is present and connected, with a means available of turning it on immediately. Therefore pipeline and cylinder should both have a pressure gauge, and pressure is an indication of contents. An oxygen meter, of the fuel cell, polarographic or paramagnetic analyser type should be used (see Chapter 16) to ensure that the oxygen supply is uncontaminated, and if other gases are used, that the rotameters give a reasonable indication of the fractional oxygen in the mixture. Note that if very low fresh gas flow is used in a circle system, the fractional oxygen delivered from the anaesthetic machine outlet is not the same as the concentration delivered to the patient's lungs, since oxygen consumption by the patient is a significant proportion of the recirculated gas. Therefore the oxygen meter should be placed close to the patient's airway rather than close to the anaesthetic machine. The human condition is such that the presence of the monitor itself may provide an undeserved feeling of security, and a low oxygen concentration alarm should be fitted. In case of oxygen supply failure, a supply failure alarm should be fitted, and ideally, the rotameters should be fitted with a device to prevent the delivery of an hypoxic mixture, either as a chain linked device on Ohmeda machines, or as hydraulic linkage as on Draeger machines (see Chapter 22). All pipelines and cylinders should be pin-indexed to prevent misconnection of gas sources.

Oxygenation of the patient

As suggested earlier, the human eye is particularly bad at detecting the adequacy of tissue oxygenation. This is no excuse for not looking at the patient, and keeping adequate parts of the patient exposed to do so. However, there is no doubt that continuous pulse oximetry is highly desirable, and it is described in Chapter 15.

Airway and ventilation

The adequacy of the airway and ventilation should be continuously monitored by observation and auscultation of the patient whenever practicable, and the use of a precordial, pretracheal, or oesophageal stethoscope has the advantage of keeping the anaesthetist in clinical and physical contact with the patient.

The correct placement of the endotracheal tube is confirmed initially by observation of equal bilateral chest movements (difficult in the obese patient) and subsequently with a stethoscope by the presence of bilateral axillary breath sounds and an absence of gastric sounds. Its correct placement or indeed the placement of a laryngeal mask and the adequacy of ventilation may be initially confirmed by the feel of manual ventilation, and subsequently by continuous measurement and display of the carbon dioxide waveform and concentration. The continual presence of carbon dioxide monitoring warns of changes to the airway, the ventilation and circulation. Most commonly available are the infrared variety, described in Chapter 16. Other devices include mass spectrometry and Raman spectrometry.

When mechanical ventilation is employed, a disconnection alarm should also be used continuously, as should measurement of the inspiratory and/or expired gas volumes. A disconnection alarm is highly desirable, and can take the form of detection of low volume, flow or pressure, or lack of a CO_2 trace. Volume detectors used to be of the Wright's respirometer type, but are manifest by electronic integration of the flow signal, which itself is measured by a differential U-tube manometer.

The continual presence of the anaesthetist is emphasised because it is only the anaesthetist or the designated, trained anaesthetic staff who have the training and experience to detect an inadequate airway. The basic clinical skills of observation of a patient's breathing, of the reservoir bag, are crucial; auscultation is a reassuring adjunct, especially if an endotracheal tube is involved.

Circulation

Cardiac rate and rhythm

The circulation should be monitored by continuous palpation or registration of the pulse and/or auscultation of the heart sounds. There should be continuous monitoring and display of the heart rate with a plethysmograph (usually incorporated into a pulse oximeter), or with an electrocardiograph (see Chapters 5 and 18). Continuity of observation is stressed here and again the basic clinical skills of palpation and auscultation are very adequate monitors. The oesophageal/precordial stethoscope is also useful, and gives a qualitative measure of force of cardiac contraction. A plethysmographic device, such as the one on a pulse oximeter (see Chapter 15), can also be useful to detect rate and rhythm. The ECG has remained an important rate monitor, despite the introduction of pulse oximeters; it is still the most sensitive detector of rhythm change. These days, software for automatic ST segment analysis is available for intraoperative cardiac ischaemia monitoring; it should be remembered, however, that ECG changes usually represent very late changes in a hypoxic episode (see Figure 11.1).

Tissue perfusion

The adequacy of tissue perfusion should be monitored continually by clinical examination and continuous monitoring with a plethysmograph or capnograph (see Chapter 16) is highly recommended. This is another example of a variable that needs the continual presence of the trained eye. The plethysmographic signal is also useful, but the signal on the pulse oximeter is electronically amplified or attenuated within certain limits, so that it is only useful outside these limits as a monitor of inadequate perfusion. The experienced clinician looks for changes in all monitors to come to a decision about a clinical situation. For example, there may be no change in the blood pressure or oxygen saturation where tissue perfusion is impaired but, in the absence of any other relevant changes, a reduction in end tidal CO_2 levels on the capnograph is an indicator of changes in blood flow to the lungs, which can be a reflection of cardiac output or of pulmonary embolus.

Blood pressure

Arterial blood pressure should be determined at appropriate intervals (see Chapter 12), usually at least every 5 minutes, and more frequently if indicated by clinical circumstances. Continuous registration of arterial blood pressure is encouraged in appropriate cases. While cardiac output is an important variable that we want to know about, it has not been as easy to measure as, for example, blood pressure, other than on the ICU, because it has in the past involved pulmonary artery catheterisation.

Contemporary methods of cardiac output measurement are relatively non-invasive (see Chapter 13). However there is a much longer history of blood pressure measurement, which represents an easily measured, clinically established variable that is directly related to cardiac output. Traditionally, it was measured manually with a double cuff device every 3 or 5 minutes; these days, there are automatically cycling devices that have a degree of accuracy that is perfectly acceptable, when compared against intra-arterial devices, except at very low pressures. Note that no circumstances are mandated on the need for intra-arterial blood pressure monitoring, this being a matter for clinical judgement.

No technique is without morbidity, and intra-arterial monitoring is no exception. Where continuous monitoring is required, then it is desirable; preconditions for this might include: poor cardiac condition, predicted large blood loss, hypotensive surgery, cardiac or vascular surgery, neurosurgery, prolonged surgery; but each case should be judged on its own merits. The requirement of the device is accurate reproduction of all the harmonics of the blood pressure waveform, with no signal distortion from inappropriate damping or phase distortion (see Chapter 12). Note also that no guideline is given on the need for central venous or pulmonary artery pressure measurement, these being a matter for clinical judgement, such as anticipated large blood loss, or poorly functioning left ventricle. The use of the pulmonary artery catheter to measure cardiac output has all but disappeared from clinical practice [van Doorn *et al.* 1994] by the emergence of the less invasive echocardiography (see Chapters 10 and 13), which uses the Doppler principle to measure blood velocity and, by deduction, the cardiac output.

Temperature

Patients can become very cold in the operating theatre, with cold air-conditioning, cold respiratory gases, and cold intravenous fluids, with a subsequent prolongation in recovery. The use of low gas flows, recirculated through a carbon dioxide absorber, heated fluids and blankets have improved this, but hypothermia is a potential problem during long operations. More dangerous is malignant hyperthermia, and patients having operations for the first time are an unknown quantity here. There are perfectly adequate and inexpensive thermistor temperature probes for oesophageal, tympanic or rectal use, and software can be incorporated into existing devices.

Depth of anaesthesia

Clinical assessment of anaesthetic depth is not easy, since all the clinical indicators are indirect, i.e. blood pressure, pulse rate. The gas analysers that measure CO_2 also measure volatile agents and a continual measure of inspired and expired vapour can be made; with a knowledge of MAC values for each agent, the expired value can be used as a measure of the adequacy of anaesthetic depth. This is particularly important where low fresh gas flows are being used, since what is dialled up on the vaporiser bears little relationship to what is in the breathing system until equilibrium is reached; the patient's absorption of the agent, especially early on in the anaesthesia, forms a large proportion of the amount of agent in the breathing system. Later on, absorption diminishes as the patient becomes saturated, and the amount in the breathing system increases with no change to vaporiser setting. More recently developed is the *BIS monitor,* which looks at changes in the EEG that are characteristic of changing depth of anaesthesia; this is described in Chapter 19.

Neuromuscular function

When neuromuscular blocking drugs are given, the use of a peripheral nerve stimulator is recommended. This recommendation is given a relatively low priority but, equally, this monitor is useful to determine the contribution of residual neuromuscular blockade in the patient who fails to resume spontaneous respiration at the end of the operation. This monitor is almost more important where the older neuromuscular blockers are used than it is where newer, shorter acting agents are used [McGrath *et al.* 2006].

Postanaesthesia care

All patients should be observed and monitored in manner appropriate to the state of their nervous system function, vital signs, and medical condition with emphasis on oxygenation, ventilation and circulation. Supplementation of clinical monitoring with quantitative methods described above for intra-anaesthetic patient care is recommended. Pulse oximetry is highly recommended.

References

Association of Anaesthetists of Great Britain and Ireland. (2007) *Recommendations for Standards of Monitoring during Anaesthesia and Recovery*, 4th edn, Association of Anaesthetists.

Barton, A. (1991) Editorial: 'Alarm Signals' over warning signs, *Anaesthesia*, **46**, pp. 809.

Chakrabati *et al.* (2009) Comparison of four different display designs of a novel anaesthetic monitoring system, the 'integrated monitor of anaesthesia (IMA)', *Br. J. Anaes.*, **103**, pp. 670–77.

Cooper, J. B., Newbower, R. S. and Kitz, R. J. (1978) An analysis of major errors and equipment, failures in anesthesia management: considerations for prevention and detection, *Anesthesiology*, **60**, pp. 34–42.

Edworthy, J. and Hellier, E. (2006) Alarms and human behaviour: implications for medical alarms, Postgraduate Educational Issue – Clinical Monitoring, *Br. J. Anaes.*, **97**, pp. 12–17.

Eichorn, J. H. (1989) Prevention of intraoperative anesthesia accidents and related severe injury through safety monitoring, *Anesthesiology*, **70**, 572–77.

Görges, M. and Staggers, N. (2008) Evaluations of physiological monitoring displays: a systematic review, *J. Clin. Monit. Comput.*, **22**, pp. 45–66.

Holland, R., Runciman, W. B. *et al.* (1993) Oesophageal intubation. The first 2000 AIMS reports, *Anaes. Int. Care*, **21**, pp. 608–610.

Kennedy, R. R. *et al.* (2009) The influence of various graphical and numeric trend display formats on the detection of simulated changes, *Anaesthesia*, **64**, pp. 1186–89.

McGrath, C. D. and Hunter, J. M. (2006) Monitoring of neuromuscular blockade. Continuing education in anaesthesia, *Critical Care & Pain*, BJA, **6**, pp. 7–12.

Nazir, T. and Beatty, P. C. W. (2000) Anaesthetists' attitudes to monitoring instrument design options, *Br. J. Anaesth.*, **85**, pp. 781–84.

O'Carroll, T. M. (1986) Survey of alarms in an ICU, *Anaesthesia*, **41**, pp. 742–44.

Pryn, S. J. and Crosse, M. M. Ventilator disconnection alarm failure, *Anaesthesia*, **44**, 1989, pp. 978–81.

Runciman, W. B., Barker, L. *et al.* (1993) The pulse oximeter: applications and limitations. The first 2000 AIMS reports, *Anaes. Int. Care*, **21**, pp. 543–50.

Runciman, W. B. (1988) Monitoring and patient safety: an overview. *Anaes. & Int. Care*, **16**, pp. 11–13.

Runciman, W. B. (1993) Risk assessment in the formulation of anaesthesia safety standards, *Eur. J. Anaesthesiol.*, **10**(suppl 7), pp. 26–32.

Schreiber, P. and Schreiber, J. (1987) *Anesthesia Systems Risk Analysis and Risk Reduction*, North American Draeger.

Sykes, M. K. (1992) Clinical measurement and clinical practice, *Anaesthesia*, **47**, pp. 425–32.

Tarassenko, L., Hann, A. and Young, D. (2006) Integrated monitoring and analysis for early warning of patient deterioration, Postgraduate Educational Issue – Clinical Monitoring, *Br. J. Anaes.*, **97**, pp. 64–68. (1993) The International Task Force on Anaesthesia Safety: International Standards for a Safe Practice of Anaesthesia, *Eur. J. Anaesthesiol.*, **10**(suppl 7), pp. 12–15.

Utting, J. E. (1987) Pitfalls in anaesthetic practice, *Br. J. Anaes.*, **59**, 877–90.

van Doorn, C. A. M., Lyons, G., Swindells, S. and Unnikrishnan Nair, R. (1994) Perforation of the right ventricle and cardiac tamponade; the use of a pulmonary artery catheter, *Br. J. of Intensive Care*, 4(Feb), pp. 66–67.

Walker, I. A., Merry, A. F., Wilson, I. H. *et al.* (2009) Global oximetry: an international anaesthesia quality improvement project, *Anaesthesia*, **64**, pp. 1051–60.

Watkinson, P. J. *et al.* (2006) A randomised controlled trial of the effect of continuous electronic physiological monitoring on adverse event rate in high risk medical and surgical patients, *Anaesthesia*, **61**, pp. 1031–39.

Webb, R. K., van der Walt, J. and Runciman, W. B. *et al.* (1993) Which monitor? The first 2000 AIMS reports, *Anaesth. Int. Care*, **21**, pp. 529–42.

Chapter 12

Blood pressure measurement

This chapter contains: history; non-invasive blood pressure measurement (palpation, auscultation, oscillotonometry, oscillometry, cuffs, plethysmographic methods); direct measurement of blood pressure (system components, transducers, strain gauges, resonance and damping); other blood pressure measurements.
The chapter has links with; Chapters 1, 3, 4, 5, 6, 10, 11 and 13.

Introduction

Blood pressure measurement occurs either non-invasively or invasively, and usually refers to systemic arterial pressure measurement, but can also refer to systemic venous or pulmonary arterial pressure measurement. In 1733 the Reverend Stephen Hales was the first person to measure the blood pressure *in vivo* in unanaesthetised horses by direct cannulation of the carotid and femoral arteries. In doing so he observed the pulsatile nature of flow in the circulation. In 1828 Poiseuille developed the mercury manometer, and used it to measure blood pressure in a dog. The mercury manometer has, of course, become the standard technique against which other techniques are compared. The earliest numerical information on blood pressure measurement came from direct rather than indirect measurement in 1856 by Faivre, using Poiseuille's device. However, in the last part of the nineteenth century, non-invasive measurement techniques were developed. In 1903, Codman and Cushing introduced the concept of routine intraoperative blood pressure measurement, which at the time was a revolutionary concept. Nowadays it is a fundamental part of minimal monitoring criteria.

Non-invasive blood pressure measurement

There are several techniques of non-invasive BP (NIBP) measurement, all of which function by occluding the pulse in a limb with a proximal cuff, then detecting its onset again distally, on lowering the cuff pressure. Detection methods include palpation, auscultation, plethysmography, oscillotonometry and oscillometry. Accuracy of all non-invasive techniques depends on cuff size in relation to the limb concerned, and over which artery the cuff is placed. Such techniques of NIBP measurement are necessarily intermittent. Much discussion has taken place on the accuracy of these devices, and the accuracy of diastolic pressure measurements needs improving, and there are ideas proposed for new non-invasive devices [Tooley and Magee 2009].

Palpation

In the absence of a stethoscope, this technique is simple and reliable. After inflating the cuff on the upper arm to a pressure of above that of systolic, the cuff is then deflated while palpating the brachial artery and the systolic pressure is measured with a mercury column at first detection of the pulse. A study by van Bergen [1954] showed that BP can be underestimated by this method by up to 25% at 120 mmHg. Factors that worsen this underestimate include bradycardia and rapid cuff deflation.

Auscultation

Auscultation depends on hearing Korotkoff sounds, first described in 1905. The sounds are described in phases, from I to V, shown in Figure 12.1. Note that there may be an *auscultatory gap* within phase II, the presence of which, in

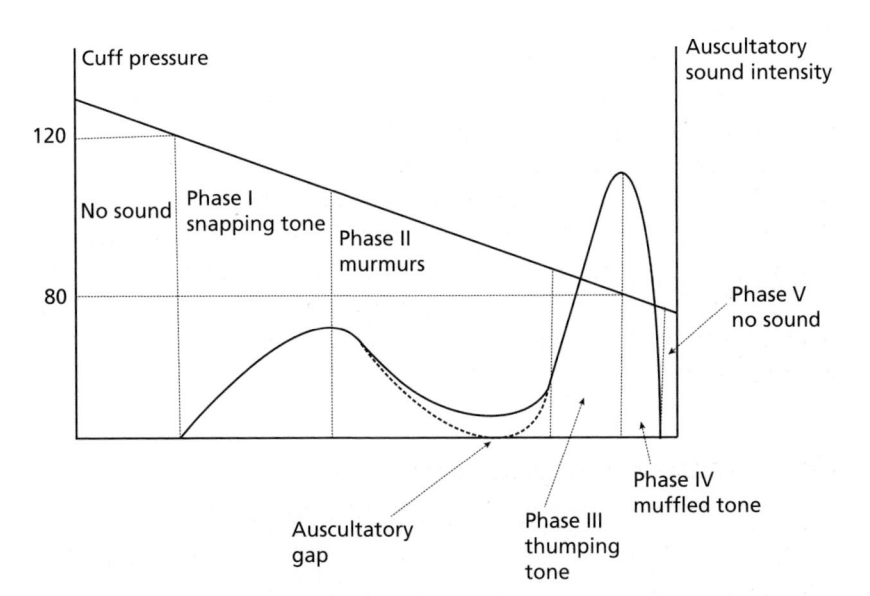

Fig. 12.1 Graphical representation of cuff deflation and intensity of sounds on auscultation when measuring blood pressure.

some hypertensive patients, indicates the need for a rough estimate of systolic BP by palpation before a more accurate measure by auscultation. The physical origin of Korotkoff sounds is uncertain, but is almost certainly due to resonance set up in tissues from turbulent flow in the artery.

Phase I is the first onset of a *tapping*, high frequency sound, which is synchronous with the pulse, which increases in intensity, which represents the onset of systolic blood pressure, and 5–10 mmHg below which there appears a palpable pulse. In phase II the frequency of the sound is lower, represented by a softening of the sound, and in some patients the sounds in this phase disappear altogether – the auscultatory gap – before the frequency increases again and the sounds then become sharper in phase III. In phase IV there is an abrupt muffling of the sounds and their complete disappearance in phase V. Diastolic blood pressure corresponds to either phase IV (American Heart Association, children) or phase V (American Heart Association, adults) or both (World Health Organisation). Phase V is closer to real diastolic than phase IV and there is better agreement between observers on phase V onset. However in some hyperdynamic states, phase V may not occur until the cuff is almost completely deflated.

One error with this method (and with other cuff methods) is using the wrong cuff size; too small a cuff overestimates the pressure, too large a cuff underestimates the pressure. Other causes of inaccuracy include lack of agreement between observers, over-rapid deflation of the cuff, and bradycardia. The studies of Pereira et al. [1985] showed that, although there were reasonable positive correlations between direct and auscultatory measurements for systolic ($r = 0.93$) and diastolic ($r = 0.79$) pressures, there was a large variation between observers. They showed that auscultation tended to overestimate at lower pressures and underestimate at higher pressures. However, the method is simple and reliable.

Oscillotonometry

Figure 12.2 shows the double cuff used in oscillotonometry, which is primarily of historical interest, but is nevertheless technically interesting because of the technologies involved. The narrower upper cuff is inflated and deflated in the fashion described above. The wider lower cuff is the means of pulse detection as the upper cuff is deflated. The most popular such device, was *von Recklinghausen's* oscillotonometer, with overlapping upper (occluding) and lower (sensing) cuffs C_o and C_s respectively. Both cuffs, as shown in Figure 12.2, are connected to a pair of aneroid barometers in a sealed box, a valve mechanism and an inflating bulb, which allows inflation and deflation of the cuffs. The aneroids are linked to each other and to a single pointer on

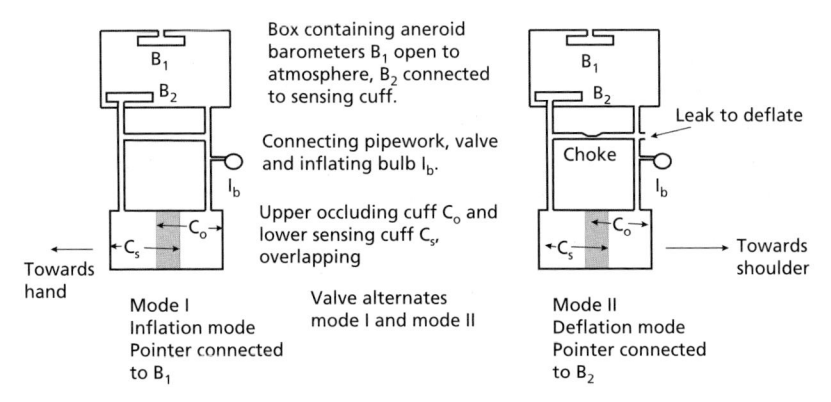

Fig. 12.2 Principles of oscillotonometer.

the dial. In mode I, on inflation of the cuffs, the absolute value of the pressure compressing the arm is read by aneroid B_1. Depression of a lever converts the device to mode II and, as the air is bled out of the system, with some delay in the lower cuff due to the choke, pulsations on the lower cuff, of increasing magnitude as systolic pressure is approached, are transmitted to aneroid B_2, which is connected, in mode II, to the pointer. Attenuation of the pulsations corresponds to diastolic blood pressure. In each case the actual value of blood pressure can be read off by releasing the lever, reconnecting aneroid B_1 to the pointer. Actual values are prone to variation depending on interpretation, particularly with regard to diastolic BP.

Oscillometry

In contrast to oscillotonometry described above, this method uses a single cuff, both to compress the artery and to detect pulsations. This is the standard contemporary method of NIBP measurement in automated devices using microprocessor technology. If the cuff is automatically inflated to some predetermined level above systolic BP, then deflated continuously in a stepwise fashion, arterial pulsations are detected on the cuff as systolic BP is reached. As mean arterial pressure is reached, these pulsations reach maximum amplitude, and as diastolic pressure is approached these pulsations diminish and disappear. This process is shown graphically in Figure 12.3. The mechanism inside such a device is shown in Figure 12.4. A single pressure transducer detects the cuff pressure itself continuously as well as the arterial pulsations. Appropriate electronic circuitry digitises and processes the signal, then systolic, mean and diastolic pressures are displayed. The original Dinamap

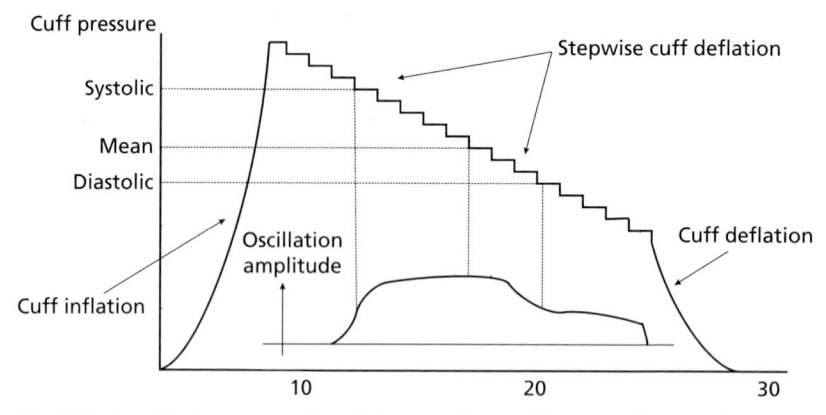

Fig. 12.3 Graphical representation of the use of an oscillometer, showing decreasing cuff pressure plotted against time in seconds on deflation with increasing transmission of oscillation and the link to measured components of blood pressure.

measured only mean arterial pressure (MAP), where cuff oscillations are at a maximum. It was found to be accurate if the compressible volume within the hydraulic components of the device was kept to a minimum. With the subsequent development of more sophisticated electronic algorithms, systolic BP (SBP), MAP and diastolic BP (DBP) became measurable. SBP is deemed to

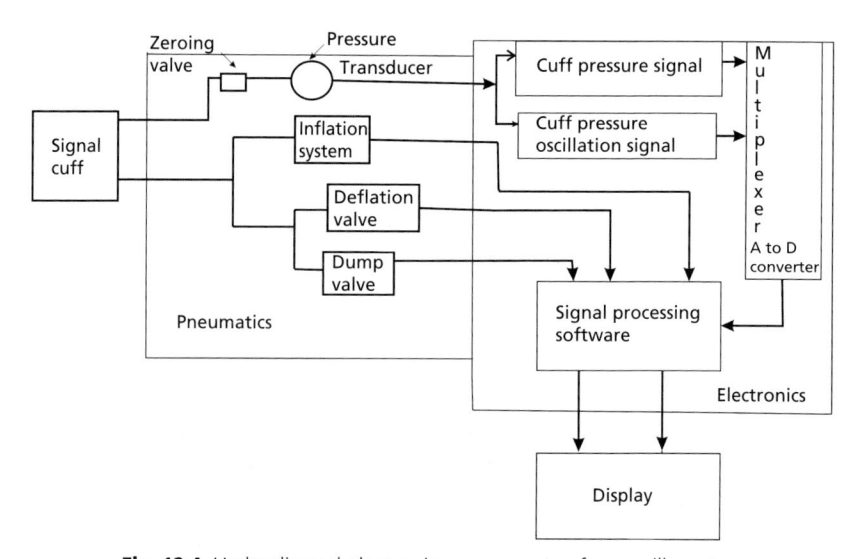

Fig. 12.4 Hydraulic and electronic components of an oscillometer.

occur where the rate of increase of the pulsations is at a maximum (rather than the onset of pulsations) and the DBP is measured where the rate of decrease is maximum (rather than the offset). It is generally more difficult to detect diastolic pressure than systolic using this technique. It is also likely that the algorithm checks the measurements with the equation

$$MAP = DBP + \frac{(SBP - DBP)}{3}.$$

This assumes that the pressure waveform is triangular in shape, with DBP at the base, SBP at the apex, with MAP at the triangle's centroid. There are also oscillometric devices that deflate the cuff continuously rather than in discrete steps, and which analyse the overall oscillation envelope.

Excessive cuff movement, rapid BP changes, abnormal pulse rhythms and interchanging cuffs can all exceed the ability of the electronic algorithms to interpret the BP components accurately. Otherwise these devices have been shown to be convenient and reasonably accurate.

Doppler ultrasound

Blood pressure can also be determined using ultrasound, by detecting blood vessel wall movement or blood flow. In either case the Doppler principle is the applicable physical principle (see Chapter 10). An ultrasonic transmitter and receiver lies against the skin beneath an occluding cuff. As the cuff is deflated and blood flow or vessel movement is detected, there is an audible Doppler frequency shift, whose character changes as the cuff pressure falls through systolic to diastolic. Analysis of frequencies by the device yields SBP and DBP. Doppler techniques have been found to yield higher systolic pressures that an old fashioned sphygmomanometric method [MacDonald *et al.* 2008].

Cuff size

This is an important factor in determining the accuracy of any NIBP measurement technique. In particular, this factor has implications for NIBP measurement in children and in obese patients. The guidelines of the American Heart Association stipulate that the width of (the inflatable bladder part of) the cuff should be 40% of the mid-circumference of the limb, and that the length of the cuff should be twice this width. A narrow cuff gives a falsely high reading and a wide cuff gives a falsely low one. A cuff reading of blood pressure may indeed not correlate with intra-arterial BP measurement across the whole range of BPs, because there is a non-linear relationship between the pressure in a cuff and its internal diameter around a limb, which is the variable important in compressing the limb concerned.

Morbidity from cuffs includes skin and underlying tissue damage and possible ulnar nerve damage, when used on the arm too close to the elbow.

Other cuff sites

Sometimes the upper arm is not the most convenient place at which to measure the blood pressure using a cuff. The calf and the ankle are two such alternative sites. It has been found that there is poor agreement between different sites for systolic pressure measurement, but better agreement for diastolic and mean pressures [Moore et al. 2008], and that the ankle is the least uncomfortable place for the cuff. If the ankle is used as a cuff site, then the additional presence of an intermittent compression device on the leg does not influence the BP reading, if the measurement is made when the compression device is not in its active phase [Parikh et al. 2007].

Plethysmographic methods

The *Finapres* uses the principle of arterial volume clamp plethysmography, developed by Peñaz and Slurer. Figure 12.5 shows the Finapres. A small, low volume cuff, which also contains an infrared light source and photo detector, so that the finger is transilluminated, is applied around a finger. The amount of infrared light absorbed depends to some extent on the blood volume in the finger at the time; this is the same technology as the pulse oximeter uses. The finger cuff is relatively non-compliant, and is attached to a small pump

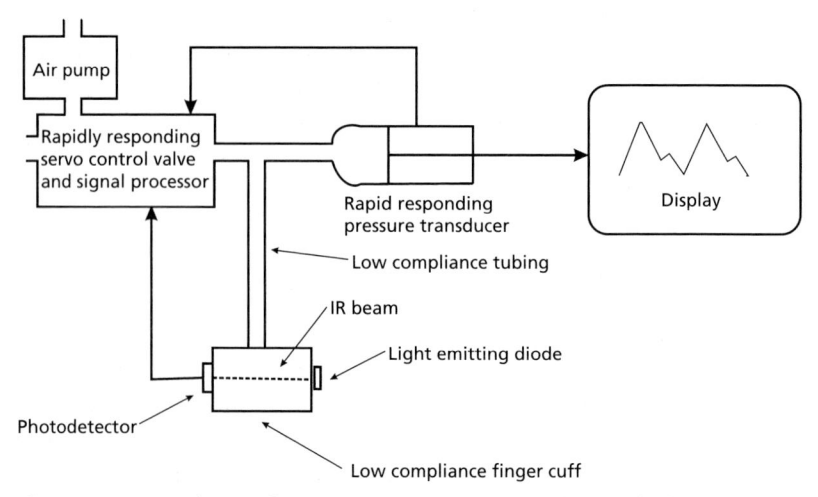

Fig. 12.5 Peñaz technique for continuous non-invasive technique for blood pressure measurement (Finapres).

and a solenoid valve, which allows rapid small amplitude pressure changes to be applied to the finger. This is linked to a servo-control mechanism with a fast response time. Initially the system determines the MAP during the 'open loop phase', and subsequently the arterial waveform is continuously followed during the 'closed loop phase'.

During the open loop phase the cuff is deflated from maximum inflation until cuff oscillations are maximum, corresponding to MAP determined by an oscillometric method. The amount of infra-red light absorbed in the finger at this point corresponds to a *set point*. During the closed loop phase, the cuff pressure is adjusted rapidly to maintain constant IR light absorption in the finger tissues. It does this in the presence of arterial pulsation and vessel distension, as deviations from MAP occur due to intra-arterial pressure rise to systolic BP, and fall to diastolic BP from the set point at MAP. Thus the device tracks arterial pressure and the output is displayed as an arterial waveform.

In order for the accuracy of the device to be maintained, all components of the compressible volume, including the non-distensible cuff, must be kept to a minimum. The whole electro-pneumatic and plethysmographic device, with associated amplifiers, is contained in a box strapped to the patient's arm. There have been variable reports of the accuracy of the Finapres under different clinical circumstances, which is probably why it has not been accepted into widespread use [Farquhar 1991]. However, its small size makes it convenient to use in mobile situations and the fact that it produces a continuous output makes it useful.

Another device that falls into this category and functions in a similar way, but without a cuff, is the 'Tensymeter' [Szmuk *et al.* 2008], a non-invasive device that is used over the radial artery. An actuator applies external pressure to the artery, and a sensor assesses the pulsations at different levels of applied pressure, and uses an algorithm to produce a continuous output of systolic, mean and diastolic pressure.

Direct measurement of blood pressure

The direct measurement of arterial blood pressure has some advantages over other blood pressure measurements: the waveform obtained from the data is continuous (as opposed to intermittent), and it can be in real-time and is potentially the most accurate of any blood pressure measurement. Direct measurement is very important in critically ill patients and in those whose cardiovascular system may be compromised. It is important in patients who require physiological or pharmacological manipulation of the blood pressure, e.g. cardiopulmonary bypass, or hypotensive anaesthesia. It also provides a convenient means for taking blood for blood gas analysis. When the waveform is obtained, a number of other indices [Hutton and Cooper 1985, Prys-Roberts 1984] can be obtained, apart from the systolic, diastolic, mean and heart rate. Myocardial contractility can be assessed from the gradient (dp/dt) of the arterial upstroke as shown in Figure 12.6.

The area beneath the curve up to the dicrotic notch gives an index of stroke volume, which is also shown in the figure. The characteristics of the diastolic decay give information about the resistance and compliance of the peripheral vascular bed. An abrupt fall in the pressure wave from its peak with no tail can be a sign of reduced peripheral vascular resistance. High peripheral resistance is shown by a shallow diastolic decay, as shown in the diagram. The pulse waveform is used as the basis of pulse contour analysis to estimate cardiac output (see Chapter 13).

Components of the monitoring system

The components of the direct blood pressure monitoring system include the hydraulic coupling, the transducer, amplification, signal processing, signal display, and power supplies to the various components.

Hydraulic coupling

Ideally, for the greatest accuracy, the pressure transducer should be inserted directly into the site to be measured. Whilst catheter based transducers do exist, this section deals with transducers that are connected to the measurement site by means of a fluid filled catheter, usually isotonic dextrose or saline. For a correct representation of the waveform, and to control damping and resonance

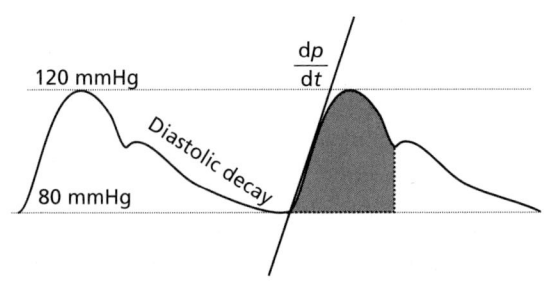

Fig. 12.6 Arterial waveform showing indices obtained: myocardial contractility (dp/dt), index of stroke volume (represented by the shaded area), and diastolic decay.

(as discussed at the end of this chapter), the catheter must be reasonably stiff, straight, and the fluid is assumed to be incompressible and must not contain any air bubbles. The catheter would terminate at the transducer dome, which is a disposable device containing a flexible membrane, which would sit in very close proximity to the diaphragm of the transducer. The pressure change at the measuring site would induce a similar change at the membrane, which would move the transducer diaphragm. The amplifier and signal processing have been discussed in Chapter 5. The electrical power to the transducer (for the bridge circuit) is normally obtained, via suitable electrical isolation, from the amplifier.

Transducers

Transducers have already been briefly mentioned in Chapters 3 and 5. This chapter deals with different transducers in more detail specifically used for direct pressure measurement. The most common type of pressure transducer consists of a diaphragm, one side of which is open to the atmosphere and the other connected to the pressure that is to be measured. Pressure, as mentioned above, causes a displacement of the diaphragm that can be measured in many ways. Strain gauges, resistive, capacitive and inductive, are the main examples, with resistive strain gauges being the most common.

Resistive strain gauges

The resistance of a wire increases with increasing strain. As the length increases and the diameter decreases, so the electrical resistance can increase; conversely the compression of the wire has the opposite effect. Different materials will have different sensitivities to the stress and strain. In addition different materials will be affected to different extents by temperature in this regard. Strain gauges are constructed to be used in a pressure transducer, as shown in Figure 12.7.

The wire is connected to a flexible base as this is bonded to the structure under measurement. Nickel and platinum wires are often used as they have appropriate sensitivities and temperature coefficients. Silicon wires can be used as they have much higher sensitivity, but are also affected by temperature change to a greater extent, with significant non-linear effects at higher temperatures.

Strain gauges therefore tend to be of two main types: crystals of silicon and metallic. A single crystal of silicon with a small amount of impurity can be used and the variation in performance with temperature can be minimised by connecting four such crystals in a Wheatstone bridge arrangement (see Chapter 4), so that all the bridge resistors are in fact the crystals themselves, all

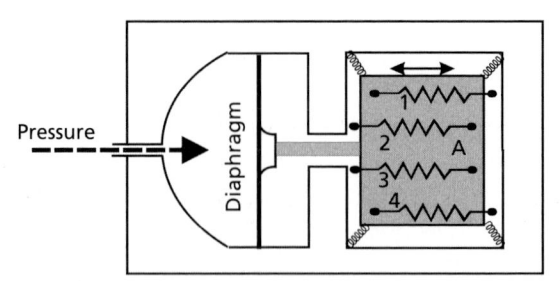

Fig. 12.7 Diagram of a typical pressure transducer which uses bonded strain gauges. The pressure moves the diaphragm back and forth, which in turn moves the plate A (which is held in place by springs at each corner as shown). With pressure in the direction of the arrow, the movement of plate A causes strain gauges 1 and 4 to be compressed, and 2 and 3 to be expanded.

contained within the transducer. Two of the strain gauges are arranged so that their resistance increases with pressure (e.g. resistances 2 and 3 in the diagram) and the other two will decrease with pressure. This bridge arrangement also has the advantage over the bridges already discussed in that the sensitivity to the pressure signal is quadrupled. The temperature will affect all the stain gauge components in the bridge equally and therefore the outputs will not change.

Inductance gauges

The inductance of a coil depends on its geometry, the magnetic permeability of the medium or core on which the coil is wound, the overlap between core and coil, and on the number of turns of the coil [Geddes *et al.* 1989]. If the core is attached to the transducer membrane, then the pressure movements will cause the core to be more or less inserted into one coil, as shown in Figure 12.8(a). The inductance will therefore change in proportion to the pressure on the diaphragm. If the inductance is part of a tuned circuit, e.g. resonated at rest at say 5 kHz, and the circuit is also either part of a Wheatstone bridge circuit, or part of an oscillator and frequency-to-voltage-converter circuit, then the blood pressure will produce a voltage proportional to it in both cases. Alternatively, two coils can be used, as shown in Figure 12.8(b). Again the transducer membrane alters the position of the core, but the core now alters the transformer action. The lower coil is supplied by an oscillator, and the top coil gives an output of the oscillator frequency, and its magnitude proportional to the pressure.

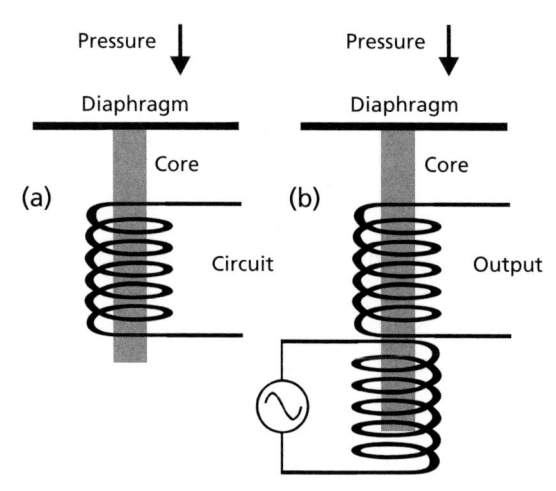

Fig. 12.8 Diagram showing the principle behind an inductance based pressure transducer. The pressure moves the diaphragm, which moves the iron core in and out of the coils(s). (a) refers to the single coil transducer and (b) refers to the two coil transducer with input and output coils.

Capacitive based gauges

If one plate of a capacitor (B in Figure 12.9) is fixed and the other plate is part of the moveable membrane (A in the figure), then the diaphragm movements due to pressure will be translated to a change in capacitance. This moveable plate should be as small and stiff as possible, to give maximum high frequency response. If the capacitor is part of a circuit with a high frequency carrier (typically 8 kHz), and the circuit is connected either to one side of a Wheatstone bridge or to the frequency-to-voltage converter circuit, then a voltage proportional to the pressure will be produced in both cases. The capacitor transducer has high stability, good temperature stability, and has a

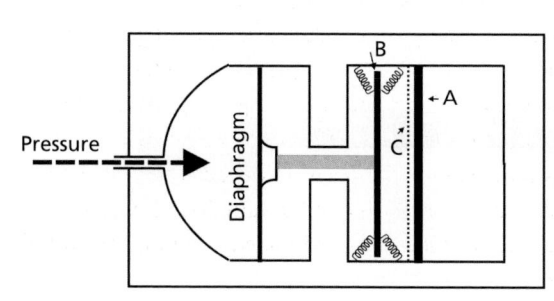

Fig. 12.9 Capacitive pressure transducer. The pressure pushes the diaphragm in and out, and this alters the position of plate B, which is connected by springs to the body of the transducer. Plate A is fixed, and so the capacitance is proportional to the pressure. C is the dielectric material.

good linear pressure range from −30 to 300 mmHg. It can also cope with high voltages from a defibrillator. It does however require more complex driving electronics than the strain gauge example.

Resonance and damping

In this section, resonance and damping will be discussed, using invasive blood pressure, as a *mechanical* example of resonance, and electrical analogues can also be derived. The mechanical measurement system is shown in Figure 12.10.

A pressure change, due to the blood pressure waveform, will cause fluid to move through the catheter and displace the diaphragm of the transducer. In this example, inertance of the transducer system (mass), resistance of the system (viscous drag or friction) and compliance of the system (elasticity or stiffness) are analogues to electrical inductance, resistance and capacitance respectively. There is an appropriate electrical circuit to the diagram in Figure 12.10, which is shown in Figure 12.11.

The 'c' subscripts represent the catheter, 't' the transducer, and 'd' the diaphragm and the driving pressure is the blood pressure. In practices, when there is a stiff catheter and a bubble free fluid, then the compliance of the catheter, and friction and inertia of the transducer are very small. A simplified model is shown in Figure 12.12. The dotted extension is the addition of air bubbles into the system. Approximate values can be estimated for the circuit components by using classical mechanical system equations [Brown et al. 1999].

$$R_c = \frac{8\eta L}{\pi r^4}, \quad L_c = \frac{\rho L}{\pi r^2}, \quad C_d = \frac{\Delta V}{\Delta P},$$

where L is the catheter length (as in Figure 12.10), η and ρ are the viscosity and density of the fluid in the catheter, r is the radius of the catheter ($d/2$ in Figure 12.10), ΔV is the volume change (related to the displacement in Figure 12.10) and ΔP is the pressure difference causing that change.

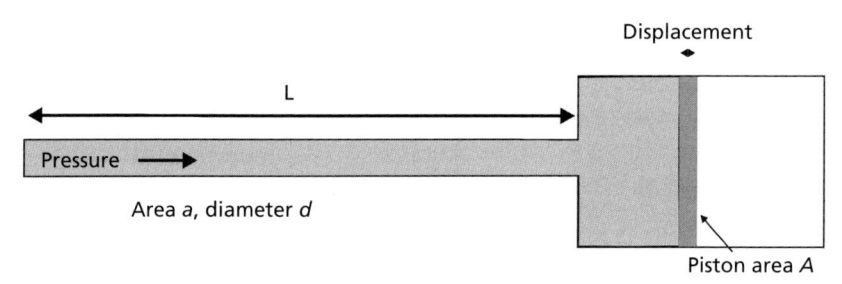

Fig. 12.10 Diagram of catheter and transducer piston.

Fig. 12.11 Electrical analogy circuit of Figure 12.10. See text for explanation.

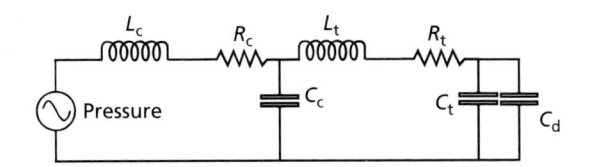

The phenomenon of *resonance* is an issue that needs to be understood in the measurement of blood pressure using arterial lines and transducers. When any physical system has vibration imposed upon it, it will vibrate at a frequency and an amplitude dependent on the relationship between its own physical properties and the externally imposed force. For a measurement system, such a phenomenon produces an unacceptable variation in response to a change in a variable being measured, and the fidelity and accuracy of the measurement is very much impaired. If the external applied force occurs at a frequency which coincides with a harmonic (a multiple) of one of the natural frequencies of the system, then the system will resonate, which means it oscillates at a characteristic *resonant* frequency at an amplitude that may increase dramatically (theoretically to an infinite amplitude) in the absence of any other forces that may prevent that. In practice, all physical systems have such frictional forces that attenuate the amplitude achieved by a system oscillating at its natural frequency. The *damping* in the system is a measure of the frictional force (hydraulic or frictional) acting on the mass and changes the oscillating response to one where the amplitude does not increase as much. If the waveform of the blood pressure is applied to the system, the fluid in the catheter will transmit the pressure to the piston and cause the diaphragm to oscillate in response. If the pressure is removed then the diaphragm will eventually return to the original position. This return can be slow or rapid and the diaphragm may oscillate depending on the relationship between the mass of the system, the elasticity and damping. These factors affect how faithfully the blood pressure signal is transferred to the pressure transducer.

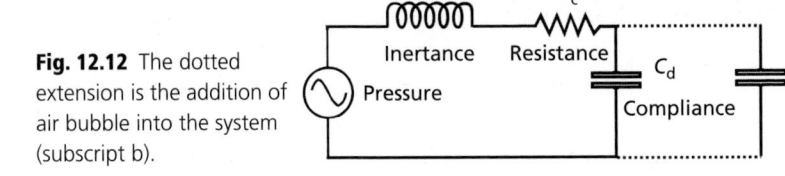

Fig. 12.12 The dotted extension is the addition of air bubble into the system (subscript b).

The following mathematical expression describes this (see also Chapter 1):

$$M\frac{d^2x}{dt^2} + k_1\frac{dx}{dt} + k_2 x = \text{sum of applied forces,}$$

where M is the mass-inertia component, k_1 is the damping constant, and k_2 is the stiffness of the system. This can be translated into the electrical equivalent:

$$L\frac{d^2q}{dt^2} + R\frac{dq}{dt} + \frac{q}{C} = \text{applied voltage.}$$

This is equivalent to

$$L\frac{di}{dt} + Ri + \frac{q}{C} = \text{applied voltage,}$$

where L is the inductance, R the resistance (the damping), and C the capacitance. q is the charge and i the current.

The response of the direct blood pressure system can be measured by triggering it with two different inputs. The first is using a *step* function and the second a driven variable frequency sine wave. The step function is a sudden change from no pressure to full pressure, instantaneously, for example by activating a continuous flush device and then suddenly releasing it. The variable frequency is obtained by putting a pressure oscillator into the system that starts at DC (zero frequency) and goes up to 200 Hz. This was discussed in the concept of dynamic calibration in Chapter 3 and shown in Figure 3.6.

If the step is applied, the diaphragm moves to its new (correct) position in a time dependent on the damping in the system. The *overshoot* (the displacement beyond the final resting level) will depend on this as well. In solving the second-order differential equation above, the solution can be described as a combination of a sinusoidal waveform inside an envelope of a decaying exponential function, as shown graphically in Figure 12.13. The factor k_1/M describes the amount of damping present in the system, and if it exceeds a certain value the oscillatory component will reduce to zero, which may adversely affect the response of the system. If it is less than a certain value, the amount of oscillation will become prolonged, and this too reduces the usefulness of a measurement system. It is normal to make this factor a non-dimensional *damping ratio*, D. When there is *critical* damping, defined as the damping ratio $D = 1$, the diaphragm will respond as quickly as possible without overshoot. Less damping results in a more rapid response of the diaphragm but there is overshoot. If there were no damping (which is not possible in practice) then the system would oscillate at the natural frequency of the system. A summary of the different type of responses encountered with different damping ratios is shown in Figure 12.13. This demonstrates that the

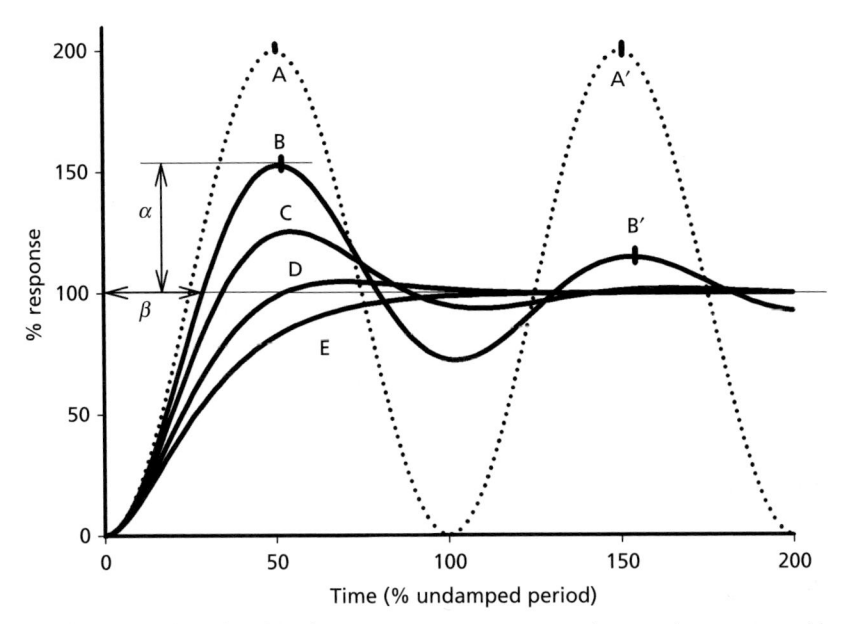

Fig. 12.13 Diagram showing the system response to a step increase in pressure, with different damping ratios. The dotted response A is with no damping ($D = 0$). B is with light damping (damping ratio of 0.2). C is with a ratio of 0.4. D is with a ratio of 0.7, E is with damping of 1. The final position is shown by the line that goes through % response $= 100$. α and β are the overshoot and rise-time of response B respectively. The undamped frequency of the system is $1/(A'-A)$ Hz and the damped frequency with damping ratio of 0.2 is $1/(B'-B)$ Hz.

more damping there is, particularly where it is greater than critical damping, the less the overshoot and the more time the system takes to reach the final value. This means that the frequency of the system becomes lower as the damping is increased. When the damping is critical, ($D = 1$), shown in curve E, then this is the fastest the diaphragm can reach the final position without overshoot.

The resonant frequency of the system: $f_o = \dfrac{1}{2\pi}\sqrt{\dfrac{k_2(\text{stiffness})}{M(\text{mass component})}}$.

The electrical equivalent can be seen to be: $f_o = \dfrac{1}{2\pi}\sqrt{\dfrac{1/C}{L}}$,

where C is the capacitance, and L the inductance. The damped natural frequency is the reciprocal of time B'–B (from Figure 10.12) and is:

$$f_D = f_o\sqrt{1 - D^2} \text{ or in the example } f_D = f_o\sqrt{1 - (0.2)^2}.$$

Therefore the damped frequency is lower than the natural undamped frequency. The rise time is related to both the resonant frequency of the system and the damping factor. To obtain a more rapid rise response without excessive overshoot, it is necessary to raise the natural frequency of the system (faster frequency = steeper rise time, shorter period). To do this a stiffer diaphragm (K_2) is required or a using a lighter mass in the system (M), from the equation above.

The second method of looking at the response of the system is to apply a sine wave pressure signal at constant amplitude to the system, but of gradually increasing frequency. The output of the pressure transducer could be similar to Figure 3.6, and the relative amplitude will follow one of the curves in Figure 12.14, depending on the damping of the system.

At low frequencies the output of the system remains at a constant amplitude, independent of frequency, indicating that the system is accurately following the input pressure waveform. However, as the frequency is increased beyond a certain value, the output waveform can increase in amplitude, the

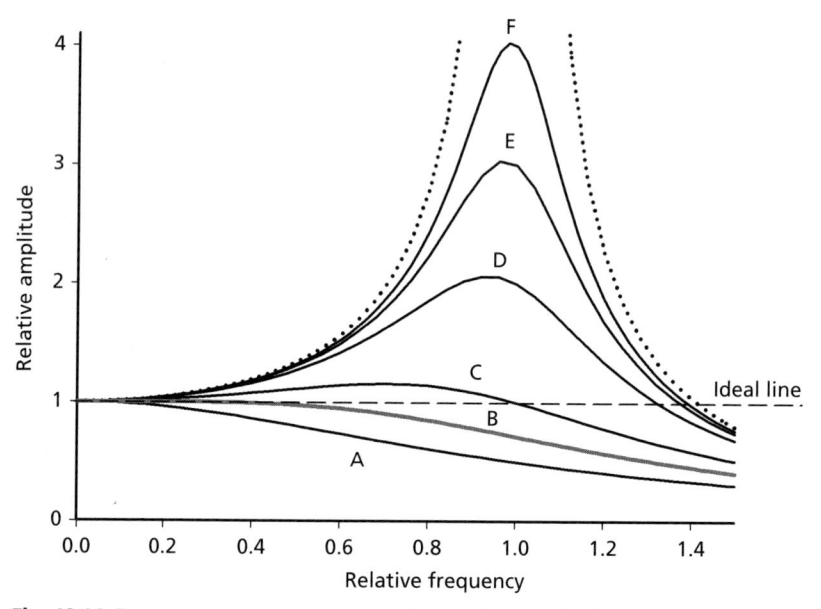

Fig. 12.14 Frequency response curves of the relative amplitude (ratio of the output wave and the input wave) against the relative frequency (the ratio of the frequency and the resonant frequency). Curve A is with the damping ratio of 1, B 0.7, C 0.5, D 0.25, E 0.17 and F is with 0.125. The dotted lines represent the system with a damping ratio approaching zero. The dashed line is the ideal frequency response with a relative amplitude equal to 1.

peak of the response occurring at the resonant frequency of the catheter–transducer system. At higher frequencies, after this resonant peak, the amplitude of the response declines towards zero.

With critical damping $D = 1$ (curve A), the amplitude of the frequency response curve, as shown in Figure 12.13, falls as the frequency is increased above 0.1 of the resonant frequency. With less damping (i.e. curves C–F), the amplitude of motion increases as the resonant frequency is approached, and then decreases. The figure suggests that with $D = 1$, the system can respond faithfully (i.e. no distortion, following the ideal response line) to sine waves of up to about 10% of the natural resonant frequency. If this resonant frequency is much higher than the highest frequency component of the pressure wave of interest, then this is not a problem. When the damping is 0.5 (curve C), then the frequency response is more uniform, but there is more overshoot. A damping factor of 0.7, shown by the response (curve B), is a special case. There is no resonant peak in the response, it follows the ideal response for about 40% of the resonant frequency, there is a 5% amplitude overshoot (see Figure 12.13 curve D), and also the phase shift is linear up to the resonant frequency. As discussed in Chapter 3, this linear phase shift means that the each sine wave frequency component in the pressure signal has a delay in the signal (e.g. due to inertia components) linearly proportional to the driving frequency. These are all important conditions for the faithful reproduction of the pressure signal by the system. With the damping at 0.7, then the shape of the blood pressure is faithfully reproduced and the step response is only slightly compromised. However, if the damping is reduced to 0.64 (*optimal damping*), then the response time shortens but the overshoot rises to 7%. This is the best compromise between speed of response and overshoot. Figure 12.15 illustrates this and shows the graph of speed (rise time) against overshoot (see Figure 12.13 as well for graphical explanation of rise time and overshoot) at different damping values. The frequency response and the phase response are good for the damping of 0.64, but not quite as good as the 0.7.

With optimal damping of 0.64, then parameters of the system can be defined as:

Step function:
- rise time for step function (β as defined in Figure 12.13) = 48% of natural frequency.
- overshoot (α in Figure 12.13) = 7% over final amplitude value.

Sine wave frequency generator response:
- resonant peak (above ideal response line) = 1.3%
- bandwidth (to 70% of ideal response line) = 0–108% resonant frequency

188 | BLOOD PRESSURE MEASUREMENT

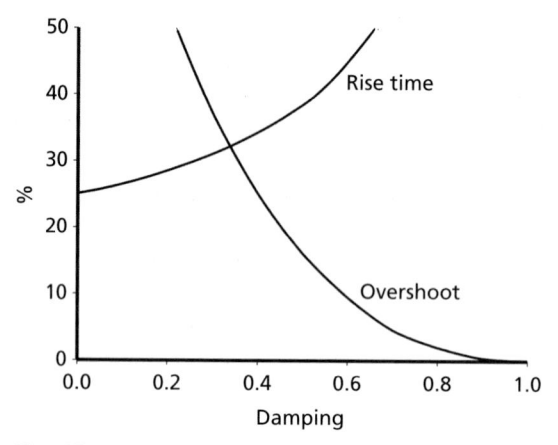

Fig. 12.15 Damping values verses rise time (% of undamped period) and overshoot (% over final value) in terms of the response to a step function. (Figure 12.13 shows a graphical description of rise-time and overshoot.)

The damping ratio can be estimated by measuring the overshoot (or rise-time) and using Figure 10.14. Alternatively, if the undamped frequency can be estimated, then the damped frequency can be measured from the step response, and using the equation: $f_d = f_o\sqrt{1 - D^2}$, where f_d, is the damped resonant frequency, f_o is the natural frequency and D is the damping factor.

Electrical analogue of the damped resonant systems

As well as practically exploring the response of the blood pressure system by mechanical means, the response can be simulated electrically using an analogue as already discussed in Figure 12.12. By choosing values of C_d and L_c, and varying R_c, the behaviour of underdamped, critically damped and overdamped systems can be explored.

For critical damping ($D = 1$), the resistance R_c is given by:

$$R_c = 2\sqrt{\frac{L_c}{C_d}}.$$

For any value of R_c, the damping is given by:

$$D = \frac{R}{2}\sqrt{\frac{C_d}{L_c}}.$$

The natural resonant frequency f_o is given by:

$f_o = \dfrac{1}{2\pi\sqrt{L_c C_d}}$ and the damped resonant frequency, f_d is given by:

$f_d = f_o\sqrt{1 - D^2}$ (which is the same as the mechanical system).

Other blood pressure measurements

Pressure itself has already been defined in Chapter 3. Blood pressure is normally measured in or around systemic arterial vessels as described in the preceding section. The generator of systemic blood pressure is the left ventricle of the heart. Figure 12.16 shows the variation of BP from heart to capillaries on right and left sides of the circulation. It is possible to measure the left ventricular pressure invasively with an appropriate catheter, a pulmonary artery catheter (see Chapter 13).

Where invasive pressure monitoring is used, the pressure waveform, pulsating between systolic and diastolic values, is transmitted down the arterial tree, normally becoming more spiky towards the periphery of the circulation. This spikiness is because of altered vessel compliance and transmitted waves towards the periphery, and needs to be taken into account when interpreting peripheral arterial traces. Accurate interpretation assumes that

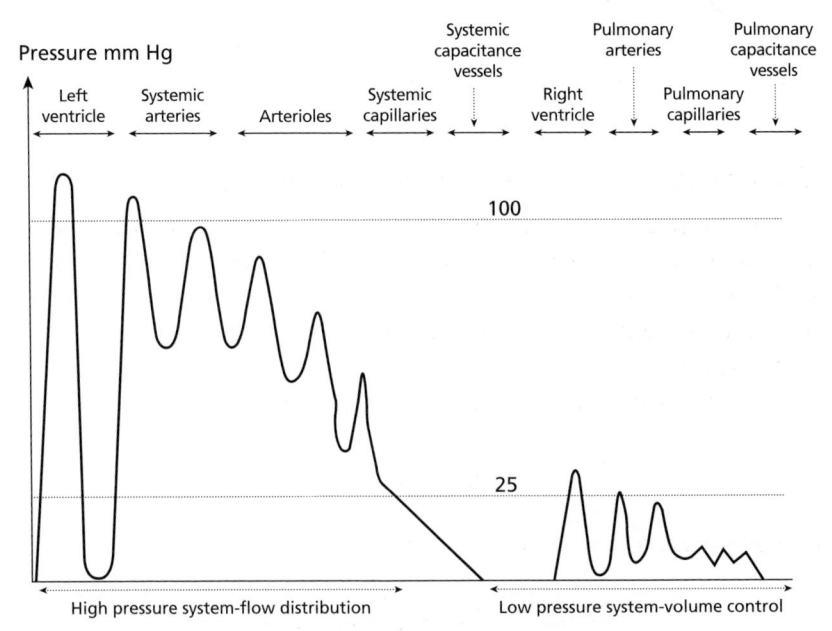

Fig. 12.16 Pressure distribution across the human circulation.

electromechanical issues, such as zeroing, calibration, and adjustment of damping have been optimised and that the clinical benefit/risk ratio of using invasive arterial monitoring in the first place has been weighed.

Capillary pressures are low, not usually measurable, and are of the same order of magnitude as oncotic pressures in the vessels and surrounding tissues, a fact that governs the transudation of fluid between vessel and tissue; oncotic pressure is the osmotic pressure exerted by the presence of large protein molecules in the vessels, which drives interstitial fluid into the capillary. This is countered by the hydrostatic pressure within the capillary, decreasing from the arteriolar end to the venular end of the capillary, driving fluid out of it.

Blood returns from the systemic circulation towards the right atrium via the central veins, whose pressure (CVP) can be measured by catheter. Normal CVP, when the patient is breathing spontaneously, is ± few mmHg, and therefore accurate zeroing of such a measurement system is important. Venous pressure waveforms have relatively small pulsatile components and a mean value is usually quoted. This means that, apart from accurate zeroing, a less stringent system of measurement is needed than for an arterial pressure measurement system, and a U-tube manometer often suffices. The CVP represents the filling pressure, not of the left side of the heart, but of the right. Nevertheless, a relationship is assumed to exist between the right atrial filling pressure and the left ventricular end diastolic pressure unless there is heart or lung disease. Under these conditions, left sided filling pressures should be measured, using a pulmonary artery catheter, in order to understand left ventricular function. In general, CVP measurements are interpreted in the light of trends in values rather than the actual values themselves. Problems with CVP measurement occur when:

- the catheter is too short and not placed in the thoracic cavity;
- the catheter is too long when it is in the right ventricle, measures ventricular pressures and may cause arrhythmias;
- the line is being used simultaneously to deliver fluids;
- compression of intrathoracic veins by positive pressure ventilation, and in particular positive end expiratory pressure is applied;
- microelectrocution risk (see Chapter 6) is now rare.

References

Brown, B. H., Smallwood, R. H., Barber, D. C., Lawford, P. V. and Hose, D. R. (1999) Chapter 18, pressure measurement, *Medical Physics and Biomedical Engineering*. IOP Publishing, Bristol.

Farquhar, I. K. (1991) Continuous direct and indirect blood pressure measurement (Finapres) in the critically ill, *Anaesthesia*, **46**, pp. 1050–55.

Geddes, L. A. and Baker, L. E. (1989) Chapter 3, *Inductive Transducers, from Principles of Applied Instrumentation*, Wiley, New York.

Hutton, P. and Cooper, G. M. (1985) *Guidelines in Clinical Anaesthesia*, Blackwell, Oxford, p. 26.

MacDonald, E., Froggatt, P., Lawrence, G. and Blair, S. (2008) Are automated blood pressure monitors accurate enough to calculate the ankle brachial pressure index? *J. Clin. Monit. Comput.*, **22**, pp. 381–84.

Moore, C. *et al.* (2008) Comparison of blood pressure measured at the arm, ankle and calf, *Anaesthesia*, **63**, pp. 1327–31.

Parikh, B. R., Simon, A. M., Kouvaras, J. N., Ciolino, R. B., Suthar, T. P. and Dorian, R. S. (2007) Influence of intermittent pneumatic compression devices on noninvasive blood pressure measurement of the ankle, *J. Clin. Monit. Comput.*, **21**, pp. 381–86.

Pereira, E., Prys-Roberts, C., Dagnino, J., Anger, C., Cooper, G. M. and Hutton, P. (1985) Auscultatory measurement of arterial blood pressure during anaesthesia: a reassessment of the Korotkoff sounds, *Eur. J. Anaes.*, **2**, pp. 11–20.

Prys-Roberts, C. (1984) Invasive monitoring of the circulation. In Saidman, L. J. and Smith, N. T. (eds), *Monitoring in Anaesthesia*, 2nd edn, Butterworth, Oxford, pp. 79–115.

Szmuk, P., Pivalizza, D. and Warters, R. D. (2008) An evaluation of the T-Line Tensymeter continuous non-invasive blood pressure device during induced hypotension, *Anaesthesia*, **63**, pp. 307–312.

Tooley, M., Magee, P. (2009) Patent no. WO/2010/061197. Method of measuring blood pressure and apparatus for performing the same.

Van Bergen, F. H. *et al.* (1954) Comparison of indirect and direct methods of measuring arterial blood pressure, *Circulation*, **10**, pp. 481–90.

Further reading

Geddes, L. A. and Baker, L. E. (1989) Chapter 16, Criteria for the faithful reproduction of an event. From principles of applied instrumentation, Wiley, New York.

Cruickshank, S. (1998) *Mathematics and Statistics in Anaesthesia*, Oxford University Press.

Chapter 13

Cardiac output measurement

> This chapter contains: the pulmonary artery catheter; echocardiography; pulse contour analysis (PiCCO, LiDCO); transthoracic electrical impedance. The chapter links with: Chapters 1, 3 and 12.

The pulmonary artery catheter

The pulmonary artery catheter was the mainstay of clinical cardiac output measurement for many years, but because of its relatively invasive nature and the lack of improvement of clinical outcome with its use, it is now seldom used in a modern clinical environment. Any perceived accuracy of the technique is now considered unnecessary in the face of the risks of its use, and with the introduction of newer non-invasive techniques. Nevertheless, it is worth describing, partly because of its historical interest, and partly because of the technologies involved.

A catheter passed into the right atrium from an easily accessible central vein can be passed through the right ventricle and out into the pulmonary arterial tree while the vascular waveforms are visualised. Figure 13.1 shows the waveforms as they appear to the user. A small balloon at the tip of the catheter allows it to be flow directed and wedged in a pulmonary arterial vessel. At this point the pulsatile waveform is lost and the tip of the catheter is looking ahead, down the pulmonary arterial tree towards the left atrium, a system with a relatively low pressure drop from one end to the other, the flow in that vessel having been brought temporarily to a standstill. Thus the *pulmonary artery occlusion pressure* (PAOP) or *pulmonary capillary wedge pressure* (PCWP) can be considered a reasonably accurate representation of left atrial pressure or left ventricular filling pressure. This assumes that there is no pulmonary vascular disease, such as pulmonary hypertension, or mitral valve disease, in which case PAOP would not be an accurate representation of left atrial pressure. If the catheter is placed in the apical region of the pulmonary vascular tree, the excess of the alveolar pressure in inspiration over pulmonary capillary pressure becomes significant, and the latter is a less accurate reflection of left atrial pressure. The balloon should not be over-inflated for fear of rupturing the

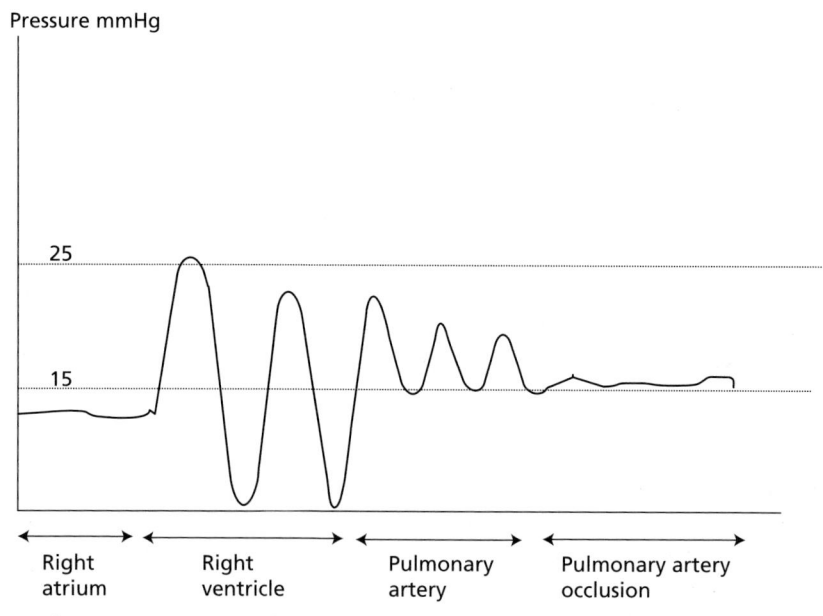

Fig. 13.1 Pressure waveform trace on insertion of a pulmonary artery catheter.

pulmonary artery, and this is one of its perceived risks that has led to less usage. Once the measurement has been made, the balloon should be deflated so that the pulmonary arterial waveform is once again visible, if necessary withdrawing the catheter a bit to achieve this; failure to do so would result in regional lack of perfusion and may result in ischaemia. The same transducer, signal processing and recording requirements are needed as for systemic arterial monitoring. If the patient is undergoing artificial ventilation the PAOP measurement must be made at end expiration, particularly if ventilation with positive end expiratory pressure (PEEP) is being applied. As with a CVP line, it is also possible to cause cardiac arrhythmias with the catheter.

A pulmonary artery catheter can also be used to measure cardiac output using the Fick principle, which states that when a dye is added at rate m to a fluid in flow, and the concentration of the dye changes from Cv before, to Ca after being added to the fluid, the flow rate, Q, of the fluid, is given by

$$Q = \frac{m}{(Ca - Cv)}.$$

In Fick's original experiment, Q was the cardiac output itself, the 'dye' was oxygen, so m was oxygen uptake and Ca and Cv were respectively arterial and venous oxygen concentration (content). In measuring cardiac output using a pulmonary artery catheter, the *dye* is typically a bolus of cold dextrose at room temperature or colder, which is therefore used as a thermal dye and the method is the *thermodilution* method. Dextrose is used rather than saline to reduce the chance of creating an electrically conducting pathway that could cause microshock. The bolus of dextrose is introduced into the circulation at a proximal part of the catheter situated in the right atrium, mixes in the circulation, and the change in temperature of the blood is measured at the catheter tip by a thermistor. The curve of temperature against time may look as in Figure 1.15 (see Chapter 1). The software associated with the catheter deduces cardiac output according to Fick's principle using the formula

$$Q = \frac{V \rho_i c_i (t_b - t_i) \times 60}{\int \rho_b c_b T \, dt},$$

where Q is the cardiac output, V is the volume of injectate, ρ_i and ρ_b are the densities of injectate and blood respectively, while c_i and c_b are the thermal conductivities of injectate and blood respectively; t_i and t_b are the temperatures of injectate and blood. The bottom line represents the area under this curve and the small amount of recirculation of the dye shown at the right hand end of the curve is ignored in the computation. Note therefore that a large *area-under-curve* represents a small cardiac output and vice versa. Variations in temperature of the pulmonary artery blood will result in some error in computing cardiac output by this method [Moise et al. 2002], and overestimation of a low cardiac output has been detected due to the increase in cardiac preload caused by the injectate bolus [Tournadre et al. 1997].

The pulmonary artery catheter has become the standard against which other less invasive methods of cardiac output measurement are compared, even though the method itself has fallen out of favour. Several factors have contributed to its demise:

- It is possible to damage the right ventricular wall, pulmonary artery wall or tricuspid or pulmonary valves with the catheter [van Doorn et al. 1994].
- Years of use of the pulmonary artery catheter in life threatening conditions such as septic shock have failed to alter outcome significantly.
- Non-invasive methods of assessing cardiac output, in particular echocardiography and pulse contour analytical methods, have been developed, which are now much more commonly used with satisfactory results and less morbidity.

It is also possible to sample mixed venous blood from the pre-alveolar pulmonary vasculature, and to measure the mixed venous oxygen saturation

(SvO_2). When compared with the arterial oxygen saturation (SaO_2), this gives useful information about oxygen usage by cells and is characteristically low in sepsis.

Echocardiography

The principles of the use of ultrasound using the Doppler principle have been described in Chapter 10. Doppler ultrasound has many applications and the qualitative and quantitative assessment of cardiac function is one. The method actually measures the velocity of red blood cells in the aorta, which, when averaged and multiplied by the cross-sectional area of the flow (e.g. of the aorta, either measured by the device itself or calculated from input data on the patient's age, height, weight and gender), gives blood flow as cardiac output.

An appropriate way to visualise the descending aorta near its origin using ultrasound is often via the oesophagus, on the end of a probe, which the patient swallows. The ultrasonic waves on the end of the probe are produced by a piezoelectric crystal in the range 2.5 to 5.0 MHz. The same transducer alternately transmits the wave for 1 μs and detects the reflected waves for 250 μs, although other formats of the method use separate transmission and receiving transducers. It is also possible to use a transducer situated on the tip of an endotracheal tube to measure cardiac output with Doppler, thus allowing continuous, intraoperative cardiac output monitoring. Alternatively, the non-invasive transthoracic or suprasternal approach is another route of examination. Whatever route is chosen, correct orientation of the ultrasound probe in relation to the descending aorta is required to ensure an accurate estimate of cardiac output.

The velocity signal from the probe is *fast Fourier transformed* and the output is represented as a velocity amplitude waveform plotted against time. This produces a characteristic triangular waveform from which further qualitative information about the circulation may be deduced, as shown in Figure 13.2. Inadequate orientation of the probe with the predominant flow direction within the aorta would give a less well defined, low amplitude waveform. The *flow time* (FT) is shown in Figure 13.2 as the distance across the base of the triangle on the time base. Since FT is influenced by heart rate, and by ventricular preload and afterload, a *corrected flow time* (FTc) is calculated to index the trace to a heart rate of 60 beats a minute, which gives a FTc of about 333 msec. A decreased FTc is seen in hypovolaemia, a decreased preload, and an increased afterload. An increased FTc is seen with a decreased afterload. The *peak velocity* is indicated on the trace by the apex of the triangle. It diminishes with

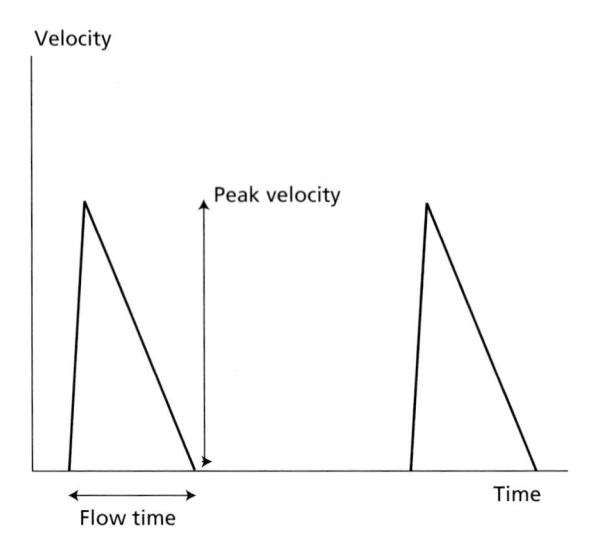

Fig. 13.2 Velocity vs. time curves obtained to measure cardiac output using echocardiography.

age, increased preload, and increased afterload. Transoesophageal Doppler monitoring can be used to guide fluid therapy [Diaper et al. 2008].

As well as measuring blood velocity and deducing cardiac output, an echocardographic examination of the heart can also look at cardiac structure and function. For example a cardiologist might be interested in looking at valvular area or dysfunction, abnormal wall movement in ischaemic parts of the heart, end diastolic and systolic volumes, determination of ejection fraction, thrombus detection, detection of air embolus and aortic dissection. It is also possible, by application of Bernoulli's theorem to deduce pressure gradient across, for example, the aortic valve by blood velocity measurement in the aortic root.

Providing the transducer can be appropriately positioned to give results in which the user can feel confident, cardiac output measurements using echocardiography can be as accurate as thermodilution methods [Legrant et al. 1993] even when used intraoperatively in the presence of cardiac disease [Ryan et al. 1992]. However one study showed that Doppler ultrasound produced significantly different results from thermodilution during all modes of ventilation encountered during anaesthesia, whereas thoracic electrical bioimpedance did not [Castor et al. 1994]. Clearly the user must be appropriately trained, and Doppler ultrasound can be used to measure blood flow in vessels other than the aorta from which a degree of atherosclerotic stenosis can be assessed.

Pulse contour analysis

In recent years, a new technique has been developed for cardiac output measurement using mathematical algorithms to analyse the pulse contour [Röding et al. 1999] obtained from an arterial waveform (as suggested in Chapter 12) [Wiseley et al. 2001]. These are less invasive than a pulmonary artery catheter, but more so than ultrasound, although the method uses invasive monitoring techniques that would already in any case be in place, such as an arterial line and a central venous line. The form of the arterial pressure waveform (the pulse contour) depends not only on stroke volume but also on vascular compliance. The concept of computing stroke volume on the pulse contour was originally described by Otto Frank in 1899 as the *Windkessel model*. To monitor cardiac output by this technique the device must be calibrated against another technique. The PiCCO and LiDCO devices both use the pulse contour technique, but have different methods of validating the result.

PiCCO

PiCCO uses an appropriately manufactured specialised arterial cannula with an integrated thermistor to analyse both the arterial pressure waveform and the blood temperature at a peripheral artery such as the radial or femoral artery [Wilde et al. 2006]. The cardiac output data derived from the pulse contour is calibrated using an intermittent thermodilution technique via a central venous catheter (rather than a pulmonary artery catheter). The cold bolus of fluid is injected into the central vein, and the temperature is measured at the arterial cannula. This is used to update the cardiac output derived from the algorithm used to interpret the pressure waveform contour. The cardiac output is given by:

$$KR \int \left[\frac{P(t)}{SVR} + C(p) \frac{dp}{dt} \right] dt,$$

where K is a calibration constant determined by thermodilution; R is heart rate; $P(t)$ is the mathematical function that describes the pressure waveform with respect to time; SVR is systemic vascular resistance; $C(p)$ is the mathematical function describing aortic compliance with respect to the pressure; and dp/dt is the rate of change of the pressure waveform curve with respect to time. The terms bracketed under the integral sign are integrated with respect to time. The displayed value is an average of a number of preceding beats.

LiDCO

This device calibrates the pulse contour derived cardiac output by using a dilution technique with lithium as the 'dye'. A small bolus of dilute lithium chloride is injected into a central or a peripheral vein. Blood is drawn from an ordinary peripheral arterial cannula at a constant rate by a small pump, and it is allowed to flow past a lithium sensitive electrode to a collection system. The lithium electrode system and blood collection system are attached to the arterial line by a T-piece and the whole tubing system is primed with saline. The electrode and its associated software plots a lithium/time/concentration curve, from whose area the cardiac output may be calculated in the usual fashion. This is the calibration value that can be used to validate the pulse contour values from the arterial waveform, which uses a different algorithm to deduce the cardiac output from the pressure waveform to the PiCCO device. The lithium delivery system at the venous end is also primed with saline to ensure prompt delivery of the bolus of lithium. It has been found that LiDCO underestimates cardiac output when the afterload is low [Yamashita et al. 2007].

Both pulse contour methods have disadvantages associated with venous and arterial cannulation, such as thrombosis or kinking of the cannulae. The accuracy of both devices depends on the calibration methods described [Johansson et al. 2007]. LiDCO cannot be used with patients on lithium therapy, nor on patients on non-depolarising neuromuscular blockers which interfere with lithium analysis. As with any method that depends on a dilution technique, cardiac shunts will also introduce error into this method of cardiac output measurement. In addition it has been found that pulse contour methods are not entirely satisfactory for monitoring stroke volume changes in response to intraoperative fluid therapy [Lahner et al. 2009]. There are other devices available that use pulse contour analysis without additional, invasive calibration procedures, such as *Vigileo* and *Heartsmart* but they have been found to be less reliable than the other calibrated pulse contour methods [Sakka 2007, Berridge 2009].

Transthoracic electrical impedance

This is included for completeness, but is not a method in common clinical use. The principle depends on the change of electrical impedance in the thorax with a number of activities such as respiration, but including the change of heart size associated with the cardiac cycle. As the method was developed, a way was found to deduce stroke volume from the change in transthoracic electrical impedance. The method has been found to have a variance of more than 20 to 48% [Tomaske 2008], which may explain why it is not in common use, although careful trials using it have found a reasonable degree of confidence in the method [Sathyaprabha 2008].

References

Berridge, J., Warring-Davies, K., Bland, J. and Quinn, A. (2009) A new, minimally invasive technique for measuring cardiac index: clinical comparison of continuous cardiac dynamic monitoring and pulmonary artery catheter methods, *Anaesthesia*, **64**, pp. 961–67.

Castor, G. *et al.* (1994) Simultaneous measurement of cardiac output by thermodilution, thoracic electrical bioimpedance and Doppler ultrasound, *Br. J. Anaesth.*, **72**, pp. 133–38.

Diaper, J., Ellenberger C., Villiger, Y. *et al.* (2008) Transoesophageal Doppler monitoring for fluid and hemodynamic treatment during lung surgery, *J. Clin. Monit. Comput.*, **22**, pp. 367–74.

Johansson, A. and Chew, M. (2007) Reliability of continuous pulse contour cardiac output measurement during hemodynamic instability, *J. Clin. Monit. Comput.*, **21**, pp. 237–42.

Lahner, D., Kabon, B., Marshalek, C. *et al.* (2009) Evaluation of stroke volume variation obtained by arterial pulse contour analysis to predict fluid responsiveness intraoperatively, *Br. J. Anaes.*, **103**, pp. 346–51.

Legrant, J. Y. *et al.* (1993) Cardiac output measurement: comparison of esophageal Doppler vs. thermodilution, *Br. J. Anaesth.*, **70**(suppl 1), p. 9.

Moise, S. F., Sinclair, C. J., Scott and D. H. T. (2002) Pulmonary artery temperature and measurement of cardiac output by thermodilution, *Anaesthesia*, **57**, pp. 562–66.

Röding, G. *et al.* (1999) Continuous cardiac output measurement: pulse contour analysis vs. thermodilution technique in cardiac surgery patients, *Br. J. Anaesth.*, **82**, pp. 525–30.

Ryan, T. *et al.* (1992) Transesophageal pulsed Doppler wave measurement of cardiac output during major vascular surgery: comparison with thermodilution technique, *Br. J. Anaesth.*, **69**, pp. 101–104.

Sakka, S., Kozieras, J., Thuemer, O. and van Hout, N. (2007) Measurement of cardiac output: a comparison between transpulmonary thermodilution and uncalibrated pulse contour analysis, *Br. J. Anaes.*, **99**, pp. 337–42.

Sathyaprabha, T. N., Pradhan, C., Rashmi, G. *et al.* (2008) Noninvasive cardiac output measurement by transthoracic electrical bioimpedence: influence of age and gender, *J. Clin. Monit. Comput.*, **22**, pp. 401–408.

Tomaske, M., Knirsch, W., Kretschmar, O. *et al.* (2008) Cardiac output measurement in children: comparison of Aesculon cardiac output monitor and thermodilution, *Br. J. Anaes.*, **100**, pp. 517–20.

Tournadre, J. P., Chassard, D. and Muchada, R. (1997) Overestimation of low cardiac output measurement by dilution, *Br. J. Anaesth.*, **79**, pp. 514–516.

van Doorn, C. A. M., Lyons, G., Swindells, S. and Unnikrishnan Nair, R. (1994) Perforation of the right ventricle and cardiac tamponade; the use of the pulmonary artery catheter, *Br. J. Intensive Care*, **4**(Feb.), pp. 66–67.

Wilde, R., Breukers R., van den Berg, P. and Jansen, J. (2006) Monitoring cardiac output using the femoral and radial arterial pressure waveform, *Anaesthesia*, **61**, pp. 743–46.

Wiseley, N. A. and Cook, L. B. (2001) Arterial flow waveforms from pulse oximetry compared with Doppler flow waveforms, *Anaesthesia*, **56**, pp. 556–61.

Woltjer, H. H. *et al.* (1996) Standardisation of non-invasive impedance cardiography for assessment of stroke volume. *Br. J. Anaesth.*, **77**, pp. 748–52.

Yamashita, K., Nishiyama, T., Yokoyama, T. *et al.* (2007) Effects of vasodilation on cardiac output measured by PulseCO, *J. Clin. Monit. Comput.*, **21**, pp. 335–39.

Chapter 14

Gas pressure, volume and flow measurement

> This chapter contains: gas pressure measurement; gas volume and flow measurement (the spirometer, the Vitalograph, Wright's respirometer, Fleisch pneumotachograph, hot wire anemometry, ultrasonic flow transducer, measurement of functional residual capacity and deadspace, the variable orifice constant pressure flowmeter).
> The chapter links with: Chapters 1, 3, 7, 12 and 22.

Introduction

The physics of pressure, flow and the gas laws have been discussed in Chapter 7 in relation to the behaviour of gas and vapour. This section will focus on the physical principles of the measurement of gas pressure, volume and flow.

Unlike a liquid, a gas is compressible and the relationship between pressure, volume and flow depends on the *resistance* to gas flow (or *impedance* if there is a frequency dependence between pressure and flow in alternating flow, see Chapter 4 for the electrical analogy of this) in conduits (bronchi, anaesthetic tubing); it also depends on the compliance of structures being filled and emptied (alveoli, reservoir bags, tubing or bellows). Normal breathing occurs by muscular expansion of the thorax, thus lowering the intrathoracic pressure, allowing air or anaesthetic gas to flow towards the alveoli down a pressure gradient from atmospheric pressure. When positive pressure ventilation occurs, gas is 'pushed' under pressure into the alveoli. Depending on the exact relationship between the ventilator and the lungs, different relationships exist between airway pressure (rather than alveolar pressure, which cannot easily be measured) and gas flow and volume.

Gas pressure measurement

Gas pressure measurement devices were traditionally in the form of an aneroid barometer, a hollow metal bellows calibrated for pressure and temperature, which contracts when the external pressure on it increases, and expands when it decreases. The movement is linked to a pointer and indicator dial (see

Figure 14.1(a)). It is often more convenient to make the device in the shape of part of a circular section, but the principle is the same. This is what the Bourdon gauge, which commonly measures pressure in gas cylinders, looks like (see Figure 14.1(b)).

The detection of movement of the diaphragm of an aneroid barometer can take several forms. The movement can either be linked via a direct mechanical linkage to a pointer, or diaphragm movement can be linked to a capacitative or inductive element in an electrical circuit, such as a Wheatstone bridge. Airway pressure during spontaneous breathing or artificial ventilation is low. The preferred units of measurement are cm H_2O and the range of values is between -20 and $+20$ cm H_2O. The aneroid barometer to measure this will therefore be of light construction, using thin copper for the bellows material. By contrast the pressure in a gas cylinder is of the order of 137 bar (about 141 000 cm H_2O, or 1.41 km height of water!) and in an anaesthetic gas pipeline it is of the order of 50 bar (52 000 cm H_2O). Therefore the aneroid barometer will be of much more robust construction.

The strain gauge transducer, one use of which is for invasive blood pressure monitoring, has evolved into a device that can also measure gas pressure. Modern semiconductor materials exhibit significant piezoresistive properties, whereby their electrical resistance changes with mechanical deformation. A piezoresistive strain gauge, stuck on to a piece of silicon can be used in conjunction with a Wheatstone bridge (see Chapter 4) as a device to measure gas pressure in a ventilator or breathing circuit. There is an hydraulic

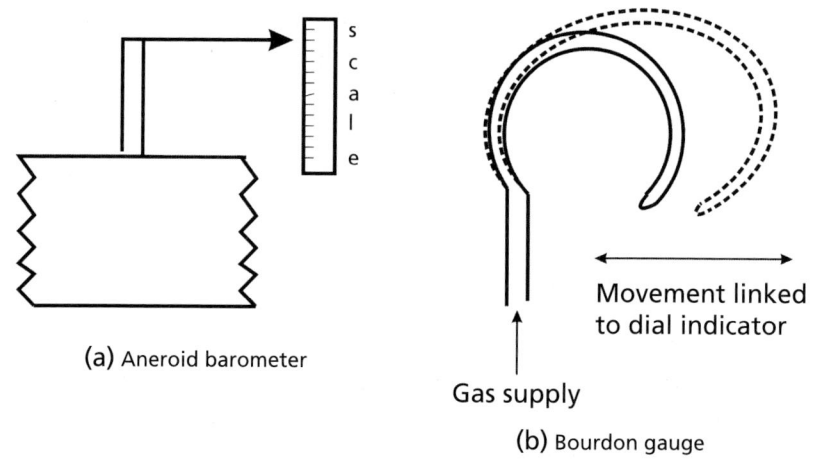

(a) Aneroid barometer

(b) Bourdon gauge

Movement linked to dial indicator

Gas supply

Fig. 14.1 Two different aneroid barometers for measuring gas pressure.

connection between the patient and the transducer composed of air, the compressibility of which limits the frequency response of the device.

It is also possible to measure gas pressure in a circuit using a static side port in the tubing that leads to a manometer, but the compressibility of gas makes this less suitable than for measuring the static pressure in a liquid.

Where a substance exists in a cylinder as a gas (i.e. above its critical temperature, e.g. oxygen), the pressure gauge effectively becomes the contents gauge for the cylinder, since the pressure and volume changes are related by Boyle's law. Where, however, the substance exists as a combination of a liquid and its associated vapour (i.e. below its critical temperature, e.g. nitrous oxide), the pressure gauge will indicate the saturated vapour pressure, which will remain constant at constant temperature (51 bar at room temperature) as long as the vapour pressure can be restored in the cylinder from the liquid reservoir. The pressure gauge in this case is therefore not useful as a contents gauge. If a nitrous oxide cylinder is continuously emptied at a fast rate, the pressure gauge would probably gradually decrease, but this is because the temperature of the contents is falling with continuing vaporisation and with it the saturated vapour pressure.

The measurement of airway pressure in a breathing system or ventilator circuit occasionally may not reflect the alveolar pressure; if there is significant upper airway obstruction, the indicated upstream airway pressure will be high, and the downstream alveolar pressure will be lower than indicated. Under these circumstances adequate lung volume filling on inspiration will not be usefully indicated by the pressure gauge. Since a ventilator consists of components such as bellows and tubing, each of which has its own compliance and flow resistance characteristics, any airway pressure measurement will be altered by these factors and indicated airway pressure may not adequately reflect alveolar pressure. This is particularly true where the frequency of ventilation is high and the volume is small, such as occurs in ventilation of neonates. Under these conditions the frequency response of the system may not be adequate.

A gas pressure gauge can form the basis of a ventilator or breathing system disconnection alarm.

Gas volume and flow measurement

The spirometer

Clinical observation of the patient often gives good qualitative information about gas volume and flow into the patient. For quantitative measurement, the *Bell spirometer*, shown in Figure 14.2, is the simplest device for measuring the various breathing volumes shown in the accompanying graph. The device

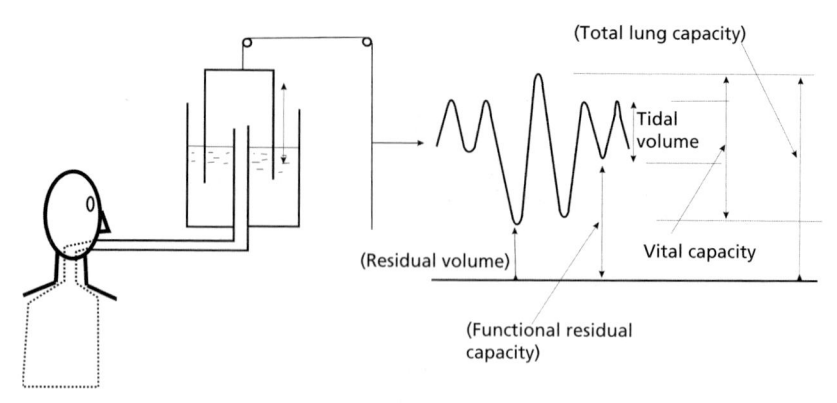

Fig. 14.2 The Bell's respirometer and the volumes it can measure. (Quantities unmeasurable by the spirometer are in brackets.)

cannot measure various residual capacities and volumes. Any device that measures flow and mathematically integrates the signal, can also be used to calculate tidal volume; *minute volume* is then *tidal volume × respiratory rate per minute*.

The Vitalograph

A Vitalograph is often used in a preoperative assessment setting (see Figure 14.3(a)) to measure *forced vital capacity (FVC)* and to deduce *forced expiratory volume in one second (FEV$_1$)*. It has even more resistance than a Bell spirometer, but is nevertheless a useful clinical tool. On taking a deep breath the patient breathes out hard into the device, breathing out gas at **B**ody **T**emperature and **P**ressure, fully **S**aturated with water vapour (BTPS). As the gas cools to room temperature, the water vapour condenses out, and the filling of the bellows moves a pen pointer, which is linked to a device, which moves a pen chart. Although the measurement is done at **A**tmospheric **T**emperature and **P**ressure with **D**ry air (ATPD), the chart is calibrated to allow final values to be computed at BTPS. The resulting graphs can be used diagnostically as shown in Figure 14.3(b).

The Wright's respirometer

Now a device of mostly historical interest, the Wright's respirometer is a turbine flowmeter, designed to allow the inspired or expired gas flow past some stationary vanes to ensure unidirectional flow, and past some moving vanes linked to a pointer, calibrated so that the amount of gas passing through

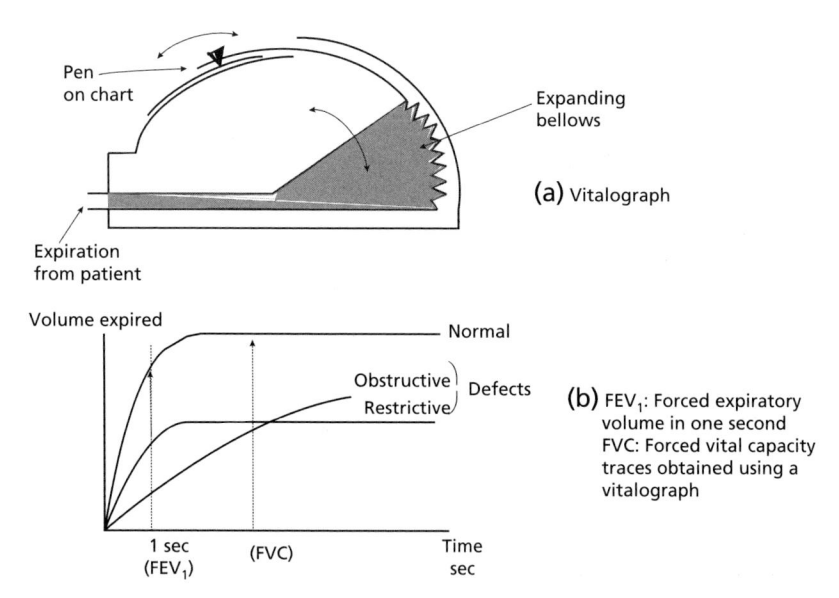

Fig. 14.3 (a)The Vitalograph; (b) graphical output from the Vitalograph.

the device is the expired volume. It is effectively a flow integrator and despite its careful engineering design, the inertia of the moving vane means that it over reads at high initial flow rates and under reads at low flow rates; it also under reads at low flow rates due to friction. Clearly there is also likely to be some inaccuracy where the expired gas contains significant amounts of nitrous oxide, whose density is 50% greater than that of air. However, despite all these potential errors, the device is accurate to within 10% of true values. It is shown diagrammatically in Figure 14.4. A more modern version of this device uses magnetic coupling of the turbine to an electronic measuring scale.

The Fleisch pneumotachograph

The principle of using differential pressure measurement to measure flow, such as the venturi device, and the associated orifice flowmeter, has been discussed in Chapter 5. An evolution of the venturi is the *Fleisch pneumotachograph*, which measures flow and is a device that depends on a linear relationship between pressure and flow during laminar flow as suggested by Hagen and Poiseuille. It achieves this by dividing the flow through the tube into a large number of parallel, small diameter tubes, thus ensuring laminar flow through each one. The pressure difference between the inlet and outlet is measured and is directly related to flow. This type of device is therefore also known as a

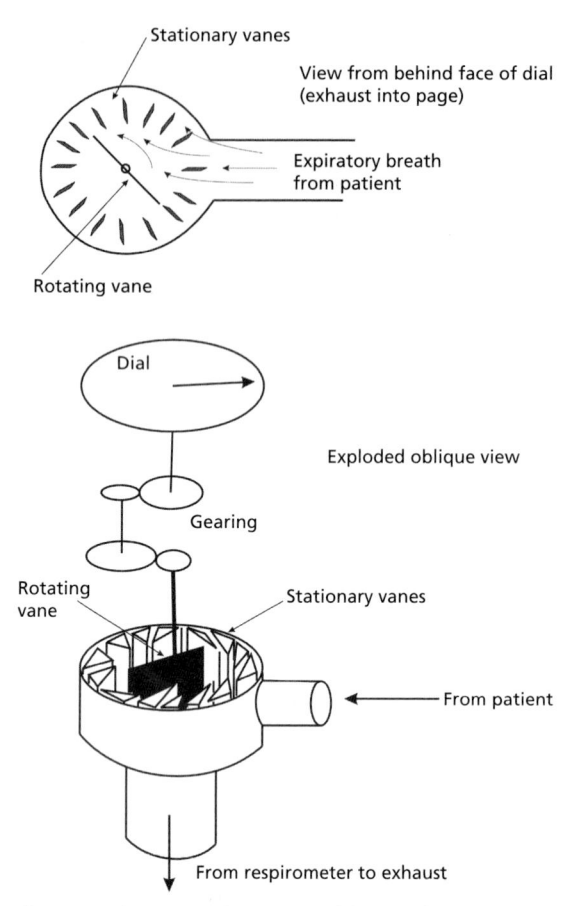

Fig. 14.4 Diagrammatic sections of the Wright's respirometer.

variable pressure drop, fixed orifice flowmeter and is shown diagrammatically in Figure 14.5. Its accuracy depends on the extent to which laminar flow has been achieved, on the precision of construction and placement of the pressure tappings, the bandwidth of response of the device and the variations in temperature and viscosity of the gas within it. There are variations on the device that, for example, use a mesh rather than a series of parallel tubes, across which the flow may not be laminar, and in which the pressure drop may not be linearly related to flow, but that are still useful flow measurement devices. For all such devices, care must be taken when measuring flows during positive pressure ventilation that there is compliance matching of tubing attachments to the pneumotachograph, and that if a gauze mesh is used, it does not

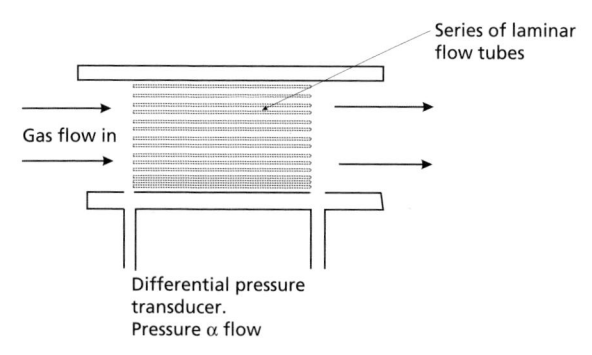

Fig. 14.5 The Fleisch pneumotachograph.

become deformed. In all these flow measurement devices the flow signal is integrated to give volume. This used to be done electronically, but nowadays the flow signal is digitised (see Chapter 4) and integrated mathematically by an algorithm in a microprocessor. Digitisation has reduced the problem of baseline drift and noise to which analogue signals were prone. Further software can use the airway pressure measurement and the volume calculation to draw a pressure/volume loop for each breath, which is a further useful addition to intraoperative airway monitoring [Bardoczky et al. 1993].

Another device that uses the principle of differential pressure measurement for flow measurement is the pitot tube, shown in Figure 14.6. One arm of a U-tube comes from the side wall of the flow tube and measures the static

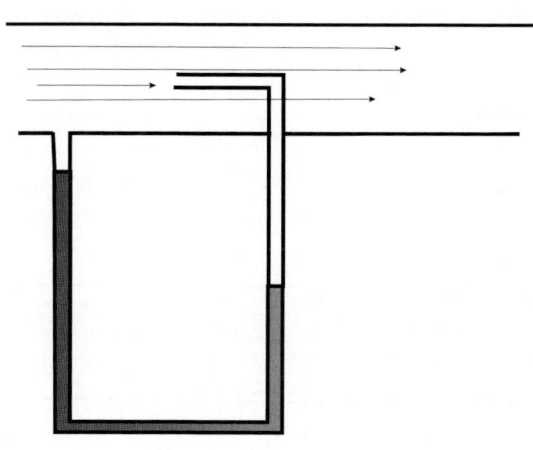

Fig. 14.6 The pitot tube.

pressure, while the other arm faces directly into the flow and measures the dynamic pressure, which is proportional to kinetic energy plus the static pressure; the difference between the two gives a (velocity)^2 term, from which flow can be deduced. The pitot tube is better known for its role in measuring aircraft velocity.

Hot wire anemometry

If a heated wire is placed in a gas stream, the degree of its cooling by the gas flow depends on gas temperature, specific heat and gas velocity (from which flow rate can be calculated). Some ICU ventilators use two anemometer wires, which are maintained at constant temperatures of 800°C and 1600°C and which form two arms of a Wheatstone bridge (see Chapter 4). Initial calibration of the device measures thermal conductivity of the gas at zero flow and accuracy varies, depending on gas composition.

Ultrasonic flow transducer

The principle of Doppler ultrasound has been described in some detail in Chapter 10 and its use in measuring blood flow has been discussed in an earlier section. Figure 14.7 shows such a device for measuring gas flow. Its principle is based on ultrasonic detection of vortex formation behind a partial obstruction to the flow. It has the advantage that it is not affected by temperature or gas composition, but below a critical flow rate vortices may not be formed to an adequate extent. Another version of an ultrasonic flowmeter has an ultrasonic pulse transmitter and receiver at both ends of a flow measurement tube; pulses are transmitted in both directions and detected at each end; with the gas flow going in one direction down the tube, there is a difference in the time for the pulses to travel in each direction, from which the gas flowrate can be deduced.

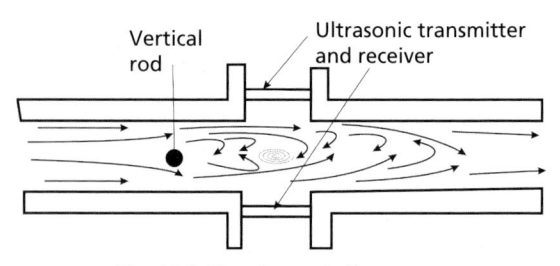

Fig. 14.7 The ultrasonic flowmeter.

Other respirometry technologies

Other technologies that have been applied to respirometry include a piezoelectric film respirometer [Moyle 1989] and the ability of polarised polyvinylidine fluoride sensors to detect temperature changes [Cyna et al. 1991].

Measurement of Functional Residual Capacity (FRC) and dead space

Although not routine in anaesthetic practice, knowledge of these techniques is included for completeness.

The body plethysmograph

This is shown in Figure 14.8. The patient sits in a sealed box and breathes through a tube attached to a pneumotachograph. He or she then makes respiratory effort against a closed shutter while changes in mouth pressure, ΔP, are recorded. In order to calculate the associated change in the gas volume within the box, which is also a change in lung volume, ΔV, a calibrating syringe is used in the box to link ΔP and ΔV. Mouth pressure changes can be thought of as intrathoracic pressure changes, since no gas flows under these conditions during panting against a closed shutter. By applying Boyle's law to the thoracic

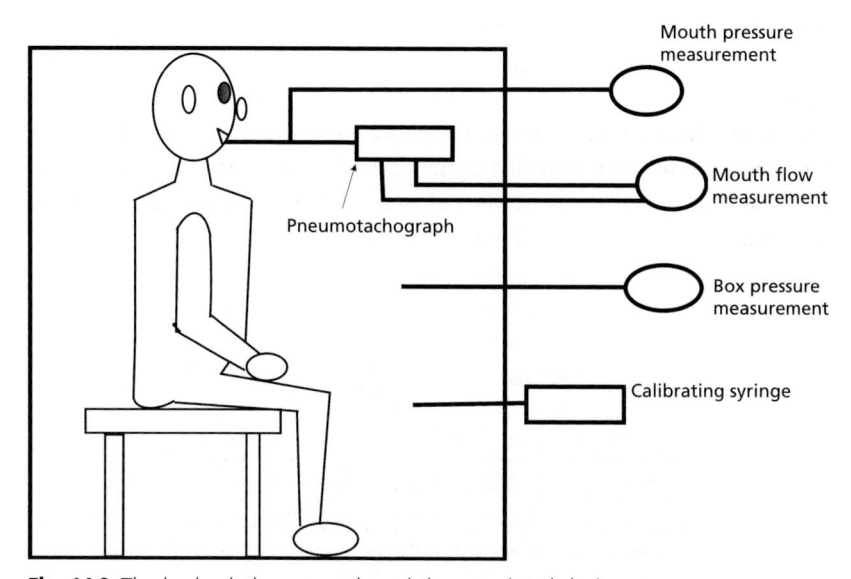

Fig. 14.8 The body plethysmograph and the associated devices to measure pressure and volume change and allow the deduction of FRC.

volume, V (the FRC), it can be said:

$$PV = (P + \Delta P)(V - \Delta V),$$

where an increase in pressure $+\Delta P$ leads to a reduction in volume $-\Delta V$ and vice versa. P is $(P_{amb} - P_{H_2O})$, ΔP and ΔV are measured or derived; hence V (FRC) can be deduced. The body plethysmograph can also be used to measure airway resistance at different lung volumes.

Dilution and washout method

One method here is the *nitrogen washout method* for measuring FRC. The patient breathes air through a circuit with a non-return valve; at the end of a normal expiration a tap is turned and the patient starts to breathe 100% oxygen; expired gas is collected in a *Douglas Bag* (see any respiratory physiology text). At the end of the nitrogen washout period (between 5 and 10 minutes in most people), the volume of expired gas and its nitrogen concentration are measured, and the nitrogen concentration in the gas is deduced. Since the initial N_2 concentration was 79%:

$$FRC = \frac{N_2 \text{ volume collected}}{0.79}.$$

Errors in this method include continued excretion of nitrogen from tissues to alveoli and trapping of alveoli where dynamic airway closure of some airways has occurred.

Using a technique involving nitrogen washout, *anatomical deadspace* can also be calculated using *Fowler's method*. At the end of expiration, air is switched to 100% oxygen as the breathing gas and on the first expiration, N_2 concentration is plotted first against time then replotted against expired volume. As residual air is expired after one oxygen breath, the initial N_2 concentration remains negligible as this is where O_2 is stored in the deadspace. The subsequent step rise in $[N_2]$ represents the point at which the N_2 containing alveoli start to empty. Some allowance is made for the front that develops between air and O_2. This is shown in Figure 14.9, which also shows how anatomical deadspace is estimated graphically.

Physiological deadspace is the functional deadspace where gas exchange does not take place, and includes the anatomical deadspace and also those areas of the lung where gas exchange should occur but does not due to poor or absent perfusion, such as in hypotension or a pulmonary embolus. In general all the CO_2 exhaled must come from the alveoli, even if it can only be observed in the mixed expired breath. If V_A, V_D and V_T are respectively the alveolar volume, deadspace volume and tidal volume, and F_{ACO2} and F_{ECO2} are the

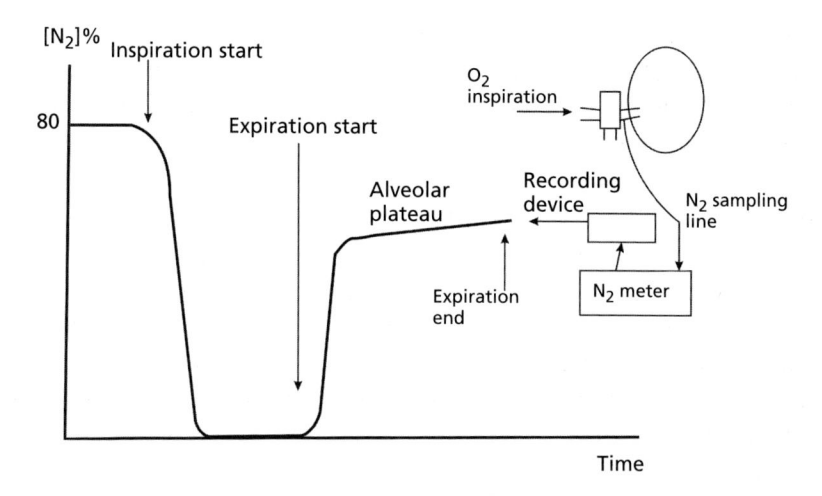

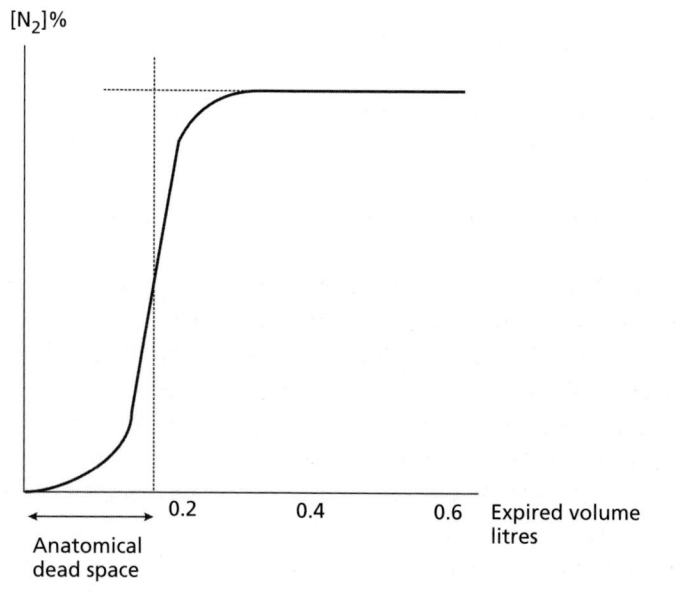

Fig. 14.9 Fowler's method for measuring anatomical deadspace; traces of exhaled nitrogen concentration plotted against time (upper curve) and against expired volume (lower curve).

concentrations of CO_2 in the alveoli and mixed expired gas respectively, it can be asserted that:

$$V_T = V_A + V_D$$

and that

$$F_{ACO2}\, V_A = F_{ECO2}\, V_T,$$

whence

$$\frac{V_D}{V_T} = \frac{F_{ACO2} - F_{ECO2}}{F_{ACO2}}.$$

If the gas concentrations on the right hand side are multiplied by ambient pressure and Dalton's law is applied, an expression can be derived in terms of gas partial pressures and this is *Bohr's equation*:

$$\frac{V_D}{V_T} = \frac{P_{ACO2} - P_{ECO2}}{P_{ACO2}}$$

The other method of measuring FRC is the *Helium dilution* technique. Helium gas is highly insoluble in blood and when inhaled, remains largely in the alveoli. A fixed volume of helium, say 10 ml at ATPD, is added to the volume, V_1, of a spirometer and the helium concentration, He_1, is measured. The spirometer is then connected to the patient at end of expiration and breathing continues. The final concentration at equilibrium of helium, He_2 is measured. Then

$$He_1\, V_1 = He_2(V_1 + FRC).$$

Hence FRC is derived.

The variable orifice, constant pressure flowmeter

An example in clinical use here is the peak expiratory flowmeter, in which the peak expiratory flow is measured, and can be as much as 500–800 l min^{-1} in healthy adults. Figure 14.10 shows the device that is commonly used. A flow of gas into the mouthpiece deflects a vane attached to a pointer, which is resisted by a spiral spring. At the same time the expired gas escapes through an annular groove round its circumference.

The Rotameter

This is another example of a variable orifice, constant pressure flowmeter, whose physical principles have been described in Chapter 7 and which is shown diagrammatically in Figure 7.3. Rotameters are standard devices on anaesthetic machines for measuring a steady gas flow being delivered to the common gas outlet, although they have been replaced on modern machines by solenoid

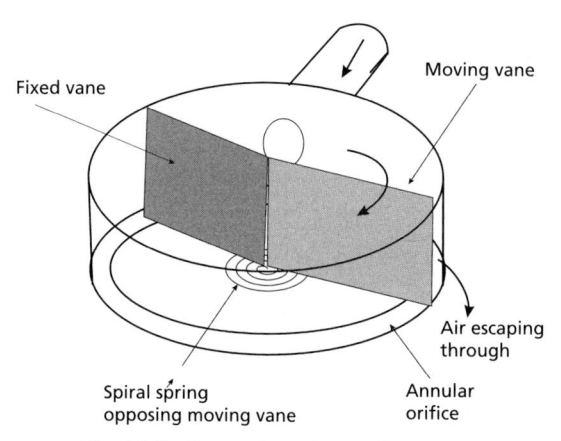

Fig. 14.10 The peak expiratory flowmeter.

flow controllers [Boaden *et al.* 1986]. Although the Rotameter was patented by Draeger in 1907 for metering carbon dioxide flow in the beer industry, it was not fitted to an anaesthetic machine until 1937. Where the bobbin is conical in shape, the flow is read from the top of the bobbin. Many Draeger bobbins are spherical and the flow is read from the centre of the sphere. By changing the taper on the Rotameter tube it is possible to change the measurement scale; it is therefore important to be aware of the non-linearity on a Rotameter scale. It is also possible to mount two Rotameters in series, one for flows up to 1 l min^{-1}, and one for higher flows.

Sources of error in a Rotameter include: failure to mount the tube vertical; the presence of dirt or electrostatic charge on the tube causing the bobbin to stick; a cracked tube.

When the gas passing through the Rotameter is pressurised, as in positive pressure ventilation or when it is used in a hyperbaric chamber, the position of the bobbin changes without any real change in the flow rate. Under pressure the density of a gas increases from, say ρ_0 to ρ_1. At a given volume flow rate a larger annular area is needed to maintain the same pressure difference and the bobbin sits high in the Rotameter tube. It is found that the actual flow rate in the flowmeter, F_A, is related to the indicated flow rate, F_I, by:

$$F_A = F_I \sqrt{(\rho_0/\rho_1)}$$
$$= F_I \sqrt{(P_0/P_1)},$$

i.e. the actual flow rate in the (pressurised) flowmeter is lower than the indicated flow rate, where P_0 and P_1 are the respective associated pressures. But, when

| O₂ | CO₂ | N₂O | O₂ | CO₂ | N₂O |

(Rotameter cracked)   (Rotameter cracked)

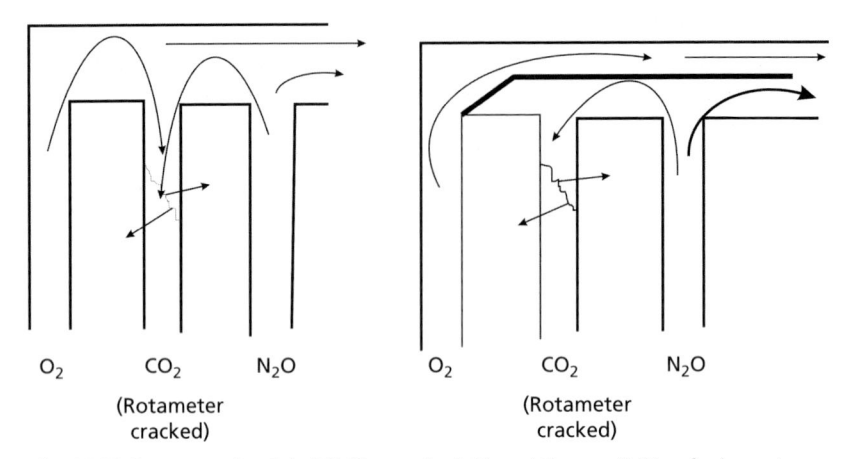

Fig. 14.11 Rotameter bank in UK (O_2 on the left) and the possibility of a hypoxic gas mixture being delivered to the patient in the event of a cracked Rotameter, unless the oxygen outlet is baffled from the other gases.

released to the lungs at atmospheric pressure, the pressure is once again P_0 and changes in flows and pressures are related to each other by Boyle's law:

$$F_0 P_0 = F_A P_1,$$

where F_0 is the actual flow rate to the lung. Hence

$$F_0 = F_A(P_1/P_0)$$
$$= F_I \sqrt{(P_1/P_0)},$$

which means that actual flow rate to the lung is higher than indicated.

A safety feature in the design of a bank of Rotameters for different gases includes baffling the oxygen tube so that it is the last gas to be mixed with the other gases. This is shown in Figure 14.10. It is to minimise the chance of delivering a hypoxic mixture to the patient in the event of a cracked Rotameter tube and despite the fact that Rotameters banks in UK are designed historically with the oxygen tube upstream of the others.

References

Bardoczky, G. I., Engelman, E. and D'Hollander, A. (1993) Continuous spirometry: an aid to monitoring ventilation during operations, Br. J. Anaesth., **71**, pp. 747–51.

Cyna, A.M. et al. (1991) AURA: a new respiratory monitor and apnoea alarm for spontaneously ventilating patients, Br. J. Anaesth., **67**, pp. 341–45.

Boaden, R. W. and Hutton, P. (1986) The digital control of anaesthetic gas flow, Anaesthesia, **41**, pp. 413–418.

Moyle, J. T. B. (1989) A Kynar piezoelectric film respirometer, Anaesthesia, **44**, 332–34.

Further reading

Clutton-Brock, T. H. and Hutton, P. (1994) Gas pressure, volume and flow measurement. In P. Hutton and C. Prys-Roberts (eds), *Monitoring in Anaesthesia and Intensive Care*, WB Saunders Co. Ltd, pp. 172–193.

Sykes, M. K., Vickers, M. D. and Hull, C. J. (1981) *Principles of Clinical Measurement*, Blackwell Scientific Publications, Chapters 12 and 15, pp. 147–158 and pp. 182–201.

Chapter 15

Pulse oximetry

This chapter contains: pulse oximetry.
The chapter links with: Chapters 5, 11 and 16.

Pulse oximetry

The pulse oximeter is a device for non-invasive, continuous measurement of oxygen saturation. As such it is arguably one of the most important intra-operative monitors at the disposal of anaesthetists, and efforts are being made to make pulse oximeters available at all operating locations throughout the world [Walker et al. 2009]. Although the device measures oxygen saturation of arterial blood, which is the physiological end point of interest, it is not a replacement for monitoring all the events which may lead to hypoxaemia; in other words it does not replace an oxygen analyser at the common gas outlet of the anaesthetic machine. Depending on the site of the probe, usually ear lobe or finger, there is a variable delay between the onset of a causative hypoxaemic event and detection of hypoxaemia by the pulse oximeter, the delay being longer the more peripherally placed is the probe. Appropriate size and design of the probe for accuracy and safety in children is important [Howell et al. 1993] and finger probes are more accurate but slower to respond than ear probes [Webb et al. 1991]. Forehead reflectance probes have been used with good results [Casati et al. 2007]. It is also true that the human eye is notoriously bad at detecting cyanosis in the range of saturations 81–85%. For additional information on Monitoring Principles see Chapter 11. It is clear, however, that in a hierarchy of monitors for anaesthesia, the pulse oximeter is indispensable.

A pulse oximeter uses two separate technologies: one is plethysmography, where reproduction of the pulsatile waveform takes place; the other is spectroscopy, where absorption of light of specific wavelengths by body tissues occurs and is analysed. The spectroscopic aspects depend on the laws of Beer and Lambert, which can be combined to state that the amount of light absorbed by a substance is proportional to the thickness of the substance sample (the path length of the light) and the concentration of the substance. The relationship is

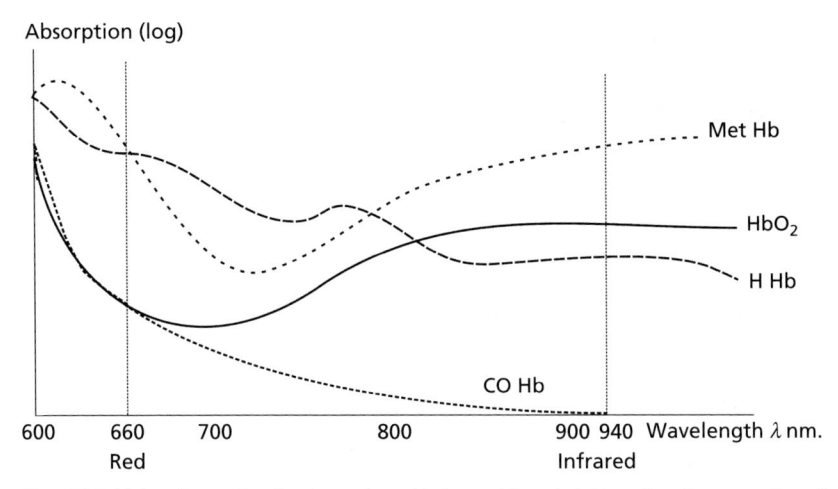

Fig. 15.1 Light absorption in the red and infra-red bandwidths of various species of haemoglobin.

a logarithmic one and can be written as:

$$adsorption = \log_{10} \frac{(intensity\ incident\ light)}{(intensity\ transmitted\ light)}$$

$$= constant \times path\ length \times concentration.$$

However there are variations between the Beer-Lambert law and the observations from pulse oximeters in humans [Awad et al. 2007].

Figure 15.1 shows the absorption spectra, in the red to infrared range, of various forms of haemoglobin, including oxyhaemoglobin (HbO$_2$) and deoxyhaemoglobin (H Hb). Pulse oximeters use light-emitting diodes (LEDs), emitting light of two different wavelengths to emit the light in from a probe, usually 660 nm (red) and 940 nm (infrared) in order to compare the spectra of oxyhaemoglobin and deoxyhaemoglobin. The LEDs have a slight temperature dependence on their absorption bands, but this does not significantly affect the accuracy of the pulse oximeter [Reynolds et al. 1991].

It can be seen from Figure 15.1 that at 660 nm, reduced (deoxygenated) haemoglobin absorbs more light than oxyhaemoglobin, while at 940 nm the converse is true. There are two wavelengths at which the curves cross over called *isosbestic points*, and a wavelength within this waveband where the curves are farthest apart. The 660 nm LED emits light close to the isosbestic point and the 940 nm LED emits light close to the point of widest separation of the curves; this gives spectrophotometric information about the total haemoglobin concentration and of the difference between oxyhaemoglobin and deoxyhaemoglobin respectively. Note that the scale is a logarithmic one, so the differences in absorbance are in fact large. The light that is not absorbed is transmitted (*transmitted = input − absorbed*), and the ratio of transmitted light at 660 nm to that at 940 nm is calculated. This ratio is used by the oximeter's software to calculate saturation. The two LEDs are switched alternately at a rate of 400 Hz, with the 'LED off' periods in between 'LED on' periods to allow a single photo detector to measure light transmission and to make allowance for the effect of ambient lighting conditions. Absorption of red and infrared light in the tissue of a finger or earlobe is not confined to arterial blood, but also occurs in all other tissues such as venous blood, muscle, nail, etc., as shown in Figure 15.2. However, only the transmission signal from pulsatile arterial blood is an 'AC'. signal, while all other constant light transmission signals are 'DC'. The amplitude of both AC and DC components from a given blood and tissue sample may well be very different at the two different wavelengths, giving raw signal levels as shown in Figure 15.3(a). Part of the software processing involves scaling the raw signal amplitudes until the DC components are equal, so that the AC components can be reasonably compared and the ratio, L, of AC signals at the two measured wavelengths can be calculated. It is then possible for the oximeter's software to look up a value of oxygen saturation, S_pO_2, corresponding to different values of L. This calibration curve is shown in Figure 15.3(b) and will have been obtained from a multiple wavelength

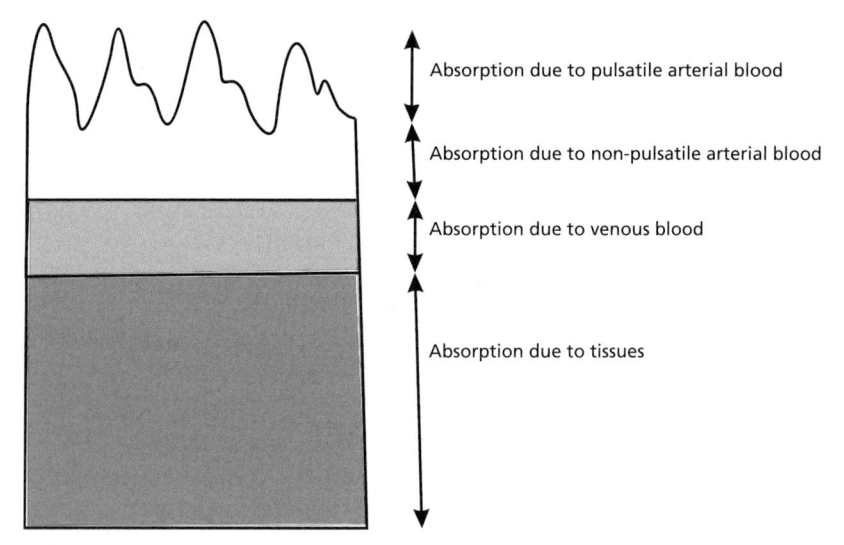

Fig. 15.2 Absorption in different tissue components by light of a pulse oximeter.

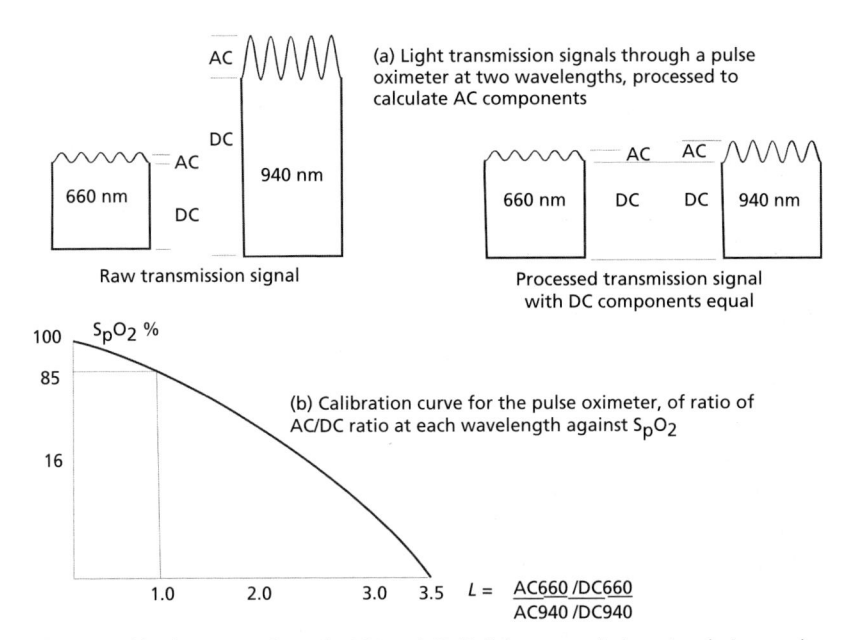

Fig. 15.3 (a) The processing of 'AC' and 'DC' light transmission signals in a pulse oximeter, normalising the 'DC' component. (b) This gives a ratio that can be plotted to give S_pO_2.

co-oximeter using formal arterial blood samples. Note that when $L = 1.0$, $S_pO_2 = 85\%$. The S_pO_2 reading on a pulse oximeter is usually an average of readings over the preceding 10 seconds or so, although this averaging period can be shortened. Errors in the readings can occur with movement and vibration [Langton and Hanning 1990], hypertension, vasoconstriction [Langton et al. 1990, Talke et al. 2006], tricuspid valve regurgitation [Stewart et al. 1991], ambient light, some IV dyes (methaemoglobin causes under reading), carboxyhaemoglobin (causes overestimate) and diathermy [Ralston et al. 1991]. Poor tissue perfusion, hypothermia [Gabrielczyk et al. 1988], hypotension and hypovolaemia also eventually cause some error [Falconer et al. 1990]. It has also been shown that the pulse oximeter under reads S_{pO_2} under conditions of low vascular resistance, such as sepsis [Secker et al. 1997]. Because the plethysmographic signal, which is often displayed alongside the S_{pO_2} value, may be subject to internal amplification, its amplitude is no guide to the adequacy of the input signal. However the width of the plethysmographic trace has been found to be consistently related to changes in systemic vascular resistance [Yang et al. 2007]. Error can also occur where there has been extrapolation by the device of values of L as in Figure 15.3 outside the calibrated range [Ridley 1988], which has been obtained from arterial blood samples of healthy volunteers breathing hypoxic gas mixtures to give S_{pO_2}s in the range 80–100%. Clearly it would be unethical to try to obtain calibration for saturations lower than this.

From the preceding paragraphs it is clear that the pulse oximeter only looks at *functional saturation*, incorporating H Hb and HbO_2, which is afforded by using two measurement wavelengths. It ignores other Hb components, which are included in a measure of *fractional saturation*, namely methaemoglobin (Met Hb) and carboxyhaemoglobin (HbCO), which can only also be measured using an additional two wavelengths. The absorption spectra of these two groups are included in Figure 15.1. The carboxyhaemoglobin absorption spectrum resembles that of HbO_2 in the red range, and its presence in arterial blood therefore contributes to an overestimate of the S_pO_2. Hence in smokers with significant levels of HbCO up to about 20%, or in victims of carbon monoxide poisoning, the pulse oximeter leads to a false sense of security. Figure 15.1 shows that Met Hb absorbs to similar extents at both wavelengths and if significant methaemoglobinaemia, which can be caused by the local anaesthetic prilocaine, nitrites and nitrates, dominates the clinical picture, the ratio L will be 1.0 and the S_pO_2 will tend to be at or near 85% (see Figure 15.3(b)).

Within the range of wavelengths 650 nm to 1000 nm, foetal haemoglobin, HbF, has the same absorption characteristics as adult Hb and therefore the pulse oximeter is accurate in neonates. HbF can, however, affect the accuracy of estimation of HbO_2 and HbCO by a four wavelength co-oximeter. Similarly, although bilirubin does not affect the accuracy of pulse oximeters, it may affect the accuracy of a co-oximeter. Skin colour does not usually affect the accuracy pulse oximeters, but some dark nail polish does.

References

Awad, A. A., Haddadin, A. S., Tantawy, H., Badr, T. M., Stout, R. G., Silverman, D. G. and Shelley, K. H. (2007) The relationship between the photoplethysmographic waveform and systemic vascular resistance, *J. Clin. Monit. Comput.*, **21**, pp. 365–72.

Casati, A., Squicciarini, G., Baciarello, M., Putzu, M., Salvadori, A. and Fanelli, G. (2007) Forehead reflectance oximetry: A clinical comparison with conventional digit sensors during laparotomic and laparoscopic abdominal surgery, *J. Clin. Monit. Comput.*, **21**, pp. 271–76.

Falconer, R. J. and Robinson, B. J. (1990) Comparison of pulse oximeters: accuracy at low arterial pressures in volunteers, *Br. J. Anaesthesia*, **65**, pp. 552–57.

Gabrielczyk, M. R. and Buist, R. J. (1988) Pulse oximetry and postoperative hypothermia. An evaluation of the Nellcor −100 in a cardiac surgical ICU, *Anaesthesia*, **43**, pp. 402–404.

Howell, S. J., Blogg, C. E. and Ashby, M. W. (1993) Howell modified sensor for pulse oximetry in children, *Anaesthesia*, **48**, pp. 1083–95.

Langton, J. A. and Hanning, C. D. (1990) Effect of motion artefact on pulse oximeters: evaluation of four instruments and finger probes, *Br. J. Anaesth.*, **65**, pp. 564–70.

Langton, J. A., Lassey, D. and Hanning C. D. (1990) Comparison of four pulse oximeters: effects of venous occlusion and cold induced vasoconstriction, *Br. J. Anaesth.*, **65**, pp. 245–47.

Ralston, A. C., Webb, R. K. and Runciman, W. B. (1991) Potential errors in pulse oximetry III. The effects of interference, dyes, dyshaemoglobins and other pigments, *Anaesthesia*, **46**, pp. 291–95.

Reynolds, K. J. *et al.* (1991) Temperature dependence of the light emitting diode and its theoretical effect on the pulse oximeter, *Br. J. Anaesth.*, **67**, pp. 638–43.

Ridley, S. A. (1988) A comparison of two pulse oximeters. Assessment of accuracy at low arterial saturations in paediatric surgical patients, *Anaesthesia*, **43**, pp. 136–40.

Secker, C. and Spiers, R. (1997) Accuracy of the pulse oximeter in patients with low systemic vascular resistance, *Anaesthesia*, **52**, pp. 127–30.

Stewart, K. G. and Rowbottom, S. J. (1991) Inaccuracy of pulse oximeters in patients with severe tricuspid regurgitation, *Anaesthesia*, **46**, pp. 668–70.

Talke, P. and Stapelfeldt, C. (2006) Effect of peripheral vasoconstriction on pulse oximetry, *J. Clin. Monit. Comput.*, **20**, pp. 305–309.

Walker, I. A., Merry, A. F., Wilson, I. H. *et al.* (2009) Global oximetry: an international anaesthesia quality improvement project, *Anaesthesia*, **64**, pp. 1051–60.

Webb, R. K., Ralston, A. C. and Runciman, W. B. (1991) Potential errors in pulse oximetry II. The effects of changes in saturation and signal quality, *Anaesthesia*, **46**, pp. 207–212.

Yang, S., Batchelder, P. B. and Raley, D. M. (2007) Effects of tissue outside of arterial blood vessels in pulse oximetry: a model of two-dimensional pulsation, *J. Clin. Monit. Comput.*, **21**, pp. 373–79.

Chapter 16

Respiratory gas analysers

This chapter contains: refractometry; infra-red spectroscopy; mass spectrometry; Raman spectroscopy; piezoelectric gas analysis; ultraviolet gas analysis; the nitrogen meter; paramagnetic oxygen analyser; polarography and fuel cells; preoperative exercise testing.
The chapter links with: Chapters 11, 5 and 17.

Introduction

The purpose of respiratory gas analysis during anaesthesia is to identify and measure the concentrations, on a breath by breath basis, of the individual gases and vapours in use. It may also be useful as a guide to cardiac function or to identify trace contaminant gases. Different techniques use different physico-chemical properties of the gas or vapour. An understanding of the physical principle underlying each method is necessary in order to recognise the value and limitations of each.

In terms of the device's ability to respond on a breath by breath basis, there are two important components: the time taken for the gas to be sampled from the anaesthetic machine or breathing system, the *delay time*; then there is the time taken for the device to measure the gas concentration, the *response time*. This is depicted in Figure 16.1. Most of the delay occurs in the delay time or *transit time* and can be reduced either by analysing the gas sample close to the airway, or by using as short and thin a sampling tube and as high a sampling flow rate to the analyser as possible [Chan *et al.* 2003]; the sampling flow rate is usually of the order of 100 to 200 ml min^{-1}. If minimal fresh gas flow rates are being used in a circle anaesthetic breathing system and the sampled gas is not returned to the breathing system, then a high gas sampling rate could represent a significant gas leak. Figure 16.1 shows a sigmoid curve of recorded gas concentration change in response to a square wave input change. The response of a gas analyser is often expressed as the time taken to produce a 90–95% response to a step or square wave input change. A square wave change in gas concentration can be produced by moving a gas sampling tube rapidly into and out of a gas stream, by bursting a small balloon within a sampling

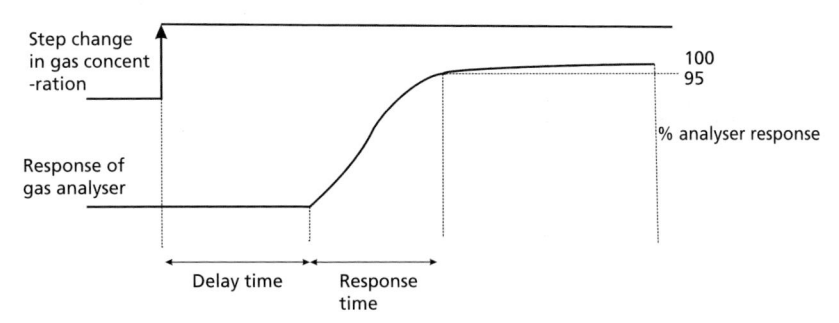

Fig. 16.1 Response of a gas analyser to a step change in gas concentration.

volume containing a gas sample, or by switching a shutter to a gas sample volume using a solenoid valve.

An important part of the use of gas analysers is zeroing and calibration since they are all prone to drift in both zero and gain. These days zeroing is often done automatically on switching on the analyser and at predetermined intervals. However the calibration – that is to say the linearity and gain of the output signal for a known input – has periodically to be checked with calibration gases. Some devices can only measure the concentration of one gas, others can estimate several gases.

Refractometry

This method uses different refractive indices of gases to identify and calibrate gases. It is included here, not because it is routinely used in anaesthesia, but because it is still a laboratory standard against which other methods are compared. The refractive index, μ, of a transparent medium is given by:

$$\mu = \frac{\text{velocity of light in a vacuum}}{\text{velocity of light in the medium}}.$$

$\mu > 1.0$ because the velocity of light in a vacuum, 3.0×10^8 m s^{-1}, is always greater than in any other medium. The delay in the passage of light through a gas depends on the number of gas molecules present; therefore the refractive index depends on its pressure and temperature. Refractometry depends on the principle that interference bands are formed when light waves from a source, which pass through two slits, are focused on to a screen. Figure 16.2 shows that when the transverse waveforms of light from the two slits arrive in phase with each other, they reinforce each other, producing a bright band of light; when they arrive 180° out of phase, they cancel each other and a dark band is produced; hence *interference fringes* are produced and the clearest fringes

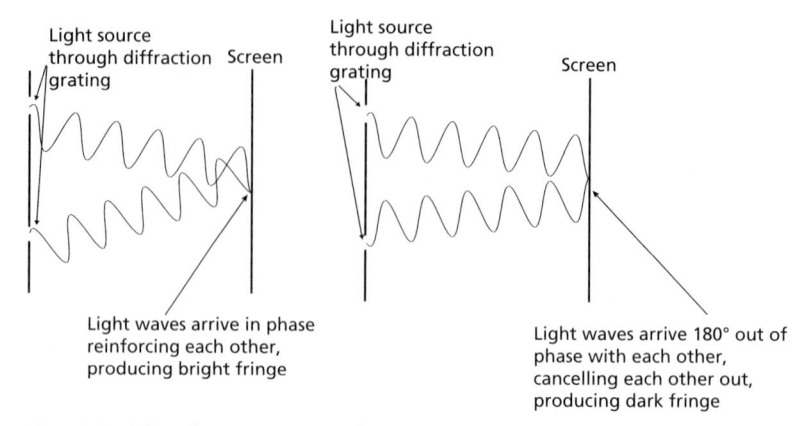

Fig. 16.2 Light refractometry. Interference fringes caused by light waves arriving in phase and 180° out of phase.

are produced when a monochromatic (single wavelength) light source is used. There is a portable device available [Cheam *et al.* 1995].

The Rayleigh refractometer

This is shown in Figure 16.3. Light from a tungsten source passes through a pair of vertical slits and a convex lens, the upper part of the pair of light beams emerging parallel from the lens; these beams are then passed through two tubes, A and B, one of which contains the gas to be analysed, the other a gas of known refractive index. The lower portion of light from the slits does not pass through the tubes containing the gases. After passing through another lens to the eyepiece, an upper and lower set of interference fringes are visible, each set being formed by virtue of interference of light beams from the slits. When tubes A and B are initially filled with the same gas, the upper fringe pattern assumes a position that can be lined up with the lower reference fringe pattern. If the gas in, say tube A, is changed to a gas to be analysed, the upper fringe pattern shifts because of the differential path lengths of the light in A and B. If one of the two glass plates, X and Y, say Y, is rotated to increase the path length through B, to match that through A, then the upper fringe pattern comes back into line with the lower one. The amount of rotation of Y, on a carefully calibrated scale, can be made to be proportional to the gas concentration in A.

Figure 16.4 shows a portable refractometer, in which the light beam from a common source is split into two beams by a prismatic glass plate, thus allowing a reference light path and one that passes through the sample. The returning light paths form an interference fringe pattern, which is viewed through an eyepiece.

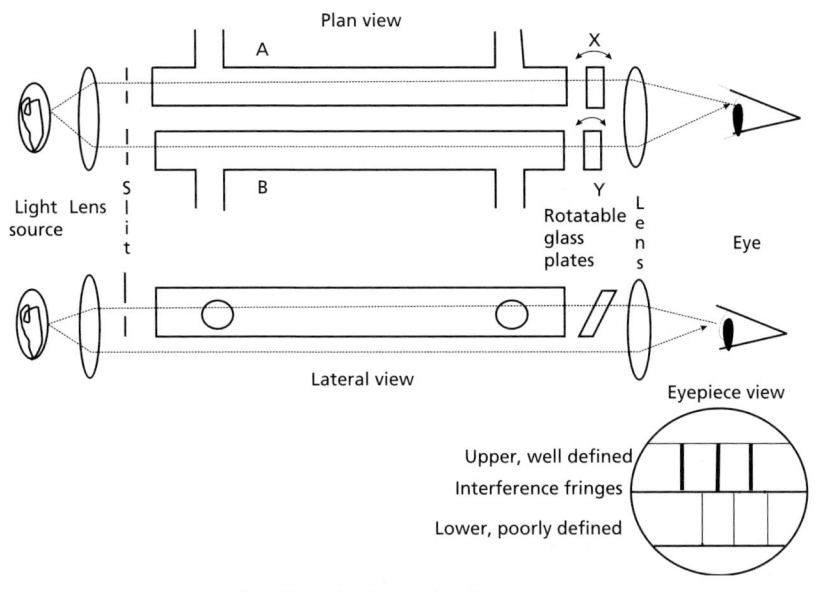

Fig. 16.3 The Rayleigh refractometer.

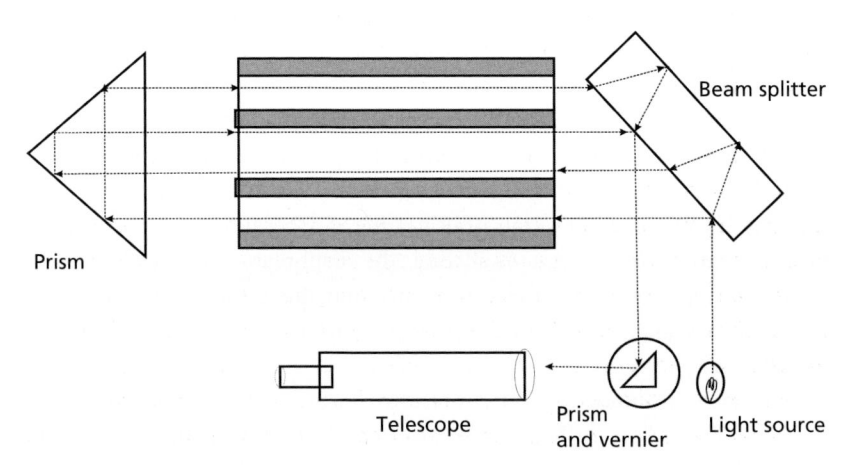

Fig. 16.4 Portable refractometer. See text for details.

Since all gases possess the physical property of refractive index, all gases can be analysed by this method. Although highly accurate, it does not lend itself to breath by breath analysis, because of its large size and the lack of ability to analyse changing gas concentrations. Nevertheless it is used as the standard method against which others are calibrated.

Infrared spectroscopy

Molecules that contain two or more different atom species (i.e. CO_2, N_2O, H_2O, but not O_2 or N_2) absorb infrared radiation because of the nature of the bond between the dissimilar atoms. This property can be used to analyse gases such as carbon dioxide, nitrous oxide and all anaesthetic vapours, but not oxygen or nitrogen. The absorption A of the IR light is related to the intensities of incident light I_i, the transmitted light I_t, the extinction coefficient ε, the path length L and the concentration C of the absorbing medium by the Beer–Lambert law:

$$A = \log_{10} \frac{I_i}{I_t} = \varepsilon L C.$$

The principle of an infrared gas analyser (*Luft* or non-dispersive type) is shown in Figure 16.5. Because CO_2 analysis is arguably as important as pulse oximetry for patient safety during anaesthesia, these devices are now commonly available in the anaesthesia workplace. Figure 16.5 shows an infrared (IR) light source emitting radiation in the range 1–15 μm wavelength. The light passes through a 'chopper' wheel rotating at 25–100 Hz, which ensures that the infrared light is only intermittently passed further down the device; this also prevents overheating of the gas sample and allows processing of the subsequent alternating signal. The light is then passed simultaneously through a pair of tubes, one of which contains a reference gas, the other of which contains the gas for analysis. Because of the differential IR absorption in the two chambers, differential amounts of IR light are transmitted to a pair of detector cells filled with air, separated by a diaphragm. The pressure in the two halves is therefore different from separate amounts of IR heating and, because the chopper wheel ensures that the IR input is oscillating, the diaphragm oscillates as well. It is usually part of an electrical capacitance and this signal is amplified and displayed as a gas concentration proportional to the number of molecules of gas present.

There are a number of sources of error in infrared spectroscopy. Major bands of absorption of IR for CO_2, N_2O and CO occur in the 4.3, 4.5 and 4.7 μm wavebands respectively, although there is considerable overlap in the 3–5 μm range. Therefore there is significant interference between the spectra of these three gases, particularly between CO_2 and N_2O in the 4.3 μm range. Another source of error is due to the phenomenon of *collision broadening*, where the absorption spectrum of one gas is actually broadened by the presence of another. This is due to transfer of interatomic bond energy between colliding N_2O and CO_2 molecules. There are electronic correction factors incorporated into the analysers to allow for different mixtures of N_2O and CO_2. It has also

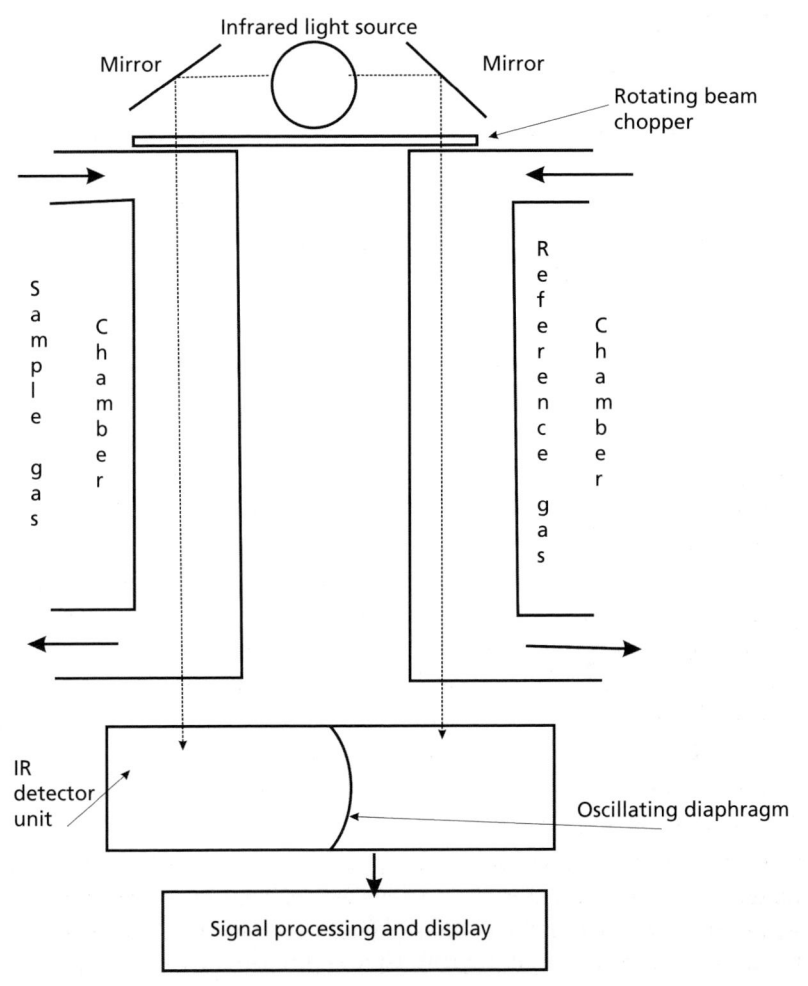

Fig. 16.5 Infrared gas analyser (Luft head).

been found that an infrared spectrometer under reads the CO_2 values in a gas mixture containing 79% helium in oxygen [Ball et al. 2003]. It has been reported that a gas mixture containing desflurane produces some error in an infrared gas analyser [Scheeren et al. 1998] as does cyclopropane [Mason et al. 1991] and alcohol [Foley et al. 1990].

A further source of error with this device is related to the fact that its output signal is proportional to the number of molecules present, which is really the partial pressure of the gas. Therefore the pressure in the sample and reference cells and the ambient pressure are all important variables against which the device must be calibrated. This includes pressure changes due to sampling through a narrow bore tube from the breathing system, as well as changes in back pressure from positive pressure ventilation. If the pressure of the measured gas mixture increases without any real change in the fractional concentration, then the partial pressure of the gas under investigation also increases. If the software of the device converts partial pressure to fractional concentration, there will be an erroneous increase in the indicated fractional concentration. Likewise if the analyser is calibrated at sea level and then used at altitude, such that there is an altered relationship between the gas partial pressure and ambient pressure, then error will result. Therefore calibration should be done under the intended working conditions and repeated relatively frequently. Another solution is to have the device calibrated for partial pressure rather than concentration, in which case error due to atmospheric pressure changes remain small. Most devices remain stable but, as well as two or three point calibration, response time should also be routinely checked. Despite long sampling lines and low sampling rates (100–200 ml min^{-1}), the 90–95% step response time should remain under 150 ms In this respect the water vapour sampled with the gas sample sometimes blocks the sampling tube, but most devices have either a water trap or a sampling line made of a material that absorbs water vapour. In this case an electronic correction is made to the result to allow for the presence of the water vapour in the expired breath sample. Alternatively water itself is a powerful absorber of IR light, and its partial pressure can be measured, usually at a different wavelength from the other constituents being analysed.

Modern IR analysers may use photocells rather than differential pressure chambers to detect IR transmission and may use narrow bandwidth filters for the different gases being sampled. For example, in the Datex-Engstrom device, N_2O is measured at 3.9 μm, CO_2 at 4.3 μm and volatile agents at 3.3 μm. It cannot distinguish volatile agents from each other unless it is 'informed' or different agents are analysed at different wavebands. Regular automatic calibration takes place using compressed vapour canisters. Note that exhaled ethyl alcohol or acetone will interfere with the readings. Any residual error due to interference between CO_2 and N_2O is taken care of electronically. The accuracy of the IR analyser is about 0.1% in a range of CO_2 of 0–10%.

Mass spectrometry

A mass spectrometer is capable of analysing a gas mixture on the physical principle of separating the components in a magnetic field according to their molecular mass and charge. Providing the constituents of the gas mixture are known, provision of appropriate detectors can be made and any gas can be analysed. Error occurs where an unknown constituent is introduced into the mixture, such as the propellant in bronchodilators, or acetone in the breath of a diabetic. Error can also occur where the molecules have the same molecular mass (see below). Although a mass spectrometer was developed for clinical use in 1958 [Fowler 1958], it is still used more often in the laboratory than in the operating room. It has been used in remote sampling, multisite and timesharing modalities.

Basic principles

The mass spectrometer has three stages, shown in Figure 16.6. In the first stage the sampled gas is drawn into a low pressure sample chamber. Only a small fraction of this gas is drawn into the second stage and through a *molecular leak*, which is in the main part of the analyser. The third stage is the electronic amplification and display of the signal.

The second stage of the analyser is virtually under a vacuum. In the ionisation chamber the gas molecules are bombarded by a transverse beam of electrons, as shown in Figure 14.6, and thereby become ionised by losing an electron. They then pass towards a plate to which a negative voltage has been applied, termed the *acceleration plate*, and through a slit in this plate, the molecular leak. From this point the path of the ions is influenced by a magnetic field. The field thus deflects and separates the ions according to their *mass:charge* ratio. Since most molecules have been ionised to carry a single positive charge, separation is mainly on the basis of mass, lightest ions being deflected the most, heaviest ions the least. The ions reach photo-voltaic detectors, where electronic registration, amplification and display of the signal occurs, proportional to the rate of arrival of ions and therefore to partial pressure of the gas.

The way in which the magnetic field is produced and ions are detected on the plates has two variants, resulting in two types of instrument. The earlier form was the *magnetic sector* type, where the magnetic field lies at right angles to the path of ions and is produced by a combination of electric and fixed magnetic fields. By varying the voltage on the acceleration plate, the velocity of the ions entering the magnetic field can be changed. This can allow the deflected ion beam to be directed across the detector plate and different components to be detected in turn. It is therefore possible to separate components of the gas sample according to mass:charge ratio.

More commonly seen these days, is the *quadrupole mass spectrometer*. This device is more compact and allows better discrimination between components.

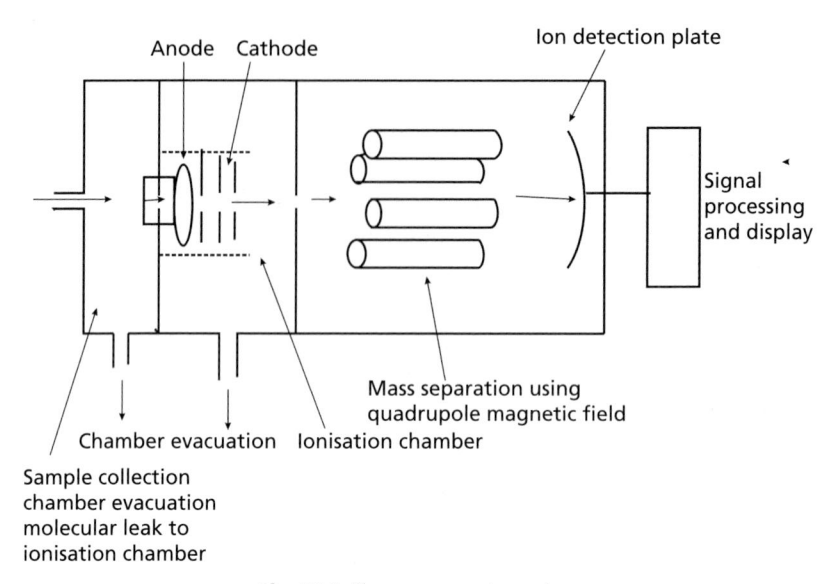

Fig. 16.6 The mass spectrometer.

The magnetic field is produced by a combination of D.C. electric and A.C. radio frequency fields. As shown in Figure 16.6, the quadrupoles consist of four rods with opposite pairs electrically connected. By careful tuning of the radiofrequency component of the magnetic field, only ions of a given mass:charge ratio proceed through the quadrupole to the detector, all other ions oscillating and colliding with the walls of the device. By a combination of changing the voltage on the acceleration plate (the cathode) and of judiciously tuning the magnetic field, a spectrum of mass:charge components can be detected and quantified. By scanning at 50 Hz it is possible to produce a continuous record of gas concentrations.

The respiratory mass spectrometer is accurate, giving good gas identification and quantification. It requires only about 20 ml min^{-1} gas sampling rate, with a 100 ms response time. However it does have some disadvantages. The second stage of the device operates under almost vacuum conditions; this requires a high quality, continuously running pump. If the device itself is at some distance from the sampling site, significant delay time may be added to the response time. Water condensation can be avoided by heating the sampling tube, but the response time for water vapour may still be longer than for other components.

Some molecule types may lose two electrons rather than one on ionisation, and appear to the magnetic field with a double charge. They then behave like an ion with half the mass, which leads to confusion in interpretation. Furthermore the ionisation process can lead to fragmentation of a molecule, so that a mass spectrum appears at the output rather than a peak. However, this anomaly is useful to distinguish the gas components with the same molecular mass, such as N_2O and CO_2 (44 Da), or N_2 and CO (28 Da). N_2O is fragmented into NO, O_2, N_2, N and O, while CO_2 is fragmented into O_2, C_2, C and O. Rather than trying to detect and distinguish both gases at 44 Da, N_2O can be detected at a subordinate peak of 30 Da and CO_2 at 12 Da. Because the fragmentation is predictable, the amplitude of the subsidiary peak can be used as a measure of the parent peak and therefore of gas concentration. Using appropriate low pass and high pass filters, obfuscating peaks in the spectra can be removed.

Raman spectroscopy

When light falls on an object, it is reflected and scattered by the object. Most of this scattering occurs without any loss of energy or change of wavelength and is known as *Rayleigh scattering*, which allows us to see the object with scattered waves in the visible spectrum. A fraction of the incident light, about 10^{-6}, is scattered with a loss of energy and a change of wavelength characteristic of the molecule off which the light is being reflected; this is *Raman scattering*, first observed in 1928 by the Indian physicist, after whom the technique is named. Raman spectroscopy has been used in industry for years as a means of identifying solids, liquids and gases, but has had to await the advent of powerful laser light sources and sensitive photocell detectors to be useful in a clinical setting for breath by breath analysis.

Figure 16.7 shows the schematic layout of a Raman spectroscope, incorporating an Argon laser source of wavelength 485 nm, high reflectance mirrors to concentrate the laser beam, a gas sampling chamber, appropriate optics, a detection system, and signal processor and display system. Table 16.1 shows characteristic wavelength or frequency changes that identify different gases. If plotted graphically, the amplitudes of the frequency shifted peaks are proportional to the gas concentrations. In Raman spectroscopy each gas is independently analysed, including CO_2, N_2O, volatile agents, O_2, N_2 and water vapour. The response time is around 100 ms and the device is therefore capable of breath by breath analysis. The sample is not altered by the process and can therefore be returned to the breathing system, which is a significant advantage in low flow anaesthesia [Lockwood *et al.* 1994], although there is some overlap between gas species [Lawson *et al.* 1993]. However the devices require a great deal of electrical power and are noisy.

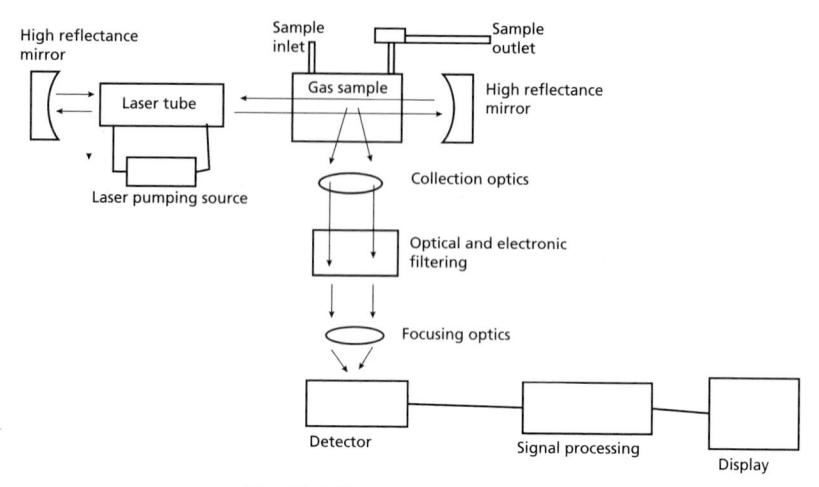

Fig. 16.7 The Raman spectroscope.

Table 16.1 Characteristic frequency shifts of substances with Raman spectroscopy

Gas	Frequency shift cm^{-1}
Isoflurane	995
Halothane	717
Enflurane	817
N_2O	1285, 2224
CO_2	1285, 1388
Oxygen	1555
Nitrogen	2331
Water	3650

Piezoelectric gas analysis

A pair of piezoelectric crystals connected to an electrical power source is made to resonate with a characteristic frequency difference. One of the crystals is coated in silicone oil, which absorbs a volatile anaesthetic agent to which it is exposed. This changes the natural frequency of the crystal, and the frequency difference between the crystals changes by an amount dependent on the amount of volatile agent absorbed. Manual identification of the agent is required, but the device is remarkably accurate and stable [Humphrey *et al.* 1991].

Ultraviolet gas analysis

UV light from a mercury lamp is the light source and the device is used to analyse halothane concentrations with an absorption spectrum in the 200 nm band.

The nitrogen meter

In contrast to the other devices described, where energy is absorbed, this device relies on electromagnetic emission from nitrogen molecules. This is the device used to measure anatomical deadspace and for assessing the regional distribution of ventilation. A sample of gas is drawn through a needle valve by a suction pump to a vacuum gas discharge tube, across which a voltage of 2000 V is applied. This ionises the gas sample, which glows and emits radiation, which can be measured by a photoelectric cell. Although the response time is short at about 20–40 ms, the output is not linear and additional electronic circuitry in the form of a logarithmic converter is necessary.

Paramagnetic oxygen analyser

Gas molecules are influenced by a magnetic field in two different ways because of their structure. Most are repelled by the field and are called *diamagnetic* molecules. Two gases, oxygen and nitric oxide, are attracted into the field and are called *paramagnetic*. This property enables oxygen concentrations to be analysed and is the result of the presence of unpaired electrons in the outer shell of an oxygen molecule (see chapter 3), which is able to generate force in a magnetic field. A paramagnetic oxygen analyser is shown in Figure 16.8 of the sort originally developed by Pauling in 1946. In this device, the two suspended glass spheres are filled with nitrogen, a weakly diamagnetic gas. The glass spheres are arranged in a dumb-bell shape, suspended between the poles of a magnet by a thread, tensioned to keep the dumb-bell in the plane of the magnetic field. Zeroing should be carried out in the carrier gas destined to have oxygen added to it at a later stage. When an oxygen containing gas mixture is drawn through the analyser, oxygen is attracted into the magnetic field, displacing the nitrogen filled spheres from it. The detection system can either be a deflection measurement or a *null deflection* type, which measures the current required in the circuitry to restore the pointer to its null point. These devices are accurate

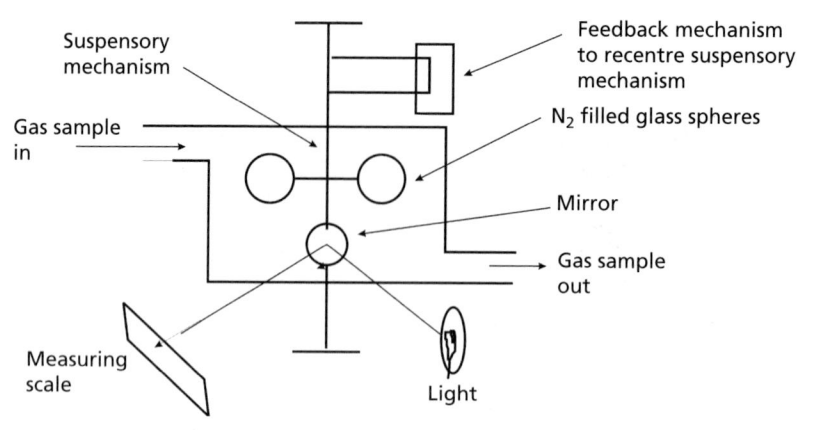

Fig. 16.8 The paragmagnetic oxygen analyser.

to within 0.1% O_2, but are adversely affected by pressurisation, vibration, water vapour and high flow rates; there is also a slow response time of up to 1 minute.

To overcome these disadvantages, a development of the device in which there is an electromagnet that produces an alternating magnetic field at a frequency of 110 Hz is shown in Figure 16.9. A reference gas enters one arm of the tube, the sample gas enters the opposite arm and the vibration set up by the alternating field is detected and measured by a pressure transducer, which detects the 20–50 µbar pressure oscillations. The oscillations are transduced into a sound signal, the amplitude of which is directly proportional to the O_2 concentration. It has been reported that a gas mixture containing desflurane interferes with the accuracy of a paramagnetic oxygen analyser [Scheeren 1998].

Polarography and fuel cells

These two techniques are combined in one section because they use the same electrochemical principle in slightly different ways.

The Clarke polarographic electrode is shown in Figure 16.10(a). It consists of a cellophane covered platinum cathode and Ag/AgCl anode in a phosphate and KCl electrolyte, between which a potential difference of −0.6 V is applied by the battery as shown. At the cathode the following reaction takes place:

$$O_2 + 2H_2O + 4e^- \Rightarrow 4OH^-.$$

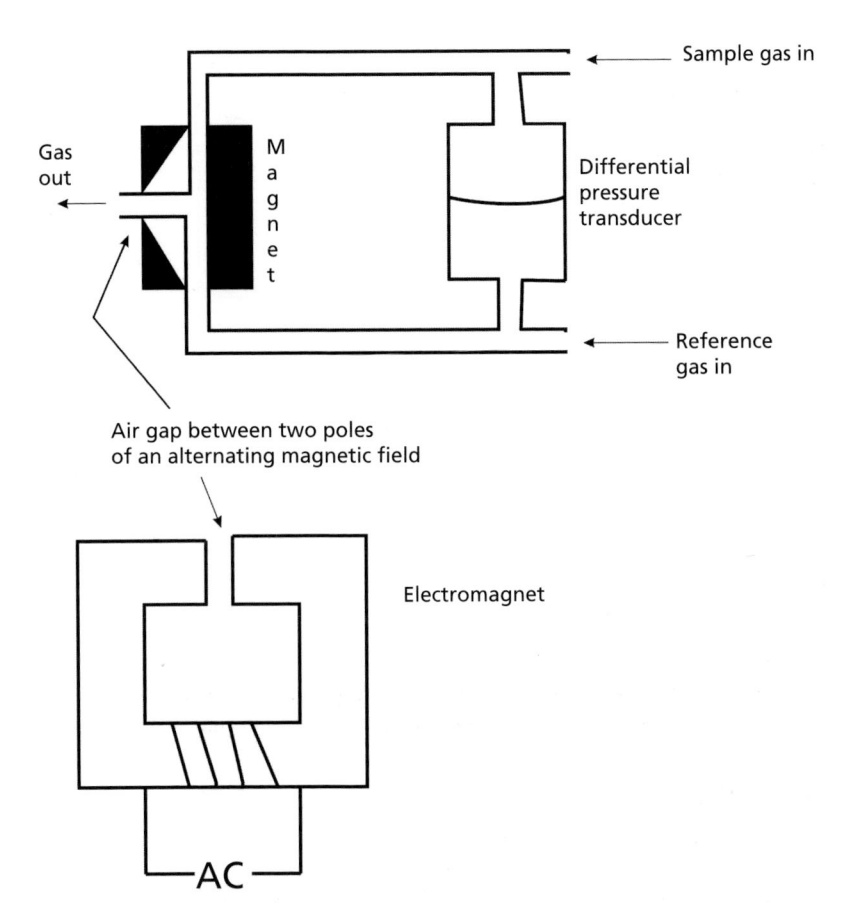

Fig. 16.9 Modern paramagnetic oxygen analyser.

Electrons are provided by the cathode, to be consumed by the O_2 in the gas or blood sample. At the anode the following reactions take place:

$$4Ag \Rightarrow 4Ag^+ + 4e^-$$
$$4Ag^+ + 4Cl^- \Rightarrow 4AgCl.$$

The current generated by the reaction represents the consumption of electrons at the cathode and is proportional to the pO_2 of the gas sample. Figure 16.10(b) shows that a plateau of current is reached for a given voltage, which represents the rate limitation of the diffusion of O_2 from the electrolyte solution. The height of the plateau is proportional to pO_2.

The problem with the polarographic electrode is the dependence on a battery and the reduction of N_2O at the cathode when it is contaminated by

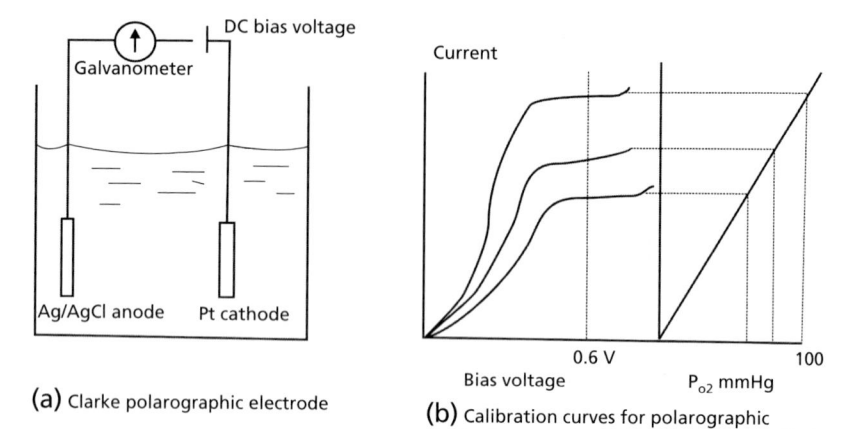

Fig. 16.10 (a) The Clarke polarographic electrode. (b) Calibration curves for the polarographic electrode.

Ag+ ions. A fuel cell is a similar device that consists of a gold cathode and a lead anode. The same reaction occurs at the cathode as in the polarographic electrode. Note that no polarising voltage is required in this circuit. Their response time is slow, making them acceptable but less suitable for breath by breath analysis than other methods [Roe et al. 1987].

Preoperative exercise testing

This section has been put in this chapter because the process involves the use of the gas analysers described above. It has become more common, both in the general community and in preoperative assessment of patients, to apply some physiological science to the process in order to classify surgical risk for a particular patient. This involves monitoring the patient's oxygen consumption and carbon dioxide output with exercise and plotting CO_2 output against O_2 consumption in ml kg^{-1} min^{-1}. These physiological variables may be calculated from *gas concentration difference (input − output) × tidal volume × respiratory rate*. Figure 16.11 shows such a plot, which generally consists of a line with two distinct slopes. At low exercise rates the slope is slightly under 1, corresponding to aerobic metabolism. As the rate of physical work increases, there is an abrupt increase of slope corresponding to a change of metabolism to aerobic plus anaerobic, and this point is termed the anaerobic threshold. Clearly the more work a patient is able to do before anaerobic metabolism starts, the greater is his or her physical capacity. A patient with an anaerobic threshold of as low as 11 ml O_2 kg^{-1} min^{-1} is deemed to be at significant

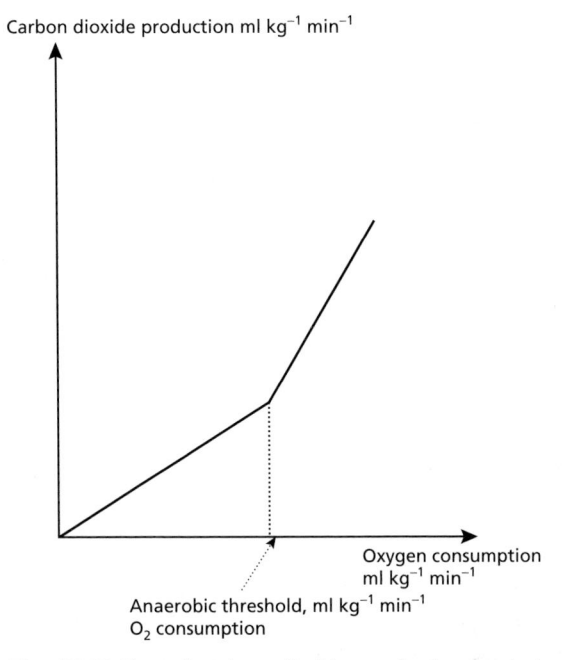

Fig. 16.11 Plot of carbon dioxide production against oxygen consumption to produce anaerobic threshold in exercise testing.

additional cardiovascular risk for major surgery (Older *et al.* 2007). Although there is some variance in the curves produced by such testing, it is found that inter-observer variance in determining the anaerobic threshold is clinically acceptable when classifying patients' surgical risk preoperatively (Sinclair *et al.* 2009).

References

Ball, J. A. S. and Grounds R. M. (2003) Calibration of three capnographs for use with helium and oxygen mixtures, *Anaesthesia*, **58**, pp. 156–60.

Chan, K. L., Chan, M. T. V. and Gin, T. (2003) Mainstream vs. sidestream capnometry for prediction of $PaCO_2$ during supine craniotomy, *Anaesthesia*, **58**, pp. 149–55.

Cheam, E. W. S. and Lockwood, G. G. (1995) The use of a portable refractometer to measure desflurane. *Anaesthesia*, **50**, pp. 607–610.

Foley, M. A. *et al.* (1990) The effect of exhaled alcohol on the performance of the Datex Capnomac. *Anaesthesia*, **45**, pp. 232–34.

Fowler, K. T. (1958) A mass spectrometer for the rapid and continuous analysis of gas mixtures in respiratory research, PhD thesis, University of London.

Humphrey, S. J. E., Luff, N. P. and White, D. C. (1991) Evaluation of the Lamtec anaesthetic agent monitor, *Anaesthesia*, **46**, pp. 478–81.

Lawson, D., Samanta, S., Magee, P. T. and Gregonis, D. E. (1993) Gas monitoring in the OR: stability and long term durability of Raman spectroscopy, *J. Clin. Monit.*, **9**, pp. 241–51.

Lockwood, G. G., London, M. J., Chakrabarti, M. K. and Whitwam, J. G. (1994) The Ohmeda Rascal II. A new gas analyser for anaesthetic use, *Anaesthesia*, **49**, pp. 44–53.

Mason, D. G. and Lloyd-Thomas, A. R. (1991) Cyclopropane and the Datex Capnomac, *Anaesthesia*, **46**, pp. 398–99.

Older, P. and Hall, A. (2007) The role of cardiopulmonary exercise testing in the preoperative evaluation of surgical patients. In Cashman, J. and Grounds M. (eds) *Recent Advances in Anaesthesia and Intensive Care*, vol 24, Cambridge University Press, pp. 1–20.

Roe, P. G., Tyler, C. K. G., Tennant, R. and Barnes, P. K. (1987) Oxygen analysers. Evaluation of five models, *Anaesthesia*, **42**, 175–81.

Scheeren, T. W. C. *et al.* (1998) Error in measurement of oxygen and carbon dioxide concentrations by the Deltatec II metabolic monitor in the presence of desflurane, *Br. J. Anaesth.*, **80**, pp. 521–24.

Sinclair, R. C. F., Danjoux, G. R., Goodridge, V. and Batterham, A. M. (2009) Determination of the anaerobic threshold in the preoperative assessment clinic: inter-observer measurement error, *Anaesthesia*, **64**, pp. 1192–95.

Further reading

Hutton, P., Hahn, C. and Clutton-Brock, T. H. (1994) Gas and vapour analysis. Chapter 13 in *Monitoring in Anaesthesia and Intensive Care*, WB Saunders Co. Ltd, pp. 194–203.

Magee, P. T. (2005) Physiological monitoring: gases. Chapter 17 in Davey, A. and Diba, A. (eds) *Ward's Anaesthetic Equipment*, 5th edn, WB Saunders Co. Ltd, pp. 355–364.

Chapter 17

Blood gas analysis

This chapter contains: the pO_2 electrode; the pH electrode; the pCO_2 electrode; derived variables; errors; temperature; transcutaneous and intravascular blood-gas analysers.
The chapter links with: Chapters 1 and 16.

A blood gas machine has electrodes to measure pH, pCO_2 and pO_2 and often measures Hb and some biochemistry as well [King et al. 2000]. Derived values from such a device include O_2 saturation, O_2 content, bicarbonate, base excess and total CO_2.

The pO₂ electrode

This is the Clarke electrode described in the previous section on gas analysers and is suitable for both respiratory and blood O_2 analysis.

The pH electrode

A *pH unit* has been defined in Chapter 1 as

$$pH = \log_{10} \frac{1}{[H^+]}.$$

In words, this can be described as 'the negative logarithm, to base ten, of the hydrogen ion concentration'.

The physical principle on which the pH electrode is based depends on the fact that when a membrane separates two solutions of different $[H^+]$, a potential difference exists across the membrane. In a pH electrode, such a membrane is usually made of glass and the development of a potential difference between the two solutions is thought to be due to the migration of H^+ into the glass matrix. If one solution consists of a standard $[H^+]$, the pH of the other solution can be estimated by measurement of the potential difference between them. The glass membrane used is selectively permeable to H^+. No current flows in this device, which does not wear out, in contrast to the Clark electrode, in which current does flow and that does need periodic replacement.

The pH measurement system is shown diagrammatically in Figure 17.1. It consists of two half cells. In one half it has an Ag/AgCl electrode and in the other a Hg/HgCl$_2$ (calomel) electrode. Each electrode maintains a fixed electrical potential. The Ag/AgCl electrode is surrounded by a buffer solution of known pH, surrounded by the pH sensitive glass. Outside the glass membrane is the test solution, usually blood, whose pH is to be measured. It is the potential difference across the glass, between these two solutions, which is variable. The blood or other solution is separated from the calomel electrode by a porous plug and a potassium chloride *salt bridge* to minimise KCl diffusion. The potential difference across the system is about 60 mV per unit of pH change at 37°C. The internal electrical resistance is high and, in order to maximise the device's accuracy, a voltmeter (or other measuring device) of very high internal resistance must be used to minimise current drawn from the system.

The pH electrode actually responds to H$^+$ activity rather than [H$^+$]. These two variables coincide at infinite dilution of solution, but might otherwise differ, because of molecular interaction between ionic species. The electrode is therefore calibrated against standard buffer solutions of pH 6.841 and 7.383.

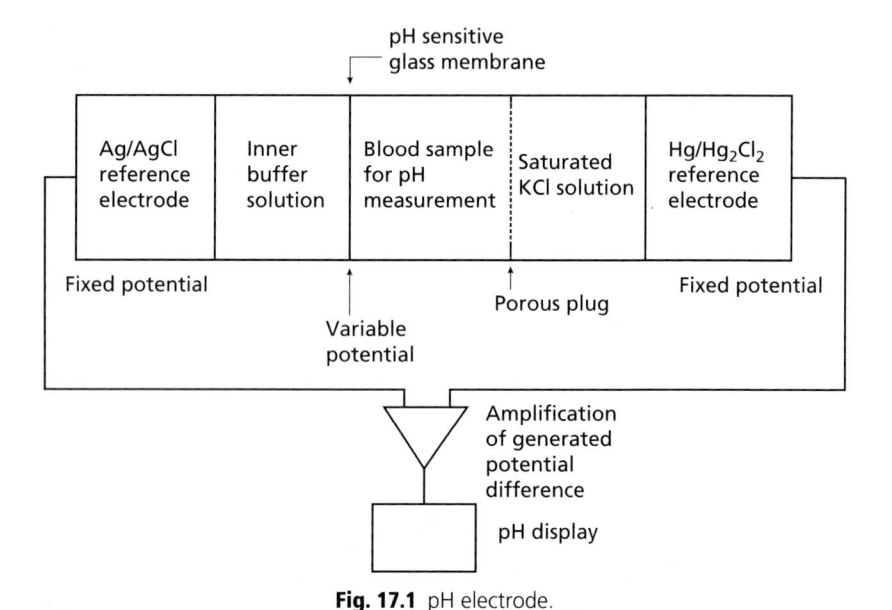

Fig. 17.1 pH electrode.

The pCO₂ electrode

There is a dynamic equilibrium between H⁺ and CO_2 by virtue of the following reactions:

$$CO_2 + H_2O \Leftrightarrow H_2CO_3 \Leftrightarrow H^+ + HCO_3^-.$$

In other words, there is a relationship between pH and [pCO₂]. The Henderson–Hasselbalch equation describes this relationship:

$$pH = pK_a + \log_{10} \frac{[HCO_3^-]}{[H_2CO_3]}$$

$$= pK_a + \frac{\log_{10}[HCO_3^-]}{\alpha \, pCO_2},$$

Where α is the Ostwald solubility coefficient, 0.003 mmol l⁻¹ mmHg⁻¹ at 37°C. pK_a, the association constant of the chemical reaction concerned is 6.1 and is the pH at which H_2CO_3 is 50% associated. Both α and pK_a vary with pH and thus lend the calculation some inaccuracy. It was Astrup, the Danish physiologist, who noticed that there was a linear relationship between pH and pCO₂.

The pCO₂ electrode is essentially a pH electrode with a difference. The sensitive glass membrane is covered with another membrane that is selectively permeable to CO_2. In the original design a layer of water was trapped between the two membranes, allowing the reaction described above to occur. A later modification had salt and bicarbonate solution in this space. A pCO₂ electrode therefore allows a pCO₂ change to generate a pH change, which is measured by the electrode.

Derived variables from a blood gas machine

Buffer base is the sum of all bases in the blood capable of buffering pH changes. These include HCO_3^-, Hb, PO_4^{3-} and the anionic parts of protein. A respiratory driven change in status, marked by, for example, the addition of CO_2, results in the formation of carbonic acid. Thus:

$$CO_2 <=> H_2CO_3 <=> H^+ + HCO_3^-$$

and H⁺ is buffered by the

$$\text{total buffer base} = HCO_3^- + Hb^- + PO_4^{3-} + \text{protein}^-.$$

The buffer base can be considered to be a limitless source of buffering capacity unless the haemoglobin component is significantly reduced The *base excess*, an indication of metabolic rather than respiratory acid-base disturbance, is defined as the amount of titratable acid required to correct a blood sample to a pH of 7.40 at a pCO_2 of 5.3 kPa and at 37°C.

Standard bicarbonate is the bicarbonate in a fully oxygenated blood sample at a pCO_2 of 5.3 kPa and at 37°C.

Errors in blood gas measurement

These include (for all electrodes):
- air bubbles in the sample
- excess heparin
- gas storage in container plastic walls
- failure to store samples at low temperature

and for the O_2 electrode only:
- metabolism in the sample
- O_2 consumption in a Clarke electrode
- material used in electrode membrane
- polarizing voltage.

and
- protein deposition on the electrode pH electrode,
- signal processing errors.

Temperature and blood gas analysis

Blood gas analysers measure blood gas variables at 37°C. A fall in body temperature means that CO_2 is more soluble in blood and that CO_2 production slows, leading to a fall in pCO_2. This shifts the O_2 dissociation curve to the left, increasing the O_2 content. However O_2 solubility also rises, decreasing pO_2 and O_2 content. pH increases with a fall in temperature.

Hypothermia itself does not alter the blood gas values at 37°C and arterial blood taken at lower temperatures should be corrected to 37°C before clinical decisions are made on the results [Rupp *et al.* 1986].

Transcutaneous blood gas analysers

pO_2 can be measured transcutaneously by applying a Clarke polarographic electrode to the skin. If the skin is 'arterialised' by heating it, a reasonable value for pO_2 can be obtained. A modified pCO_2 electrode is used to measure transcutaneous pCO_2. A common O_2/CO_2 permeable membrane and electrolyte solution is used, ensuring a common pH for both measurements. Actual measured values do not correlate well with arterial values, but trends are clinically useful and the device is still used in neonatology.

Intravascular blood gas analysers

These are microminiaturised versions of the pH, pCO_2 and pO_2 electrodes described above.

References

King, R. and Campbell, A. (2000) Performance of the Radiometer OSM 3 and ABL 505 blood gas analysers for the determination of Na, K, and Hb concentrations, *Anaesthesia*, **55**, pp. 65–69.

Rupp, S. M. and Severinghaus, J. (1986) Hypothermia. Chapter 57 in Miller, R. (ed.), *Anesthesia*, 2nd edn, Churchill Livingstone, p. 2000.

Further reading

Sykes, M. K. (1982) The determination of pH. Chapter 9 in *Scientific Foundations, Anaesthesia*, 3rd edn, Heinemann, pp. 108–114.

Sykes, M. K., Vickers, M. D. and Hull, C. J. (1981) pH and blood-gas analysis. Chapter 18 in *Principles of Clinical Measurement*, Blackwell Scientific Publications, pp. 242–63.

Clutton-Brock, T. H., Venkatash, B. (1994) Blood gas monitoring. Chapter 16 in *Monitoring in Anaesthesia and Intensive Care*, Hutton, P. and Prys-Roberts, C. (eds) WB Saunders Co. Ltd, pp. 242–55.

Al-Shaikh, B. and Stacey, S. (2002) *Essentials of Anaesthetic Equipment*, 2nd edn, Churchill-Livingstone, London, pp. 164–66.

Chapter 18

Electrophysiology and stimulation

This chapter contains: the electrocardiogram, artefacts and interference; the electroencephalogram, data collection and analysis; evoked response, auditory evoked responses; monitoring neuromuscular blockade and the electromyogram, nerve stimulators, electrodes, responses.
The chapter links with: Chapters 3, 4 and 19.

Introduction

This chapter covers the processing and application of electrical signals from the body, in particularly the *electroencephalogram* (EEG), the *electrocardiogram* (ECG), and the *electromyogram* (EMG). The EEG and ECG will be considered in their monitoring capacity. The EMG will be discussed along with simulation and neuromuscular blockade and monitoring.

The electrocardiogram

The electrocardiogram (ECG) is a surface reflection of the propagation of electrical depolarisation and repolarisation over the various contractile chambers of the heart. Depolarisation is the trigger for releasing the stored contractile energy in the cardiac muscle. Each chamber also produces electrical action and polarising recovery potentials associated with the mechanical contribution of the recovery. The ECG can be divided into two major components: one associated with the propagation of excitation and recovery of the atria; the other with these events occurring in the ventricles. Excitation of the atria gives rise to the P wave, after which the atrial contractions propel blood into the ventricles. An atrial recovery wave exists, but it is rarely seen, as it is obscured by ventricular excitation, which is signalled by the QRS wave. During the later part of the QRS wave, ventricular contraction commences. Recovery of the ventricles is preceded by the T wave. The ECG labels, i.e. PQRST, are shown in Figure 18.1.

To localise the direction of excitation and recovery of the heart chambers (and also to estimate the extent of cardiac injury), a variety of electrode arrangements can be used. The electrodes (which are normally disposable

THE ELECTROCARDIOGRAM | 245

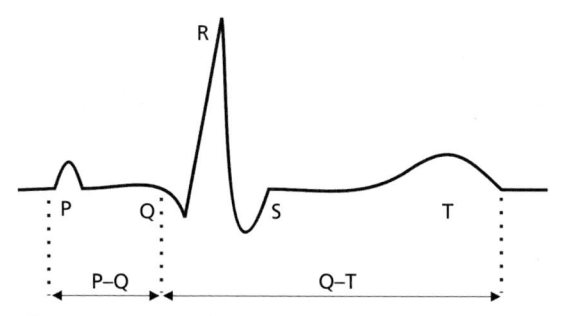

Fig. 18.1 Diagram of an ECG showing the standard labels of various parts of the waveform.

silver–silver chloride as described in Chapter 5) are positioned on easily located anatomical landmarks such as the right arm (RA), the left arm (LA), and the left leg (LL), with the right leg usually providing the reference or common. The standard (1, II, III), augmented (aVR, aVL, aVF) and precordial (V) leads are routinely recorded by electrocardiographers. It is possible to locate the direction of excitation and recovery by considering that the direction of the event (excitation or recovery) is at right angles to the isoelectric lead (i.e. the lead with equal forces in the positive and negative). This can be demonstrated by forming an equilateral triangle (Einthoven's triangle) such as in Figure 18.2(a). This gives the direction of the leads, for example lead I is RA to LA. The amplitude of lead I (0.87 mV in this example) is drawn and a right angle (perpendicular) is dropped down from this as shown in Figure 18.2(b). Similarly the lead II voltage of 0.87 mV is drawn along the lead II axis and the

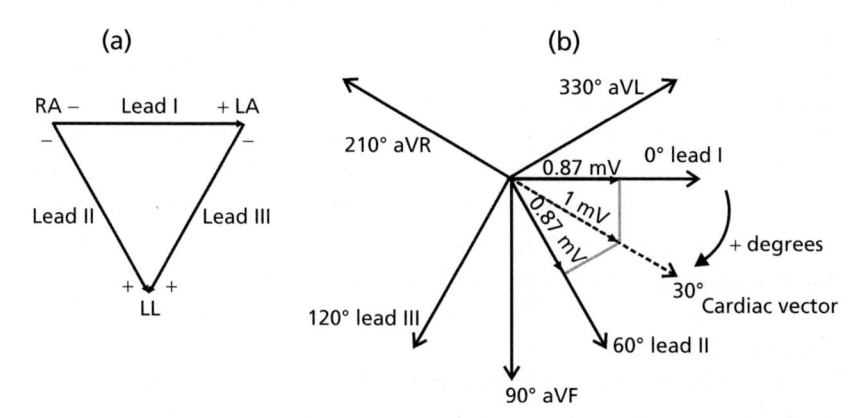

Fig. 18.2 Calculation of cardiac vectors.

perpendicular drawn. This will interact with the lead I perpendicular and this dotted line will be the cardiac direction or *vector*. The vector will be 1 mV, 30 degrees, and lead III will have zero amplitude, indicating that the vector is at right angles to this lead.

Indices obtained from the ECG

The bandwidth required for full 12 lead ECG analysis is from 0.05 to 100 Hz. Real time monitoring in anaesthesia or intensive care generally requires three leads, both arms and left leg. Lead II (RA, LL) lies close to the normal cardiac electrical axis permitting the best (maximum amplitude) P and R waves, and can be very useful for diagnosing certain arrhythmias. The ECG can be recorded graphically, usually on a time scale of 0.04 s mm^{-1} (25 mm s^{-1}) but also the ECG can be digitised, using a sampling rate of typically 1 kHz, and *algorithms* (computer programs) are used to calculate the position of various parts of the waveform (e.g. R wave).

Many algorithms can be developed to calculate R–R intervals (i.e. reciprocal of heart rate), but a simple and effective one is to first *differentiate* the ECG waveform [Coleman *et al.* 1979] by taking the digitised waveform and creating an approximation of the differential in another part of the memory. This is demonstrated in Figure 18.3. Sample A in the raw signal is subtracted from B, and this forms the first sample in the (approximation) differentiated signal as shown. The next sample will be C – B, and so on.

If the whole raw waveform is treated in this manner, then the (approximated) differentiated wave will result, as indicated in Figure 18.4. As shown in the diagram, this method has the advantage of removing baseline wander due to respiration, removing the T wave (the large wave at the end of the raw signal) and enhancing the R wave. Once the position of the R wave has been

Fig. 18.3 Diagram showing principle of digital differentiation of a small signal. The raw signal consists of five samples at amplitudes A–E. The approximation of the differentiated signal is created by taking the differences of the raw samples, i.e B−A, C−B, D−C, and E−D.

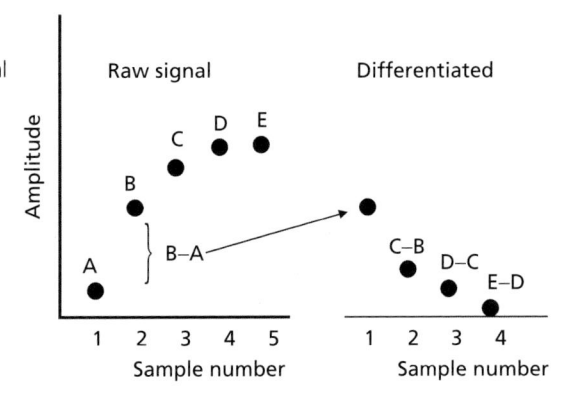

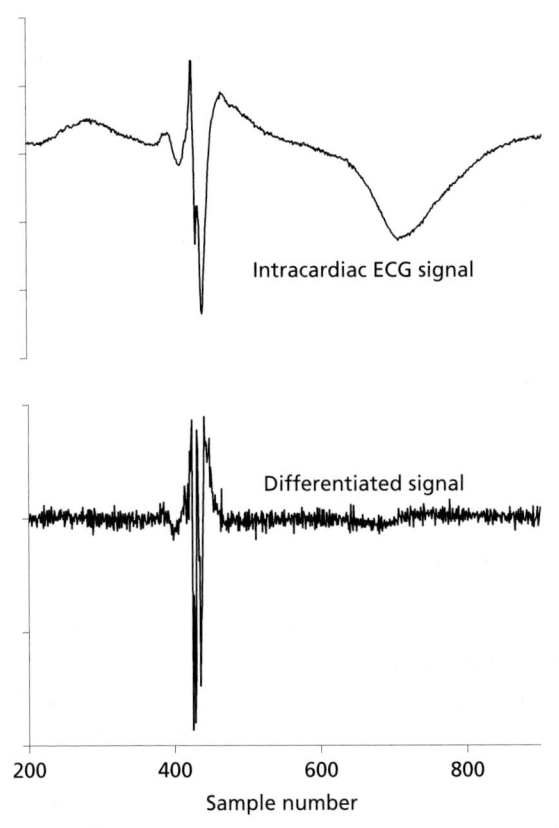

Fig. 18.4 An intracardiac ECG signal sampled at 1 kHz and the approximation to the differentiated signal.

detected, then the number of samples till the next R wave will be related to the R–R time interval. For example if there are 1000 samples between detected R waves, and the sampling rate is 1 kHz, then the R–R time interval will be 1 s and the heart rate will be 60 beats per min (BPM). A digital *refractory* period can be built into the algorithm so that the next R wave will not be searched for until a certain time has passed, e.g. 0.5 s. This enables artefacts to not be counted as a genuine R wave. Obviously R–R intervals could be manually counted by hand using squares on a hard copy.

Other indices can be derived from the ECG such as S–T segments and this can provide prompt detection of myocardial ischaemia. Normally the detection is if the ST line is greater than 0.1 mV below the isoelectric line. As in the detection of R waves, a suitable algorithm used on the digitised waveform can produce this measurement automatically.

Artefacts and interference

Although the amplitude of the ECG is larger than that of the EEG, it is still a small potential (0.1–6 mV) and artefacts and interference can cause problems in interpretation. Mains interference can be minimised by ensuring good electrode contact and a high common mode rejection ratio in the differential amplifier as discussed in Chapter 5. Filters can be used with success to produce a cleaner signal, although the filter parameters must be chosen with care. Mains filters can only be used in the monitoring mode (rate detection). In the diagnostic mode of ECG analysis, the bandwidth must be 0.05–100 Hz and the mains filter would remove some of the frequency components of the ECG signal, so should not be used. Low pass filters set to 100 Hz can be used to remove the higher frequency components of the EMG signals.

The electroencephalogram

The waves that can be recorded from the surface of the head are the summated result of postsynaptic potentials, which are the responses generated by the pyramidal cells to rhythmic discharges from the thalamic nuclei. The frequencies and amplitude of the thalamic discharges, and of the EEG itself, are determined by the excitatory and inhibitory connections between the thalamic cells. During activation, inputs to the reticular formations abolish the rhythmic discharges in the thalamic nuclei and cause the EEG to become desynchronised, i.e. not one frequency alone but a mixture of low amplitude frequencies. During de-activation single high frequencies appear. The EEG is made up of a range of frequencies between DC and around 40 Hz. In the frequency domain, the EEG is often divided into frequency bands, which are described as delta (< 4 Hz), theta (4–7 Hz), alpha (8–13 Hz) and beta (13–40 Hz). Normally alpha activity can be recorded from the surface of the brain only during relaxation with the eyes closed and this is the de-activation state. If the subject carries out mental activities or eye opening, the waveform becomes *desynchronised* and at a higher frequency but lower amplitude. The amplitude of the normal EEG is between 10 and 50 μV, but the range can be between 1 and 100 μV. The EEG is affected by sleep and anaesthesia. Anaesthetic drugs, whether intravenous or volatile affect the EEG. Low drug doses cause an increase in amplitude of the signal and an increase in frequency, so that the dominant frequency is in the beta range. Higher doses cause the EEG frequency to decrease in a dose dependent manner, with the amplitude remaining high. Very high doses cause very low frequency, low amplitude signals, mixed with bursts of high amplitude and higher frequency. There are also bursts of no activity as well, and this is called *burst suppression*.

Other drugs such as opioids also change the EEG and there are other factors affecting it such as oxygenation levels, pCO_2, temperature and noxious stimulation. Reduced cerebral oxygenation causes slowing of the EEG, as does hypocapnia, and hypothermia. Noxious stimulation, unless the patient is heavily analgesed causes an increase in EEG frequency. Because all these things affect the EEG, if processed properly, it can be a useful monitor in theatre, especially for monitoring the depth of anaesthesia and as an awareness monitor.

EEG data collection

The EEG is normally collected by using small silver–silver chloride electrodes (see Chapter 5). These need careful application using a suitable preparation paste to minimise the electrical impedance of the electrode skin interface. This will reduce interference and maximise the signal transfer to the amplifier (see Chapter 5). For secure fixing, the electrodes are fixed in an electrolyte gel and secured typically by collodion. For monitoring purposes, 2–6 sites plus a *ground* or reference are used compared with 21 sites or more for routine diagnostic EEGs in the clinical neurophysiological laboratory. There is a standardised electrode location method used in clinical neurophysiology called the 10–20 system [Jasper 1958], in which the 21 electrodes are fixed at reproducible locations. The system is shown in Figure 18.5, and it employs skull landmarks as reference points to locate the electrodes. There are 19 scalp electrodes and 2 earlobe ones. The optimal sites for monitoring used are those least likely to pick up unwanted potentials of biological or external origin. Bilateral centroparietal placements such as C_3–P_3 and C_4–P_4 [Prior 1987, Maynard 1984] (see Figure 18.5) are preferable to those including frontal or temporal electrode placements, because of the risk of picking up slow wave activity due to eye movements or sweating and faster components from scalp or facial EMG. The electrodes are connected to an isolated differential amplifier, which minimises 50 Hz and other external electrical interference being amplified. After amplification, the signal is band-pass filtered, normally to between 0.1 and 32 Hz. Depending on the signal processing methods used to quantify the EEG signal, the filtered signal can be processed by analogue electronics or converted by an ADC (as described in Chapter 5) and processed digitally.

Electroencephalogram analysis

The EEG, an example of which is shown in Figure 18.6, is a complex (comprising many different frequencies) time domain waveform and, for monitoring purposes, it needs to be data reduced and is usually quantified to a single number. Most analysis techniques examine *epochs* (short time

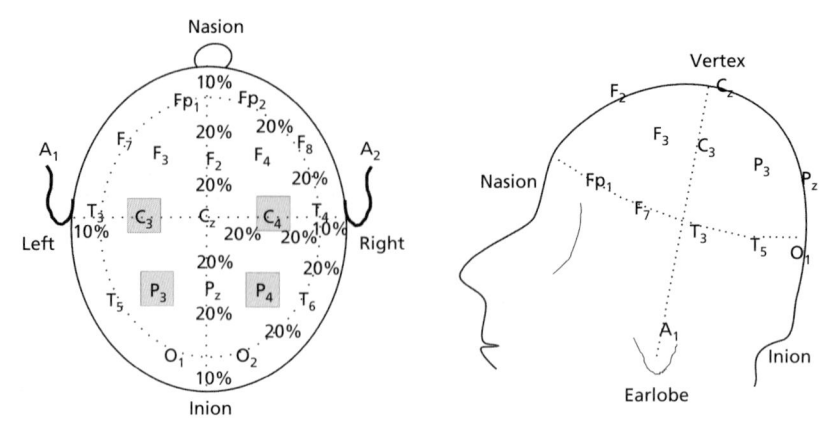

Fig. 18.5 The 10–20 system for electrode placement. Fp is frontal pole, F is frontal, C is central, P is parietal, O is occipital, T is temporal. Even numbered electrodes are on the right and odd on the left. The midline between the nasion and inion is defined as subscript Z. The measurements are based upon the line drawn from the nasion to the inion. 10% along this line, there is a circumference drawn around which the first ten electrodes are located. 30% along this mid-line is the frontal electrode, then the central electrode after another 20%. The rest of the electrodes follow a similar pattern – most of the inner electrodes are spaced 20% apart. The boxed electrodes are typical monitoring locations referred to in the text.

periods) of EEG, usually in the range 2–16 s (there are mathematical reasons why this range is chosen). Analysis can be in time (time domain) or frequency (frequency domain).

Time domain analysis

This analysis, which is discussed in Chapter 3, is where the amplitude or power of a signal is considered against time. These analyses include maximum amplitude or zero crossing frequency (number of times the EEG crosses the zero voltage line) per epoch, as demonstrated in Figure 18.7.

Frequency domain analysis

The EEG can also be examined in the frequency domain, and this can reveal hidden information or clearer patterns than can be seen in the time domain. The EEG, as in time domain analysis, is broken down into epochs. It is then converted (by Fourier analysis) into many different sine waves, all at individual amplitudes and starting points (phase) as already demonstrated in Chapter 3, Figures 3.3 and 3.4. The frequency resolution of the frequency

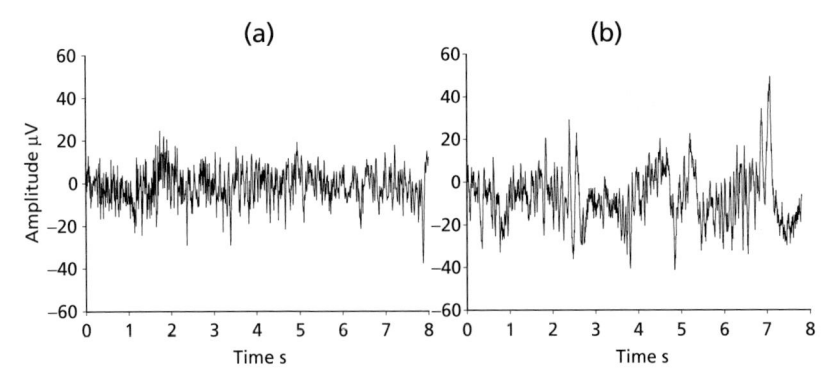

Fig. 18.6 Example of two 8 s EEG epochs. The signal is collected from C_3–P_3, and (a) is from an awake patients, receiving a very low dose of propofol, and (b) is from an unresponsive patient receiving a high infusion of propofol [Forrest et al. 1994].

analysis is determined by the length of the epoch where epoch^{-1} is the frequency resolution. For example if the epoch length was 4 s, then the resolution of the spectrum would be 0.25 Hz.

Quantification of the frequency spectrum in anaesthesia has been well studied using indices such as the *median power frequency* (MPF) and the *spectral edge frequency* (SEF). A spectrum is produced of each from the EEG. From this spectrum are obtained single indices such as the MPF and SEF. The MPF is

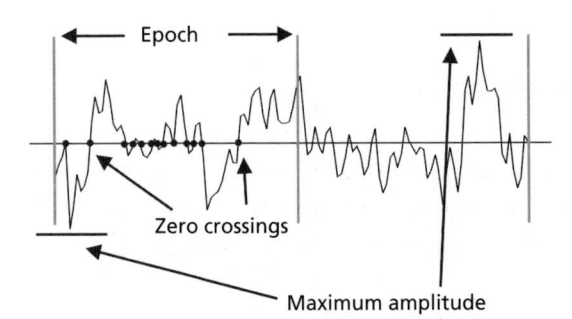

Fig. 18.7 Diagram illustrating simple time domain analysis of the EEG waveform. In this example, the zero crossings frequency in the first epoch is 12. The maximum amplitude is the maximum amplitude value (positive or negative) in each epoch. The number of turns (i.e. change of direction of the signal) in each epoch could be another measure, as could the integrated voltage per epoch.

the frequency value where 50% of the power is below and 50% above that value and the SEF is where 95% of the power is below. This is demonstrated in Figure 18.8.

The MPF and SEF have been studied during single drug anaesthesia and correlate well with drug levels and clinical effect [Schwilden *et al.* 1989, Forrest *et al.* 1994]. The problem with these indices is that they tend to have different values with different drugs and also have the problem of the *biphasic effect*. This is where many anaesthetics agents initially increase the EEG frequencies and therefore the MPF. As the dose of the anaesthetic is increased, the EEG frequency (and therefore the MPF) will decrease. There is therefore an MPF value for awake and the same MPF value will be for some level of anaesthesia as well. This is shown in Figure 18.9 where the value of 9.7 Hz is both the MPF for all awake, drug free patients and the value where, by coincidence, 50% of the patients will be anaesthetised [Forrest *et al.* 1994].

Evoked responses

The cortical potentials so far described (EEG) are *spontaneous*, but another group of cortical potentials can be considered as *evoked*. This is where a short

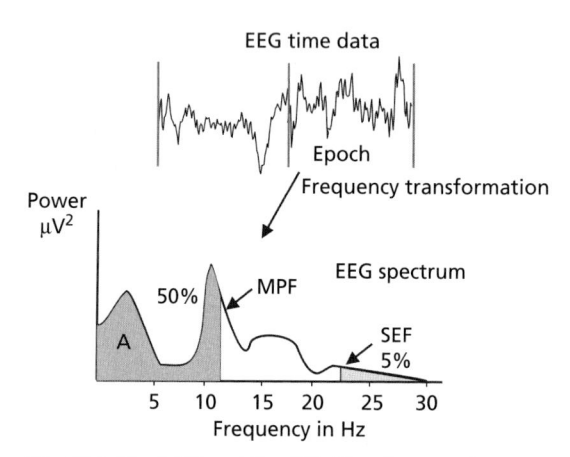

Fig. 18.8 The MPF and the SEF. The diagram shows an epoch of EEG that is transformed to a spectrum. The MPF and SEF are obtained from this. The MPF is the frequency value where 50% of the power of the spectrum is below it (shown as the shaded part A). The SEF is the frequency value where 5% of the power of the spectrum is above it, or 95% below.

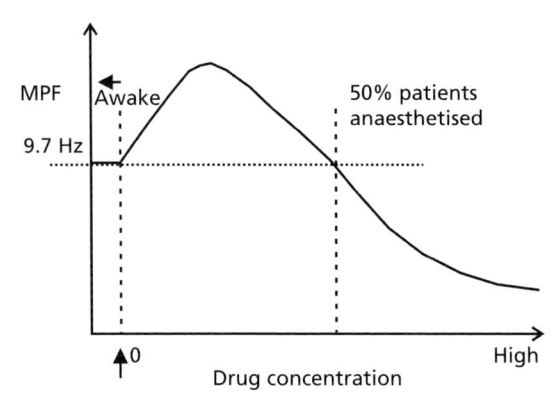

Fig. 18.9 The biphasic response of the MPF. The diagram shows the MPF at different concentrations of an anaesthetic drug (propofol). Awake patients with no anaesthetic have an MPF of 9.7 Hz but this value is also when 50% of them will be unconscious at a later stage in the research procedure.

external stimulus is given and the response to that is measured. The stimulus can be auditory, visual or somatosensory (electric shock or mechanical). The response is normally lower amplitude than the background EEG, and therefore submerged in the EEG. This is demonstrated in Figure 18.10(b), which shows an *auditory evoked response* (AER) submerged in the EEG. This auditory evoked response consists of a number of labelled waves as shown in Figure 18.10(a) but each modality (i.e. visual, auditory, etc.), although having a similar wave, has different labels. It is the *latency* and amplitude of each wave that is of interest. The latency is the time delay of the wave from the stimulus (at time zero).

As the AER is submerged, it cannot be visualised directly. Usually *coherent* averaging is used to reduce the *random* background EEG to an average voltage lower than the evoked response; this allows the evoked potential to be obtained and visualised. In a steady state clinical or physiological condition, the stimulus will normally elicit a repeatable response from the cortical electrodes. The EEG will also be present, and will overwhelm the response. The averaging technique is demonstrated in Figure 18.11. The stimulus is given at time 0 and the EEG following it for 100 ms (sweep 1) is stored in computer memory. This sweep contains both the EEG and the AER. After enough time has elapsed for the response to fade away (i.e. greater than 100 ms in this example) another stimulus is given and the EEG (sweep 2) is stored in another memory. The two submerged evoked responses will be very similar, but the background EEG will dominate and appear as a random signal, i.e. the EEGs will be very different from each other. If, for example, 500 stimuli are given sequentially, and all of the *sweeps* (or time periods) of the EEG are added together (in time), the amplitude of the evoked response will be multiplied by 500 times (because it is in the same time frame each time), but the EEG will have largely cancelled out, due to its random nature. In the example, a time point at 15 ms after

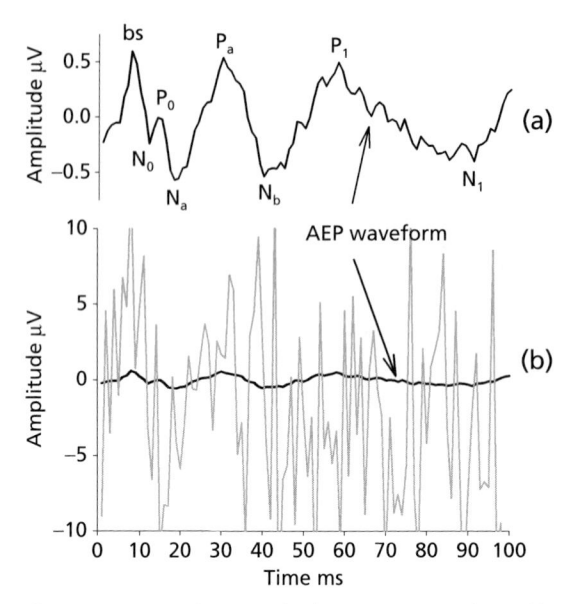

Fig. 18.10 An auditory evoked response waveform (a), which is a magnified version of the waveform in (b), which shows it submerged in the spontaneous EEG background signal. bs = brainstem wave voltage. P_n are positive waves with subscripts 0, a, b, 1, etc. N_n are negative waves with subscripts 0, a, b, 1, etc.

each stimulus is shown. The EEG values at this time will be part random EEG and part AER. The AER component at 15 ms can be considered -0.1 μV. The amplitude values will be (*the EEG value — the AER value*): 20−0.1, −20−0.1, −15−0.1, 12−0.1, 3−0.1 μV for sweeps 1, 2, 3, 4 and 5 respectively. The average of this simple example will be the summation of these divided by the number of sweeps. This is $(0 − 0.5)/5$, which is 0.1, or the value of the AER only. The EEG values have been *averaged* out. If all time points are considered in a similar manner, the EEG will tend to zero and the complete AER will appear.

Different waves in the evoked response come from different parts of the brain. For example, the first 10 ms of the auditory response come from the brainstem and the 10–50 ms come from the auditory cortex. All three modalities auditory [Thornton *et al.* 1998], visual [Uhl *et al.* 1980] and somatosensory [Clark *et al.* 1973] are affected by anaesthesia; the visual responses are the largest responses, but they have not been studied extensively because of the problem of the patient seeing the chequerboard stimulus (a chequerboard in which the black and white squares reverse at a certain

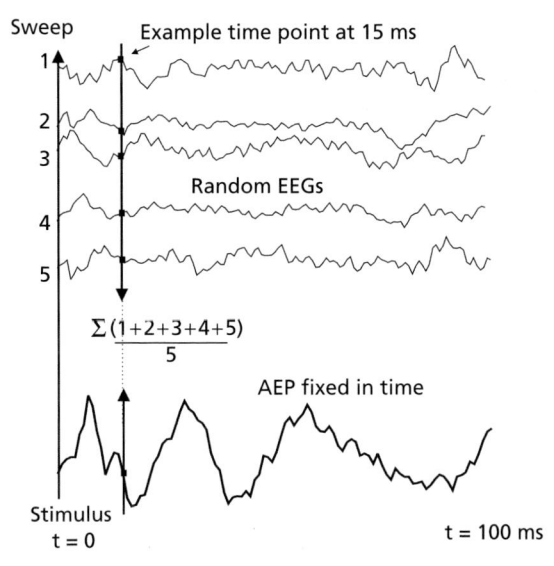

Fig. 18.11 The principle of averaging to obtain the AER. Many sweeps of the random EEG are averaged (5 in this example) to obtain the AEP (lower trace). Amplitude values at a time of 15 ms are added together and averaged. The EEG background will tend to zero, but the AER will be reinforced.

frequency) during the operation. Somatosensory responses tend to be affected by both anaesthetic drugs and muscle relaxants. The auditory responses seem to be the best modality for anaesthesia monitoring – the brainstem part of the response is robust to anaesthesia, but the 10–80 ms seem to change in a dose related manner. Electrodes are normally connected as in EEG monitoring, but in specific locations, usually vertex-mastoid gives the optimum response for auditory responses, but vertex–inion placement prevents interference from myogenic responses such as the post auricular response, which can overwhelm the AER [Tooley *et al.* 2004].

Obtaining the evoked response

To obtain the AER, the EEG signal is fed, via a suitable head box, to a patient isolated differential amplifier and then to a signal averaging device. The principle of operation is shown in Figure 18.12. In this example, a frequency generator is running at 10 Hz. This means that a stimulus pulse is generated every 100 ms. This pulse is supplied to an auditory stimulus generator that

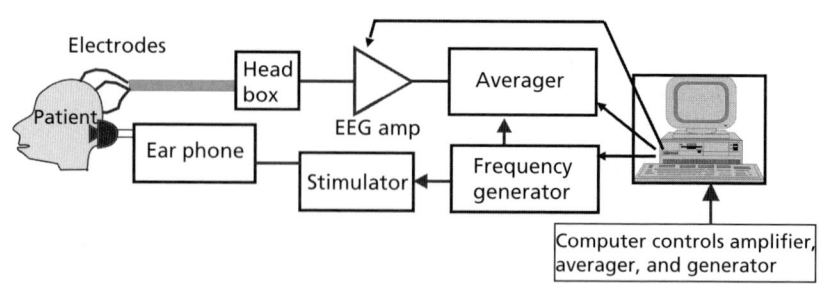

Fig. 18.12 Diagram of an averaging for the auditory evoked response. The frequency generator outputs a pulse every, e.g. 100 ms, and this synchronises the averager and presents a click or tone to the patient.

supplies a click or tone to a patient earphone. The pulse also synchronises the averager so that each new sweep is added to the previous sweep in the computer memory.

Monitoring neuromuscular blockade and the electromyogram

In clinical nerve stimulation, three different types of stimulation are used: single twitch, tetanic and the train of four. In stimulation, a supramaximal stimulus is used. This is where the stimulation intensity, usually provided by electrical current stimulation, is measured 25% above that for a maximal response.

Stimulations

Single supramaximal current impulses can be applied at frequencies of between 0.1 and 1 Hz, the lower frequency of 0.1 Hz allows greater biochemical recovery in the *off* time so the mechanical responses do not *fade*, although any tests take longer with the lower frequency. The frequency used to induce a tetanic (total) contraction is 50 Hz for 5 s. There is no fade during this stimulation if a depolarising relaxant is used. If a non-depolarising relaxant is used, this can cause fade during this response. After the tetanic stimulus is switched off, there is an enhancement of the response to lower frequency stimulations, such as the supramaximal single twitch stimulus or the *train-of-four*, called *post-tetanic facilitation* (PTF). In *train-of-four* (TOF) stimulation, supramaximal stimuli are applied at intervals of 0.5 s (2 Hz) and this can be repeated every 12 s. The ratio of the amplitude of the fourth to the first pulse is called the TOF ratio and this is inversely proportional to the degree of blockade (provided by a non-depolarising relaxant). The PTF can be a useful response to determine the degree of blockade when it is greater than 100%. When no response is obtained to a single stimulus, the number of responses detectable after the tetanic response can be a useful measure [Viby Mogensen *et al.* 1981].

The nerve stimulator

The stimulator used for the various stimulations must be a constant current device, which means that the current delivered is independent of the electrode impedance. It must maintain a constant set value, even though the electrode impedance may change during a session, or differ between patients. Normally current is inversely proportional to the electrode impedance, so special circuits comprising operational amplifiers and capacitors need to be used. Ideally the constant current should be the same value whether the electrode impedance is zero or near infinity, but in practice the range of impedance across which the current is constant is from near zero to around 20 kΩ. The current should be in the range 10 to 50 mA and a square wave of pulse width 0.2–0.3 ms at a frequency between 0.1 and 1 Hz. The timing circuits should be able to give TOF, single twitch, double burst (two 50 Hz periods of 40 ms, separated by 750 ms off period) and titanic stimulations. Nerve stimulators are also used to carry out local anaesthetic nerve blocks. An insulated needle is used as the active electrode, with the inactive electrode applied to the skin. The current will be much less than previously indicated, a value of 0.5–1 mA, which allows the needle to approach the nerve without damaging the nerve prior to injection. It is important that a lower threshold is detected with the stimulator to confirm that the nerve itself has not been entered by the needle.

Electrodes

The electrodes used are normally, silver–silver chloride, mentioned previously, and tend to be pre-gelled. They can be EEG type electrodes, which can be reused. The site should be cleaned and prepared. The median and ulnar nerves are the two most often used for stimulation, although the facial and common peroneal nerves are also used. Two electrodes must be used to complete the circuit.

Measuring the response

Although any motor nerve may be suitable, it has to be accessible and suitable for use with surface electrodes so that the response can be measured. Therefore only a limited number of nerves have been used: the ulnar, the median, the posterior tibial and the facial. Most often the ulnar nerve is stimulated and the evoked response is evaluated visually, mechanically, electrically or clinically.

Mechanical responses

In this case, the thumb acts on a force–displacement transducer. The transducer, and the rest of the hand and arm are fixed. Movement of the thumb pulls or pushes the transducer, shown in Figure 18.13, where the force applied is converted to an electrical signal. This can be amplified and supplied to a chart recorder or a computer. The arm must be rigidly fixed and the transducer should be placed correctly in relation to the muscle. Other mechanical methods include the measurement of acceleration by using a piezoelectric transducer fastened to, for example, the thumb [Jensen *et al.* 1988]. Force is mass multiplied by acceleration and so if the mass is constant, then the force is proportional to the signal from the transducer.

Electrical responses

Electrical responses to the stimulators include the EMG. This signal can be measured via surface or needle electrodes and is a summation of the many motor unit potentials. The EMG recorded is the result of the stimulation and is

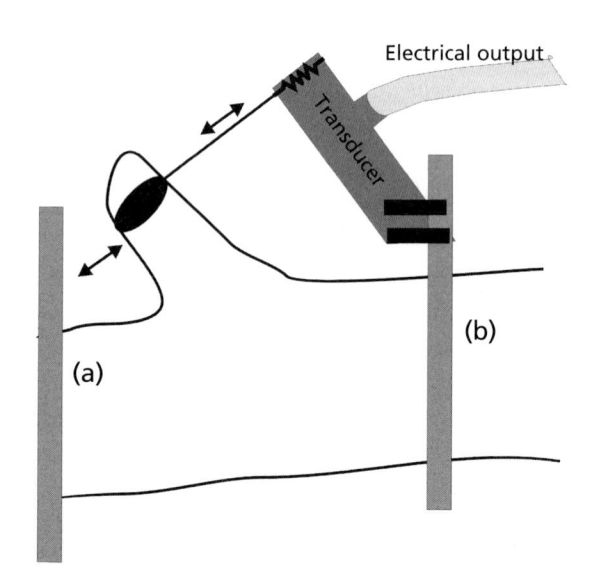

Fig. 18.13 An example of the measurement of the mechanical evoked muscle response. A and B are clamps to keep the hand and transducer fixed. The thumb is connected to a strain-gauge transducer. The movement of the thumb compresses or extends this and via the transducer electronics produces an electrical output.

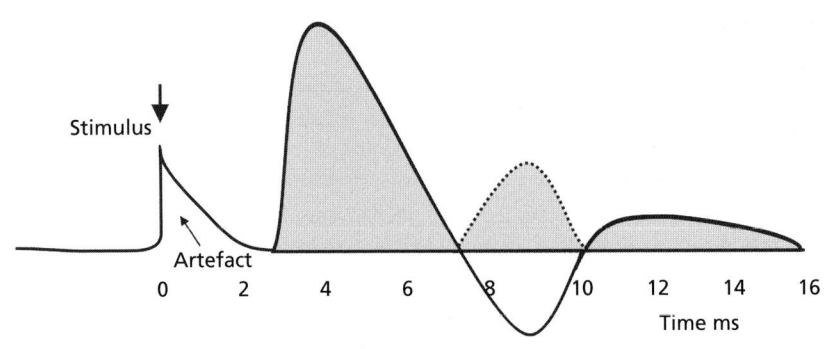

Fig. 18.14 The evoked EMG. The stimulus is at time zero, and the stimulus artefact lasts for about 3 ms This is excluded from the quantification of the wave. The wave from 3 ms till 16 ms is first rectified, and this is shown by the shaded areas (the dotted curve replaces the negative part of the waveform. The rectified waveform is then integrated by measuring the total shaded area.

therefore *evoked*. Three electrodes (positive, negative and reference) supplied to a differential amplifier (as discussed in Chapter 5) will be needed to sense the signal. The EMG is recorded in a similar manner to the EEG, but it is of much larger amplitude and the bandwidth of the signal is much higher (typically 0–4 kHz). The EMG can be processed and filtered and an algorithm can be used to provide quantification output from a computer. A simple measure could be the integrated evoked EMG [Carter *et al.* 1986]. This is where the EMG activity is rectified (i.e. the negative values are multiplied by −1) and then the area summated. This is shown in Figure 18.14. The signal is normally integrated from after the stimulus artefact until about 13 ms after the stimulus.

The electrical signals can be simple to set up, mechanical transducers are eliminated, and the electrode site (for example the arm) can be out of sight. The EMG is able to detect changes that persist into the recovery period that the mechanical responses can not [Harper *et al.* 1986].

References

Coleman, J. D. and Bolton, M. P. (1979) Microprocessor detection of electrocardiogram R-waves, *J. Med. Eng. Technol.*, **3**(5), pp. 235–41.

Carter, J .A., Arnold, R., Yate, P. M. and Flynn, P. J. (1986) Assessment of the Datex Relaxagraph during anaesthesia and atracurium-induced neuromuscular blockade, *Br. J. Anaesth.*, **58**(12), pp. 1447–52.

Forrest, F. C., Tooley, M. A., Saunders, P. R. and Prys-Roberts, C. (1994) Propofol infusion and the suppression of consciousness – the electroencephalogram and dose requirements, *Br. J. Anaesth*, **72**, 35–41

Harper, N. J. N., Bradshaw, E. G. and Healy, T. E. J. (1986) Evoked electromyographic and mechanical responses of the adductor pollicis compared with the onset of neuromuscular blockade by atracurium or alcuronium during antagonism by neostigmine, *Br. J. Anaesth.*, **58**, 1278–84.

Jasper, H. H., (1958) The ten twenty electrode system of the international federation, *Electroencep.Clin. Neurophysiol.*, **10**, 371–375.

Jensen, E., Viby-Mogensen, J. and Bang, U. (1988) The accelerograph: a new neuromuscular transmission monitor, *Acta Anaesthesiol. Scand.*, **32**, 45–48.

Prior, P. F. (1987) The EEG and detection of responsiveness during anaesthesia and coma. In Rosen, M. and Lunn, J. N. (eds) *Consciousness, Anaesthesia and Pain in General Anaesthesia*, Butterworths, London, p. 37.

Thornton, C. and Sharpe, R. A. (1998) Evoked responses in anaesthesia, *Br. J. Anaesth.*, **80**, 771–78.

Tooley, M. A., Stapleton, C. L. Greenslade G. L. and Prys-Roberts, C. (2004) The auditory evoked response during propofol and alfentanil anaesthesia, *Br. J. Anaesth.*, **92**, 25–32.

Uhl, R. R., Squires, K. C., Bruce, D. L. and Starr, A. (1980) Effect of halothane anaesthesia on the human cortical visual evoked response, *Anesthesiology*, **53**, 273–6.

Viby Mogenson, J., Howardy-Hansen, P., Chraemer-Jorgensen, B., Ording, H., Engbaek, J. and Neilson, A. (1981) Post-tetanic count (PTC): a new method of evaluating an intense nondepolarising neuromuscular blockade, *Anesthesiology*, **55**, 458–61.

Further reading

Sykes, M. K., Vickers, M. D. and Hull, C. J. (1991) Monitoring the nervous system. In *Principles of Measurement and Monitoring in Anaesthesia and Intensive Care*, Blackwell Scientific Publications, Oxford.

Chapter 19

Monitoring depth of anaesthesia

This chapter contains: the cerebral function monitor; the compressed spectral array monitor; bispectral index monitors; entropy monitors; Narcotrend monitors; patient state analyser; auditory evoked response based monitors; respiratory sinus arrhythmia based monitors; emerging trends.
The chapter links with: Chapters 5 and 18.

Introduction

There are, and have been, many monitors designed to monitor depth of anaesthesia and to give an indication of awareness during surgery, which use electrical signals obtained from the human body. Some have been designed as just research devices, some have been available commercially, but have been withdrawn, and some are still available. Most, but not all, are based on the spontaneous EEG and the AER. Some have been designed to use properties of the ECG. Although useful, all of the discussed monitors have some shortcomings, and not all are 100% sensitive and specific to discriminate between consciousness and unconsciousness, and none correlate exactly with clinical states and levels of anaesthesia.

Depth of anaesthesia monitors using the EEG

Cerebral function monitor

The design of the commercial monitor, the *Cerebral Function Monitor* (CFM) was based on simple time domain measures already discussed [Maynard *et al.* 1969]. The CFM took the EEG from a single pair of parietal electrodes. The signal was amplified and passed through a band-pass filter and differentiator, which had the effect of accentuating the gain of the higher end of the 2–15 Hz pass band. The output of this specialised filter was integrated to produce a voltage output, which varied with time. It was plotted on a logarithmic scale. The trace on the paper gave an indication of the power of the EEG and the width of the line gave an indication of the signal's variability. A schematic of an example of a CFM trace is shown in Figure 19.1(a). The CFM although useful did have its problems [Sechzer 1977]. When used to monitor depth of

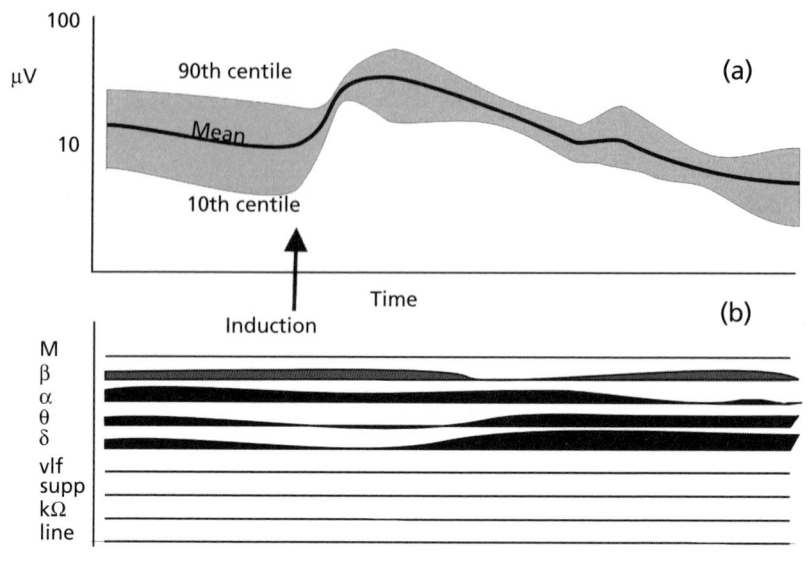

Fig. 19.1 Schematic of the CFM trace (a) and the CFAM traces (a) + (b). In trace (b), M = muscle, β, α, θ, and δ are the traditional EEG bands, vlf = very low frequency, supp = burst suppression, kΩ = electrode impedance, line = mains interference. The figure shows a simulated example of the expected traces during induction of anaesthesia [Maynard 1984].

anaesthesia, the machine was shown to be unreliable, especially when using inhalational agents. The response is biphasic, as has already been discussed in chapter 18. Also burst suppression, as already discussed, is smoothed out by the action of the filtering in the CFM, so effectively the burst suppression can artificially elevate the readings producing a paradoxical rise in cerebral function [Sinha 2007] The machine was further developed into the *Cerebral Function Analysing Monitor* (CFAM) [Maynard 1984]. This machine produced two chart recorder outputs, as shown in Figure 19.1. There was a chart similar to the CFM trace, and also a chart that produced frequency domain data consisting of the EEG displayed as traditional EEG frequency bands.

Compressed spectral array monitor

The first use of a monitor that used frequency analysis for anaesthesia was using the *Compressed Spectral Array* (CSA) [Bickford 1972], and a simplified example is shown in Figure 19.2.

Although this technique displays information in real time with the added advantage of historical information as well, the display was often difficult for

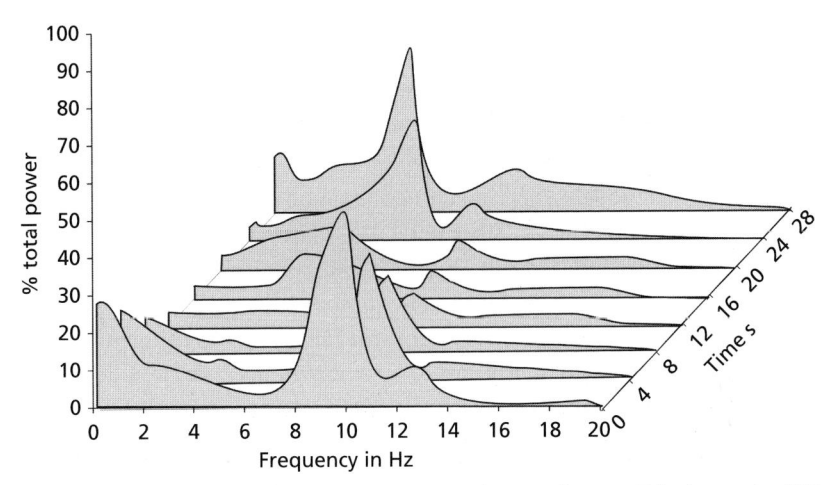

Fig. 19.2 A simplified version of the compressed spectral array. This shows the EEG power spectrum obtained from 4 s epochs obtained at different time points. Depending on the mode of display (i.e. computer screen or chart recorder), the most recent spectrum can be either at 0 or 28 s. Historical spectra are shown at time 4 s, 8 s, etc. This display gives the impression of a subject being anaesthetised (24 and 28 s) and awake (4 and 0 s).

the untrained eye to interpret. Frequently the goal of quantification of the data was not achieved. Alternative forms of the CSA were the *Density Spectral Array* (DSA), where the amplitude information was conveyed by the brightness of the trace, frequency on the ordinate and time on the abscissa [Fleming 1979].

Bispectral index monitors

The BIS monitor is the most extensively studied monitor in the anaesthetic literature, and so it will be described in some detail. The monitor specifically aims to use a new index called the *Bispectral Index* (BIS), which attempts to eliminate the biphasic effect mentioned previously and get around the fact that different anaesthetic agents affect the EEG in different ways. This commercial monitor (the BIS monitor, marketed by Aspect Medical Inc) uses a complex algorithm that produces an index between 0 (very unconscious) and 100 (very awake), with 40–60 the recommended values for general anaesthesia. The BIS number is designed to be drug independent, i.e. light sedation should give the same number irrespective of what drug is producing that effect [Johansen *et al.* 2000]. The analysis involved in the algorithm is based on the frequency and time domain measures already described (frequency analysis = beta ratio, time domain analysis = burst suppression indices) but it is dominated by bispectral analysis [Rampil 1998]. Bispectral analysis is a *higher-order* (more complex than simple Fourier transformation) spectral analysis based on the phase relationships of individual frequency components to each other. Information *hidden* in the frequency spectrum may be uncovered if the bi-spectrum is included in the analysis algorithm. This can be explained by considering, in a thought experiment, two situations that have different brain states, such as awake and anaesthetised, but the simple frequency spectra are identical. Is this example the EEG spectra alone cannot discriminate between the two clinical states (see Figure 19.3, which shows the spectrum of two EEG processes, EEG 1 and EEG 2). The spectrum of EEG 1, which represents clinical state 1, can be considered to be produced by the summation of four dominant frequencies, all having independent phases (i.e. they are not coupled together). The spectrum of EEG 2 or clinical state 2, can be considered to be produced by the multiplication of two different frequency generators. The effect of multiplication will be to produce a spectrum consisting of the frequency sum of the two generators, the frequency difference, and double each generator. The phases of these frequencies will be coupled together. It is plausible that the two different clinical states will give these different mathematical situations. For example, awake EEG is non-synchronised EEG that could be produced by independent generators as above and the anaesthetised EEG is often more synchronised. Therefore the two spectra, EEG 1 and 2 will have been produced by different brain processes (e.g. awake and anaesthetised), but have identical simple spectra morphologies. If the bi-spectra of EEG 1 and 2 were calculated, they would be very different because of the very different phase relationships, and may be useful as the dominant component in an algorithm to discriminate between clinical states. The BIS monitor actually uses the relationship between frequency values known as the *bicoherence*; this measure approaches zero when there is no relationship between components (as in the awake state) and approaches one where there is a fixed relationship (as in the deeply anaesthetised state mentioned).

Although the BIS machine is very useful in clinical practice and seems to predict awareness [Myles *et al.* 2004], there are a number of situations where it is not so useful, such as using nitrous oxide related anaesthesia, opioid dominant anaesthesia, and when interference from other devices is present [Costa *et al.* 2000, Dahaba 2005].

Entropy

The concept of *entropy* for physiological signals is derived from the concept of thermodynamic entropy for physical systems. In the context of signals, the entropy is about the irregularity, complexity or unpredictability of the

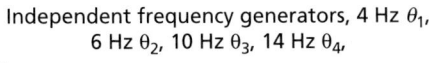

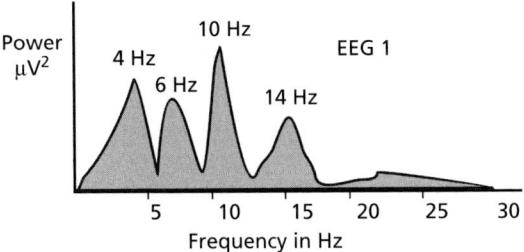

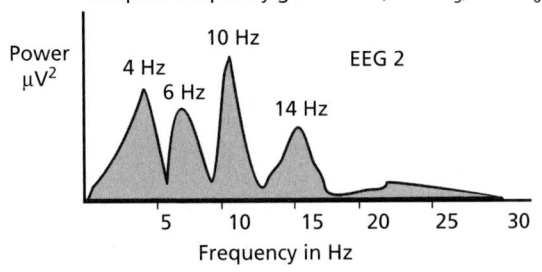

Fig. 19.3 Two identical EEG spectra, EEG 1 and EEG 2, which are produced by different brain processes, such as awake and anaesthesia. The frequency generators sum together and produce the surface EEG. The generators in EEG 1 are not synchronised or coupled together, but the generators in EEG 2 are coupled. EEG 1 is produced by four separate generators, 4, 6, 10, and 14 Hz. These produce a spectrum dominated by these frequencies, as shown. EEG 2 is produced by the multiplication of two generators 7 and 3 Hz. This might be the process that occurs under anaesthesia. The multiplication of 7 and 3 Hz will produce the sum (10 Hz), the difference (4 Hz), double 7 Hz (14 Hz) and double 3 Hz (6 Hz). In this case all the frequencies are coupled together, because of their common origin, and so the phase relationships will be very different from EEG 1.

EEG signal. In a simple example, shown in Figure 19.4, a signal that has one state, i.e. a pure sine wave, shown in Figure 19.4(a) has an entropy of zero, whereas white noise, i.e. a random unpredictable signal such as described in Figure 19.4(b), has an entropy approaching one. An awake EEG is much more complex and unpredictable than an anaesthetised EEG, which is much more sine wave like. This means that the entropy value of the EEG should change with

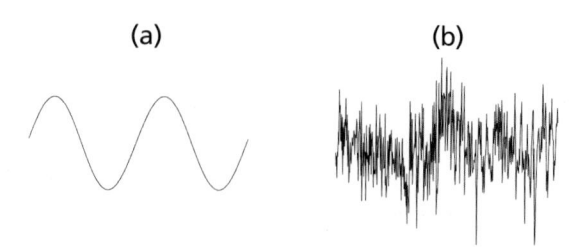

Fig. 19.4 Diagram showing two states of a signal: (a) represents a simple signal that has an entropy of zero, and (b) is a complex, white noise like signal that has an entropy approaching one.

clinical state. The commercially available entropy module calculates entropy over time windows of variable duration and produces two entropy values, State Entropy (SE) and Response Entropy (RE) [Vietio et al. 2004]. SE is an index between 0 and 91 using the frequency range 0.8 to 32 Hz and RE is an index between 0 and 100 using the frequency range 0.8 to 47 Hz. The higher index number represents awake values. RE includes higher frequencies than SE, and therefore frontal EMG is included more. The theory is that patients waking up or actually responding to noxious stimuli will have more frontal EMG and therefore RE will be different from SE in these situations. The two indices are displayed and their similarities and differences are supposed to facilitate the monitoring process. Studies have reported promising results and similarities to the BIS monitor [Bonhomme 2006] but others also indicate caution [Takamatsu et al. 2006].

Narcotrend

The Narcotrend (MonitorTechnik, Bad Bramstedt, Germany) is a monitor that uses an algorithm based on pattern recognition of the EEG. Standard ECG electrodes are used for this device. The EEG signal is amplified and analysed by multivariate statistical analysis to classify the signal into fifteeen different stages, A–F where: A is awake, C_0 is light anaesthesia, D_0 is general anaesthesia and F_1 (general anaesthesia with increasing burst suppression), depending on the clinical state of the patient. These letters were initially based on the classical Kruger anaesthetic clinical states, and visual analysis of the EEG. The fifteen stages are then converted into an index of 1–100 where, for example, A is 95–100, C_0 is 75–79, D_0 is 57–64 and F_1 is 1–4 [Kreuer et al. 2004].

Patient state analyser

The patient state analyser (PSI) uses a scale of 0–100, with 100 being awake. The PSI algorithm uses four channels of qualitative frequency analysis of the EEG and uses this information to estimate depth of anaesthesia, together with the changes that occur between the hemispheres and anterior and posterior regions of the cortex [Drover *et al.* 2002].

Depth of anaesthesia monitors using the AER

Auditory evoked response monitors

There have been several versions of commercial AER (also called Auditory Evoked Potentials, AEP) monitors produced, which have been based on the early cortical responses of the AER between 10–80 ms, which are also termed the mid latency auditory evoked response. There have been two different streams of these monitors. One stream has been marketed and produced by Danmeter (A/S Odense, Denmark) and includes the A-line monitor, the A-line ARX and the AEP/2 monitor, which have all have been validated in various clinical trials [Bruhn *et al.* 2006, Litvan 2002]. All of these are no longer commercially available. The other stream is the system produced by Glasgow University in the UK called the aepEX system, now marketed by MDM (Essex, UK), [Nishiyama 2009]. All of the monitors use similar principles, but have variations on the sophistication of the processing used. The A-line and the aepEX are produced using coherent averaging, already described in chapter 18, but differ in their method of producing an index from the evoked response waveform. The more recent versions of the Danmeter monitors used autoregressive modelling to produce a faster average, and the AEP/2 monitor used a combination of spontaneous EEG and AER to produce an index.

Depth of anaesthesia monitors using the ECG

Respiratory sinus arrhythmia based monitors

Monitors have been developed to measure depth of anaesthesia that do not use the EEG, but that are based on the ECG. They were based on the concept of *Heart Rate Variability* (HRV) [Pomfrett 1999]. HRV is where the heart rate, as measured by the reciprocal of the R–R ECG interval, is measured. This rate is normally plotted against uneven time points and then extrapolated into a continuous waveform with equal time points. The waveform is converted to the frequency domain and it has been found that the spectrum of this represents depth of anaesthesia. Two methods were proposed and used in commercial devices (although they are no longer available). One of the devices, Fathom

(Amtec Medical Systems) used HRV coupled with the respiratory cycle. In this case a chest belt was used, or an encoding from a Weight Respirometer. The other device (Anemon-1) just used HRV. Although the method seems to have gone out of favour, there are still recent studies on its use as a possible measure of anaesthesia [Jeanne 2009].

Emerging devices

Emerging new devices use *fuzzy logic* analysis using the EEG (the Cerebral state function monitor by Danmeter), non-linear complex analysis (symbolic dynamics) of the EEG for the index of consciousness monitor (Morpheus Medical, Spain), and changes in both spontaneous and evoked skin conductance as a measure of depth of anaesthesia (stress detector monitor, Med-Storm innovation AS).

References

Bickford, R. G., Billinger, T. W., Fleming, N. I. and Stewart, L. (1972) The compressed spectral array. A pictorial EEG, *Proc. San Diego Biomedical Conf. Symp. 11*, Academic Press, London, pp. 365–70.

Bonhomme, V., Deflandre, E. and Hans, P. (2006) Correlation and agreement between bispectral index and state entropy of the electroencephalogram during propofol anaesthesia, *Br. J. Anaesth.*, **97**, pp. 340–46.

Bruhn, J., Myles, P. S., Sneyd, R. and Struys, M. M. R. F. (2006) Depth of anaesthesia monitoring: what's available, what's validated and what's next? *Br. J. Anaesth.*, **97**(1), pp. 85–94.

Carter, J. A., Arnold, R., Yate, P. M. and Flynn, P. J. (1986) Assessment of the Datex relaxagraph during anaesthesia and atracurium-induced neuromuscular blockade, *Br. J. Anaesth.*, **58**(12), pp. 1447–52.

Clark, D. L. and Rosner, B. S. (1973) Neurophysiological effects of general anaesthetics, *Anaesthesiology*, **38**, pp. 564–82.

Coste, C., Guignard, B., Menigaux, C. and Chauvin, M. (2000) Nitrous oxide prevents movement during orotracheal intubation without affecting BIS value, *Anesthesia and Analgesia*, **91**(1), pp. 130–35.

Dahaba, A. A. (2005) Different conditions that could result in the Bispectral index indicating an incorrect hypnotic state, *Anesth. Analg.*, **101**, pp. 765–73.

Drover, D. R., Lemmens, H. J., Pierce, E. T., Plourde, G., Loyd, G., Ornstein, E., Prichep, L. S., Chabot, R. J. and Gugino, L. (2002) Patient State Index: titration of delivery and recovery from propofol, alfentanil, and nitrous oxide anesthesia, *Anesthesiology*, **97**(1), pp. 82–9.

Fleming, R. A. and Smith, N. T. (1979) An inexpensive device for analyzing and monitoring the electroencephalogram, *Anesthesiology*, **50**(5), pp. 456–60.

Jeanne, M., Logierb, R., De Jonckheereb, J. and Tavernier, B. (2009) Heart rate variability during total intravenous anesthesia: Effects of nociception and analgesia, *Autonomic Neuroscience: Basic and Clinical*, **147**(1), pp. 91–96.

Johansen, J. W. and Sebel, P. S. (2000) Development and clinical application of the electroencephalographic bispectrum monitoring, *Anesthesiology*, **93**(5), pp. 1336–44.

Kreuer, S., Bruhn, J., Larsen, R., Bialas, P. and Wilhelm, W. (2004) Comparability of Narcotrend index and bispectral index during propofol anaesthesia, *Br. J. Anesth.*, **93**(2), pp. 235–40.

Litvan, H., Jensen, E. W., Revuelta, M., Henneberg, S. W., Paniagua, P., Campos, J. M., Martinez, P., Caminal, P. and Villar Landeira, J. M. (2002) Comparison of auditory evoked potentials and the A-line ARX index for monitoring the hypnotic level during sevoflurane and propofol induction, *Acta Anaesthesiol. Scand.*, **46**, pp. 245–51.

Maynard, D. E. and Jenkinson, J. L. (1984) The cerebral function analysing monitor, *Anaesthesia*, **39**, 678–90.

Maynard, D. E., Prior, P. F. and Scott, D. F. (1969) Device for continuous monitoring of cerebral activity in resuscitated patients, *Br. Med. J.*, **4**, pp. 545–46.

Myers, P. S., Leslie, K., McNeil, J., Forbes, A. and Chan, M. T. V. (2004) Bispectral index monitoring to prevent awareness during anaesthesia: The B-aware randomized controlled trial, *The Lancet*, **363**, pp. 1757–64.

Nishiyama, T. (2009) Comparison of the two different auditory evoked potentials index monitors in propofol-fentanyl-nitrous oxide anesthesia, *Journal of Clinical Anesthesia*, **21**(8), pp. 551–54.

Pomfrett, C. J. D. (1999) Heart rate variability, BIS and depth of anaesthesia, *Br. J. Anaesth.*, **82**, pp. 659–62.

Rampil, I. J. (1998) A primer for EEG signal processing, *Anesthesiology*, **89**, pp. 980–1002.

Schwilden, H. (1989) Use of median EEG frequency and pharmacokinetics in determining the depth of anaesthesia. In Jones, J. G. (ed.) *Clinical Anaesthesiology. Depth of Anaesthesia*, Bailliere Tindall, vol. 3, no. 3, pp. 603–22.

Sechzer, P. H. Cerebral function monitor: evaluation in anaesthetic critical care, (1977) *Current Therapeutic Research*, **22**, pp. 335–47.

Sinha, P. K. (2007) Monitoring devices for measuring the depth of anaesthesia–an overview, *Indian Journal of Anaesthesia*, **51**(5), pp. 365–81.

Takamatsu, I., Ozaki, M. and Kazama, T. (2006) Entropy indices vs. the bispectral index for estimating nociception during sevoflurane anaesthesia, *Br. J. Anaesth.*, **96**, pp. 620–26.

Viertio-Oja, H., Maja, V., Sarkela, M. Talja, P., Tenkanen, N., Tolvanen-Laakso, H., Paloheimo, M., Vakkuri, A., Yli-Hankala, A. and Merilainen, P. (2004) Description of the Entropy algorithm as applied to the Datex-Ohmeda S/5 Entropy module, *Acta Anaesth. Scand.*, **48**, pp. 154–61.

Chapter 20

Pacemakers and defibrillators

This chapter contains: pacemakers, codes, rate responsive, interference, tachyarrhythmias; defibrillators circuits; defibrillator waveforms; biphasic defibrillators, rectilinear and truncated exponential waveforms and circuits. The chapter links with: Chapters 4, 12 and 21.

Introduction

Cardiac pacemakers and defibrillators are used to stimulate cardiac muscle directly. The pacemaker corrects for abnormalities in the heart rate (this can be fast or slow). Defibrillators are used to restore a fibrillating or tachycardic heart, to sinus rhythm. These are normally external, battery or mains powered, but can be internal devices, which are called *Implantable Cardiac Devices* (ICDs).

Pacemakers

Pacemakers for bradycardia

Pacemakers that deal with bradycardia will be considered first. Normally a slow or irregular heart rhythm is caused by three types of heart block:

- First degree, where the delay at the AV junction is increased beyond the normal 0.2 s;
- Second degree, where a proportion of the depolarisation wave fails to pass through the AV junction;
- Complete block, where none of the depolarisation waves pass through the AV junction, and ventricular electrical activity is independent of supraventricular activity.

In all these cases, the ventricles will beat at a slower or irregular rate. Dizziness or loss of consciousness may occur.

Pacemaker components

The simplest pacemaker consists of three major components: batteries, the pulse generator, and the electrode leads. The pulse generator is required to provide a rectangular pulse. Typical parameters are the duration of 1 ms, a voltage of 5 V and capable of delivering a current of 10 mA. The power needed per second (if the pacemaker is on all the time) would be $I^2R = 50$ mW, for an electrode tissue resistance of 500 Ω. If the pacemaker is operating at 1 Hz (60 beats per minute), then the average power consumption would be 50 μW, as the pulse width is 1 ms (the pacemaker is on for 1/1000 of a second, and so the power consumption will be divided by 1000). A typical small battery has a capacity of 1 A h, so that this battery could supply the average current (10 μA) for about 11 years. The circuitry would also absorb power so that the battery life would drop to around 5 years. The batteries used are now commonly lithium iodide. The output pulse is applied to the tissue via an electrode. The electrode tip, which can screw in (or more unusually, is sown in), can be made of platinum, silver, stainless steel, titanium as well as various alloys.

Unipolar and bipolar leads

The electrode leads can be *unipolar*, or *bipolar*. Both type of leads can be used to *pace* (stimulate the heart) or to *sense* (receive signals from the heart). Unipolar is where one of the electrodes of the circuit is in the heart and the other part of the circuit is the pacemaker metal container. Bipolar leads are where both electrodes are contained in the chamber to be paced, about 15–30 mm apart. The unipolar leads are simpler to construct and, because there is only one lead, it can be thicker than the bipolar equivalent (two leads in the same space), and therefore will be stronger. They will be more sensitive than bipolar leads, as they are picking up signals from a wider area, but unipolar leads can also pick up far-field interference such as diathermy, domestic mains and EMG (away from the immediate sensing area). The unipolar leads can also unintentionally stimulate other areas of the heart. Bipolar leads are more selective and have a better signal-to-noise ratio. They are also more selective where they pace. However, the leads are more complex and expensive to produce.

Simple pacemakers

The simplest type of pacemaker produces a constant stream of pulses at 70 beats per min (or whatever rate is set on the pacemaker). The heart rate is constant and will not vary in response to the patient's activity. If some heart beats do occur naturally, then competition between the heart's own pacemaker and the artificial pacemaker's pulses will occur, which will waste power (as this pacemaker beat is not needed) and can cause the heart to revert to a potentially dangerous rhythm such as tachycardia or fibrillation. The alternative to fixed rate pacing is to allow the pacemaker's electrodes to sense the activity in either the atrium or the ventricle. The pacemaker then produces a pulse in response to the detected atrial or ventricular waveform, which will be used to stimulate the ventricle. Atrial triggered devices produce a pulse *triggered* by the P wave of the ECG. Ventricular paced devices can be either *triggered* by the R wave or *demand* led where a pulse is only produced when there is no R wave sensed. The latter types are also labelled inhibit mode pacemakers. A block diagram of an advanced pacemaker is shown in Figure 20.1.

Pacemaker codes

There is a lettering code to define pacemakers; either three letters or, for more complex modes, five letters. The first letter indicates which chamber of the heart is paced: A for atria, V for ventricle and D for dual (both). The second letter indicates the chamber to be sensed, with the same letters. Additionally, O indicates that the function is turned off and not used. The third letter indicates the pacing mode: I for *inhibit* or sometimes called demand mode, T for *triggered* or also called *synchronised* mode. The fourth letter indicates programmable functionality and rate responsiveness. The fifth is for anti-tachyarrhythmic functions.

In general use, five forms of pacing are likely to be used clinically, and these give an example of the use of the codes. The codes are:

- VVI. This is where a single lead is placed in the ventricle, for pacing and sensing. If the natural ventricular rate is above the pacemaker set rate, then pacing is inhibited. If the natural ventricular rate is below, then pacing occurs;

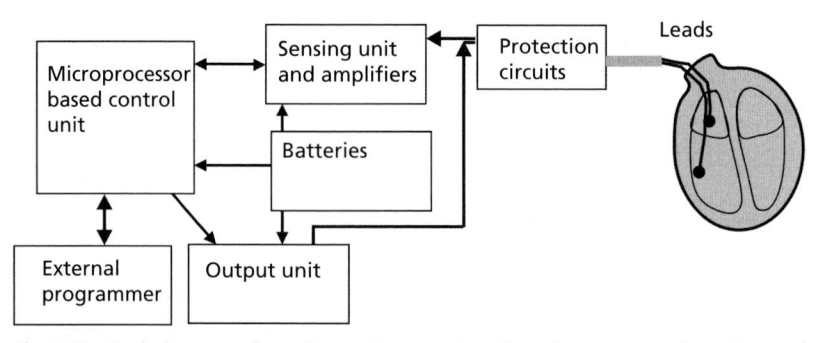

Fig. 20.1 Block diagram of an advanced pacemaker. The microprocessor based control unit can equally be a hard-wired electronics unit – it would enjoy less flexibility in this case. The external programmer communicates with the pacemaker via telemetry, using radio waves.

- AAI. This is where a single lead is in the atrium, and operation is in a similar way to the VVI mode.
- DDD. This is where leads are present in both the atrium and the ventricle. Both chambers (i.e. dual) can be paced and sensed. If activity is sensed in either chamber, then the pacing pulses will be inhibited, but if no ventricular activity follows atrial activity, then ventricular pacing will be triggered. This is the dual response of either inhibition or triggering.
- AOO/VOO/DOO. These are the simplest pacing modes where no sensing occurs (and also of course no response to the sensing). This mode can be turned on temporarily using, for example an external magnet for activation, when the sensing of external noise could lead to inappropriate lack of pacing, as described in the next section. This mode is also called asynchronous pacing.
- VDD. In this mode, only the ventricle is paced, but both chambers are sensed. Normally only one lead is used with electrodes in both chambers. This mode allows physiological pacing described in the next section. This mode is not often used but is sometimes a simpler system alternative to DDD pacing.

Physiological and rate responsive pacemakers

Some pacemakers are *physiological*, in that sensing and pacing can be in both chambers. These devices tend to be microprocessor controlled. The pacemaker attempts to use any atrial activity to help pace the ventricles (and atria if appropriate) to produce heart-rate increases to cope with physiological demands on the body. *Rate-adaptive* pacemakers make use of other *clues* to control heart-rate, when atrial information is not available to the pacemaker. For example, sensors in this type of pacemaker, such as electromagnetic or piezoelectric accelerometers, can directly detect other physiological activity. Also indirect measures can be used that rely on measurements from the body, such as respiration, intrathoracic impedance, central venous temperature, and mixed venous oxygen saturation. All these indirect measures will change in proportion to the physiological demands on the body.

Pacemaker interference

In the complex pacemakers, the design must ensure that under fault conditions the pulse generator will revert to a fixed pacing rate. Also interference, such as diathermy [Goodman 1993] and drug induced muscle movement, must be detected by the pacemaker and the device must revert to a (safe) fixed rate.

Magnets are also used to put the pacemaker in a safe mode, so that when the magnet is passed over the device (by the patient or the clinician), a detector senses the magnet, and reverts the pacemaker to a fixed rate.

Tachyarrhythmia pacemakers

Pacemakers can also be used to change tachyarrthymias or fast heart rates. When the heart rate is detected by the pacemaker as *too fast* by a suitable algorithm, then a specially designed sequence of pulses is fed into the ventricles. These can revert the rhythm to a normal sinus one. Also an electric shock can revert the rhythm back to sinus, as is used in defibrillators. Clearly pathological tachycardia must be distinguished from physiological appropriate (sinus) tachycardia [Davies *et al.* 1986].

Defibrillators

Defibrillators are devices that are used to apply a large electric shock to the heart. They normally are stand alone external devices and have three main uses:

- elective cardioversion, a synchronous shock that is applied across the chest to correct for atrial fibrillation, atrial flutter or ventricular tachycardia;
- emergency defibrillation in cases of ventricular fibrillation;
- direct defibrillation during heart surgery.

Also the devices can be implanted in the body, as *Implantable Cardiac Defibrillators* (ICD). ICDs supply typically 25 J directly to the heart, and used complex algorithms automatically to detect ventricular fibrillation or ventricular tachycardia [Smith *et al.* 1999; Compton *et al.* 1995].

Both types of defibrillator are used to restore a normal sinus rhythm to a heart that is not contracting in a coordinated fashion, i.e. in fibrillation. The cause of fibrillation is commonly ischaemia of the heart conducting tissue but can also be due to electric shock (of a lower voltage and different frequency than the defibrillation voltage, e.g. domestic mains voltage), drugs, electrolyte disorders, hypoxia and hypothermia. The defibrillation shock aims to stop the heart totally and then the heart should restart in an orderly manner, returning to sinus rhythm.

Defibrillator waveforms

The voltage and shape of the waveform needed to defibrillate the heart is critical. Early external defibrillators used domestic AC mains with a step-up transformer to obtain the correct voltage needed. These did not work very well because of the instantaneous power needed from the domestic mains is typically 125 kW! (50 A needed across the thorax of 50 Ω, and power is I^2R). The maximum power from a normal mains socket is about 15 kW, which is enough to defibrillate a small child or animal only. Defibrillators now use the charge stored in a capacitor. However, just discharging a capacitor through the thorax would result in a decaying exponential, from an initial high peak current with a very long tail. There is evidence that the high peak currents cause damage to the heart and the long tails associated with capacitor discharge could reverse the defibrillation process [Schuder 1996, Peleska 1963]. A waveform that has been used successfully in external devices is the critically damped sine wave current waveform as shown in Figure 20.2 which has a duration of around 10 ms The current flow from one electrode to the other is all in one direction. This waveform has now been superseded by biphasic waveforms, which are waveforms that initially are positive, causing current to flow in one direction, and then part way through the procedure, the current waveform goes negative, and so the current flow changes to the opposite direction. Biphasic waveforms have been shown to be more efficient and require less energy, and have a higher success rate for first shock [Bardy *et al.* 1996]. Typical values quoted are 130 joules for biphasic against 200 joules for monophasic and initial shock conversion of 97.6% for biphasic against 85.2% for monophasic [Schneider 2000]. All new external defibrillators in the UK now use biphasic waveforms. However the circuits of monophasic defibrillators are useful to discuss, as many of the principles involved are also used in the newer biphasic devices. Also some monophasic devices are still in use in the UK and elsewhere and they are still very effective.

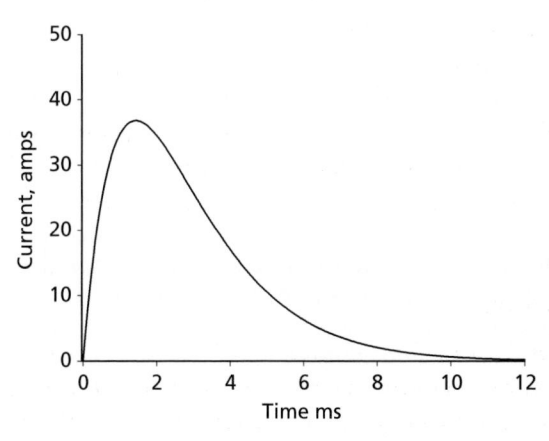

Fig. 20.2 Critically damped waveform commonly used in the output circuits of monophasic defibrillators. The values shown are typical ones.

Monophasic defibrillator circuits

The waveform for a monophasic defibrillator is produced using a circuit shown in Figure 20.3(a). The inductance in the circuit produces a resistor–inductance–capacitance (RLC) circuit and the ratio of these electrical components produces a degree of resonance and damping (as discussed at the end of Chapter 12). Changing the values of the resistance, inductance and capacitance can produce many different waveforms, from a critically damped waveform (in Figure 20.2) to an underdamped waveform shown in Figure 20.4. Biphasic defibrillators produce waveforms that could be produced by a similar circuit to that in Figure 20.3(a), but with different values of RLC circuits that produce the underdamped waveform, but in current practice the waveform is produced using more sophisticated electronic circuits as discussed in the next section.

The voltage required for defibrillation is typically 2500 V to produce a current of 50 A into 50 Ω. The duration of for the majority of the discharge is about 4 ms and so the time constant formed by the resistor and capacitor would be 4 ms. If R is 50 Ω, then C will be 80 μF. The energy stored when charge to 2500 V will be 0.5 (CV^2), which is 250 J. The delivered energy would be about 10% less. The inductance used is calculated using differential equations, and is not trivial, but a typical value would be 40 mH.

The circuit described is powered by domestic mains voltage. Defibrillators can also be battery driven. The circuit shown in Figure 20.3(a) is modified in

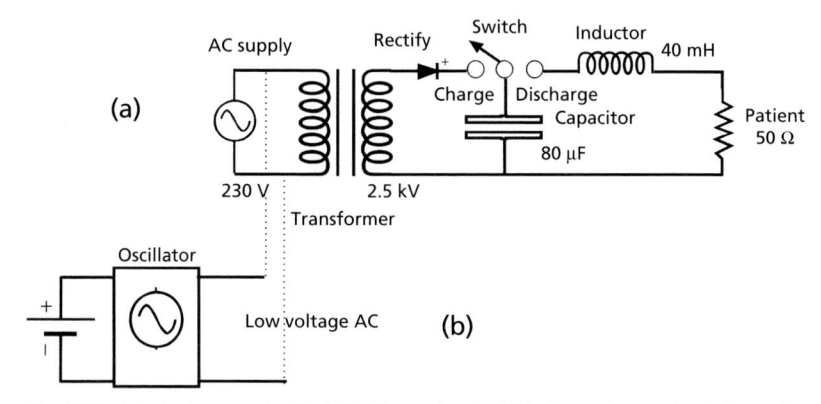

Fig. 20.3 A typical monophasic defibrillator circuit. (a) is the mains version. The mains voltage is transformed up to a very high voltage, e.g. 5000 V. This voltage is rectified by a diode that only passes the positive part of the alternating cycle to the capacitor. When the switch is in the charge position, the capacitor is charged up to the supplied voltage. When the switch is passed over to discharge, it is discharged through the inductor and patient. (b) is the battery version. This part replaces the 230 V AC supply shown in (a).

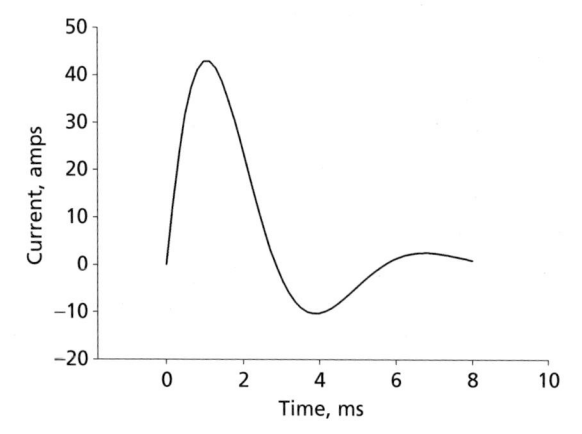

Fig. 20.4 Underdamped biphasic waveform.

that a battery powers a high frequency oscillator as shown in Figure 20.3(b). This sine wave oscillator produces an AC output similar to the battery voltage and this is supplied to a step-up transformer as before (but the step-up in this case is much higher).

Biphasic waveforms

There are now two types of biphasic waveforms in current use. One waveform is the truncated exponential biphasic waveform, shown in Figure 20.5. This waveform is again generated by the discharge of a capacitor, but the process is now controlled by electronic switches. In this process, the charged capacitor is discharged through the patient, but now there is no inductance equivalent in the circuit, so there is a normal exponential decay. After typically 4 ms, the

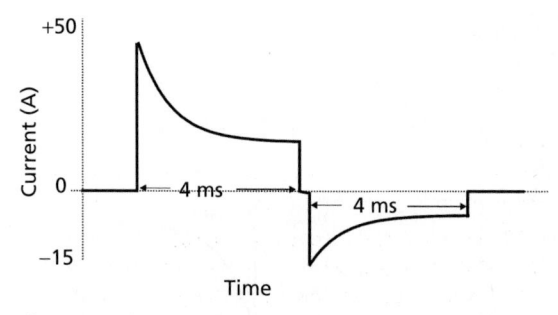

Fig. 20.5 Diagram showing a biphasic waveform, the truncated exponential.

connection to the patient is broken and the capacitor connections are reversed by the equivalent of a dual pole electronic switch, as shown in Figure 20.6, and the patient reconnected. The discharge is again programmed to be about 4 ms, but in the opposite direction. The other type of waveform used is the rectilinear biphasic waveform [Mittal 1999], but there are many permutations of this. The circuitry required to make this waveform is more complex, but the waveform is typically a positive square wave of around 6 ms duration, followed immediately by a negative truncated exponential as before, of around 4 ms The square wave is typically produced by rapid charging and discharging of the capacitor, as can be seen by the triangular nature of the top of the square wave. An example of the waveform is shown in Figure 20.7.

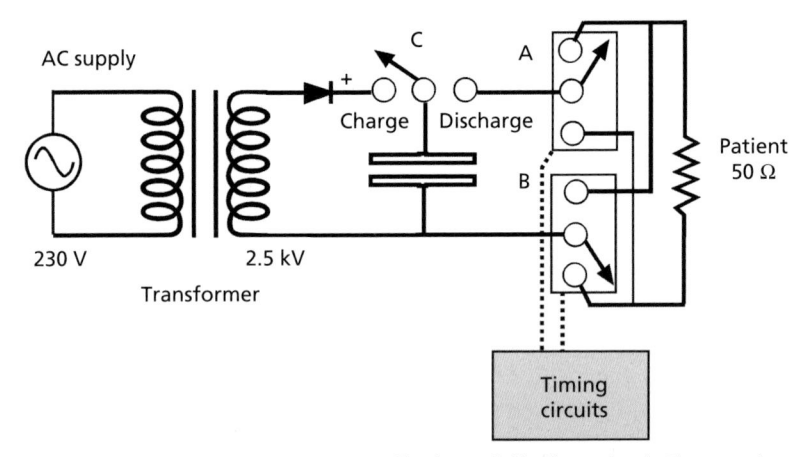

Fig. 20.6 A typical truncated exponential biphasic defibrillator circuit. The capacitor is charged up as described in the monophasic system. When the switch C is passed over to discharge, it is discharged through the patient. After a certain time, the timing circuits switch over the electronic switches A and B, so the polarity is reversed.

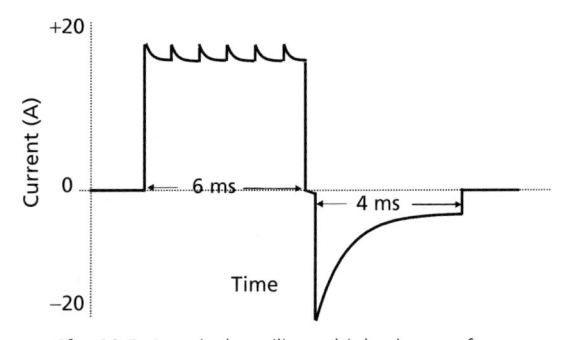

Fig. 20.7 A typical rectilinear biphasic waveform.

References

Bardy, G. H., Marchlinski, F. E., Sharma, A. D., Worley, S. J., Luceri, R. M., Yee, R., Halperin, B. D., Fellows, C. L., Ahern, T. S., Chilson, D. A., Packer, D. L., Wilber, D. J. Mattioni, T. A., Reddy, R., Kronmal, R. A. and Lazzara, R. (1996) Multicenter comparison of truncated biphasic shocks and standard damped sine wave monophasic shocks for transthoracic ventricular defibrillation, Circulation, 94, pp. 2507–2514.

Compton, A. J., Bolouri, H. and Nathan, A. W. (1995) Arrhythmia recognition strategies and hardware decisions for the implantable cardioverter-defibrillator – a review, *Med. Eng. Phys.*, **17**(2), pp. 96–103.

Davies, W., Wainwright, R., Tooley, M. and Lloyd, D. (1986) Detection of pathological tachycardia by analysis of electrogram morphology, *Pacing and Clinical Electrophysiology*, **9**, pp. 200–208.

Goodman, N. W. (1993) Diathermy and failure of cardiac pacemakers, *Anaesthesia*, **48**(9), pp. 824–5.

Mittal, S., Ayati, S., Stein, K. M., Knight, B. P., Morady, F., Schwartzman, D., Cavlovich, D., Platia, E. V., Calkins, H., Tchou, P. J, Miller, J. M., Wharton, J. M., Sung, R. J., Slotwiner, D. J., Markowitz, S. M. and Lerman, B. B. (1999) Comparison of a novel rectilinear biphasic waveform with a damped sine wave monophasic waveform for transthoracic ventricular defibrillation, *J. Am. Coll. Cardiol.*, **34**, pp. 1595–601.

Peleska, B, (1963) Cardiac arrhythmias following condenser discharges and their dependence upon strength of current and phase of cardiac cycle, *Circ. Res.*, **13**, pp. 21–32.

Schneider, T., Martens, P. R., Paschen, H., Kuisma, M, Wolcke, B., Gliner, B. E., Russell, J. K., Weaver, W. D., Bossaert, L. and Chamberlain, D. (2000) Multicenter, randomized, controlled trial of 150-J biphasic shocks compared with 200- to 360-J monophasic shocks in the resuscitation of out-of-hospital cardiac arrest victims, *Circulation*, **102**, pp. 1780–7.

Schuder, J.C., Rahmoeller, G.A. and Stoeckle, H. (1966) Transthoracic ventricular defibrillation with triangular and trapezoidal waveforms, *Circ. Res.*, **19**, pp. 689–94.

Smith, W. M. and Ideker, R. E. (1999) Automatic implantable cardioverter-defibrillators, *Ann. Rev. Biomed. Eng.*, **1**, pp. 331–46.

Further reading

Babbs, C. F. and Bourland, J. D. (1992) Defibrillators. Chapter 3 in Cook A. M. and Webster J. G. (eds) *Therapeutic Medical Devices, Application and Design*, Prentice-Hall, New Jersey, pp. 66–85.

Brown, B. H., Smallwood, R. H., Barber, D. C., Lawford, P. V. and Hose, D. R. (1999) Chapter 22, Cardiac electrical systems. In *Medical Physics and Biomedical Engineering*, IOP Publishing, Bristol, pp. 680–692.

Citron, P., Duffin, E. G., Fetter, J., Goodman, L. and Shearon, L.W. (1992) Chapter 2 in Cook, A.M. and Webster, J.G. (eds) *Cardiac Pacing: in Therapeutic Medical Devices, Application and Design*, Prentice-Hall, New Jersey, pp. 29–65.

Chapter 21

Surgical diathermy

> This chapter contains: current density; surgical effects; diathermy waveforms; electrodes, monopolar and bipolar; safety; interference.
> The chapter links with: Chapters 4, 6 and 20

Introduction

As discussed in Chapter 4, when a voltage is applied across a conductor, a current will flow. If the voltage is applied across the body via suitable electrodes the body becomes part of the circuit (Figure 21.1) and a current will also flow, the magnitude depending on the properties of the tissues in its path, particularly the resistance. This current can cause heating or other physiological effects, depending on the *frequency* of the driving voltage. The effects of the domestic mains current flowing through the body was discussed in Chapter 6, but different effects occur as the frequency of the voltage is increased. As the frequency goes up, the heating increases but the tissue stimulation decreases and, at frequencies above 100 kHz (i.e. radio frequencies), the effect is entirely heating. This heating effect in the body by electric current is called *diathermy*, but the location, concentration and how this heat is used is dependant on the electrode design and the current concentration or *current density* at any point in the circuit.

Current density

For a certain applied voltage, the *average* current throughout the circuit will be the same. The current density is the current per unit area, and so if the material in which the current passes is smaller, the heating effect increases. The resistance of the material is proportional to its size, so as the material becomes smaller then its resistance gets larger. The heating power is the product of the current squared and the resistance ($power = I^2 \times R$).

Surgical diathermy (or electrosurgery) is where either one or both of the electrodes are very small, and it is used to cut and coagulate tissue. The smaller electrode can be made into a pointed surgical tool and localised heating will occur at the tip of the instrument. The smaller and more pointed the instrument

SURGICAL EFFECTS AND ASSOCIATED WAVEFORMS | 281

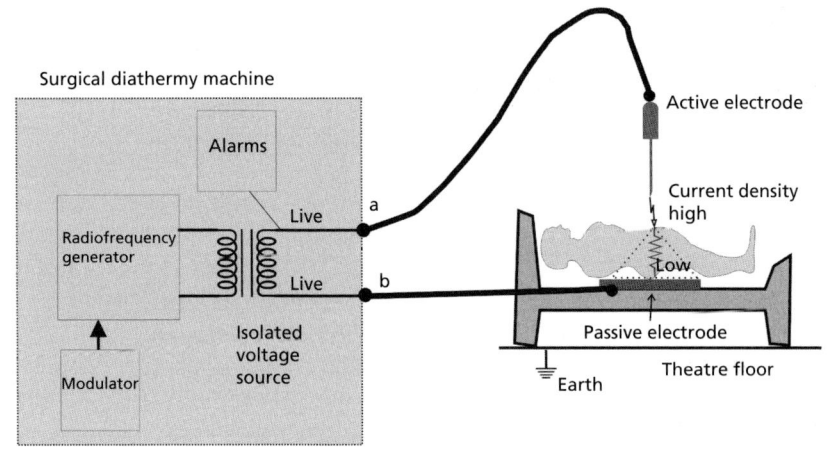

Fig. 21.1 The set up for an isolated surgical diathermy system in the monopolar mode, showing the passive and active electrodes. **a** and **b** are the electrode connections, active and passive respectively. [Figure adapted from Tooley [2004] with permission.]

is, the greater the current density will be at the tip. This electrode is classified as the *active* or *live* one. The current densities around this electrode can be as much as 10 A cm^{-2}, and the total heating power typically around 200 W.

The other electrode in the circuit is much larger and situated on another part of the body, as shown in the Figure 21.1. This is the *passive* electrode (also called dispersive, neutral, plate, patient or indifferent) and, because of its large surface area, the current density is low, and minimal heating occurs. The current density is high only around the active electrode and quickly becomes lower in the rest of the body, as demonstrated also in the figure.

Surgical effects and associated waveforms

Cutting

Rapid heating of tissue over a very small area is produced by a fine intensity arc from the pointed active electrode. This causes boiling of intracellular fluid. The tissue is cut without heat dissipation in the surrounding tissue. The most efficient cutting waveform is a continuous sine wave (see Figure 21.2(a), at a voltage between 250 and 3000 V, depending on the application, the electrode and power required. Smaller voltages may be used if the tip of the electrode is pristine (the sharper the tip, the higher the current density and heating).

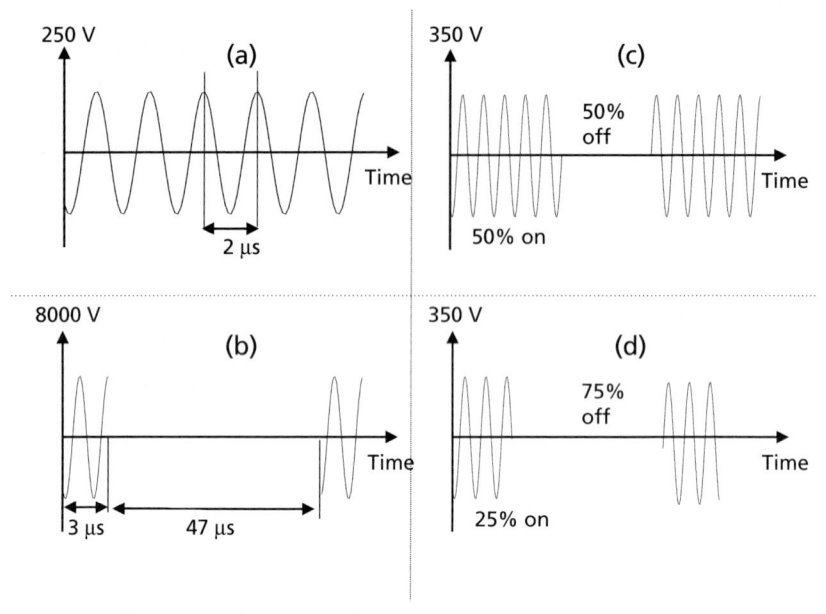

Fig. 21.2 Different waveforms for different types of surgery. A is a continuous wave at 500 kHz at 250 V peak, suitable for cutting. B is a waveform suitable for coagulation. The time scales are expanded for clarity. C shows a 50% *on* waveform, and D shows a 25% *on* waveform. [Figure adapted from Tooley [2004] with permission.]

Coagulation

The continuous waveform produces excellent cutting, but the edges bleed freely and hence the coagulation effect is poor. Coagulation can be provided by both *fulguration* (literally lightning!) and *desiccation* (drying).

Fulguration

Fulguration is the destructive charring of the tissue by arcing, and is normally carried out using a flat spatula or ball shaped electrode. Blunted instruments need higher current densities for arcing, which is obtained using higher voltages, up to 9 kV. The instrument would be positioned 2–4 mm above the surgical site. The waveforms used are not necessary continuous, and the wave can be *active* for as little as 6% of the complete cycle. For example, the 500 kHz wave could be *on* (an active voltage) for 3 μs and then *off* for 47 μs, as shown in Figure 21.2(b).

Desiccation

Coagulation by desiccation is the heating of tissue due to contact between the active electrode and tissue. The temperature of the tissue is increased at the point of contact and intracellular water is slowly steamed off and the tissue desiccates. The waveform shape is not critical here, and the power and voltage requirements are lower.

Blended modes: cutting and coagulation

Modes can be *blended* as well, so that a mixture of reasonable cutting and coagulation is provided. The waves can be 50% on/50% off (Figure 21.2(c)), or 25% on/75% off (Figure 21.2(d)), and any interim range. The shorter the on-phase of the waveform, the more useful the coagulation, and vice versa (Figure 21.2(c, d)). These ratios are controlled by the *modulator*, as shown in Figure 21.1.

Electrodes

The electrode systems used in surgical diathermy can be monopolar or bipolar. Each has different characteristics and uses.

Monopolar electrodes

The monopolar system is the one in Figure 21.1. The power available in this mode is high, and produces efficient cutting. The active electrode is available in an assortment of tips: loops to cut and fulgurate, needles to cut and coagulate, ball electrodes for desiccation and fulguration, and blades to cut and coagulate. However, as the patient forms a major part of the electrical circuit, every precaution for safety must be taken. This mode should not be used in vicinities where the target tissue is connected to other adjoining tissue via small delicate structures. These small structures could experience high current densities and be damaged due to excessive heat transfer. If a metal prosthesis, or a pacemaker, forms part of the circuit, then this can cause problems. The device could become a major part of the circuit and some or most of the current could flow through it. The device will heat up, and any tissues joining sharp extremities of the device could become areas of high current density with the associated damage.

Bipolar electrodes

In the bipolar arrangement both electrodes are small, forming each end of the diathermy forceps, in which high current densities are produced. The intense heating effects are the same at each electrode, and the body does not form part of the circuit. The bipolar surgical diathermy has a localised, precise effect on the tissue and suits delicate surgery such as ophthalmic surgery. The technique has good coagulation effect but less cutting ability due to the low power available.

Safety of diathermy

Older diathermy machines used to have the passive electrode earthed to keep the patient earthed at zero voltage. As discussed in the Chapter 6, directly earthing of the patient is not advisable for electrical safety reasons. Modern machines have outputs isolated from earth, as shown in Figure 21.1, to minimise leakage currents from diathermy circuits to earth. To complete an electrical circuit for the maximum diathermy current to flow, the circuit must be completed from **a** (in Figure 21.1) to the active electrode, via the patient resistance, to the passive electrode, back to output **b** (in Figure 21.1). In theory, if the passive electrode were to become disconnected, no current would flow. If an earthed object (such as a drip stand) were to touch the patient, then this should have no effect, as it is not part of the return current path. However, there is also the problem of stray capacitance. This has already been mentioned in the Chapter 6, where stray capacitance exists between conductors causing leakage currents, but also stray capacitance can occur between conducting objects.

Stray capacitance

As discussed in Chapter 4, capacitors allow the flow of alternating current. They can be considered as frequency dependant resistances, where the resistance deceases as the frequency increases. Capacitors can occur, for example, as the passive electrode and the ground of the theatre (which can be indirectly connected to earth). This 'capacitor' will have a low capacitance value but pass very small currents at domestic mains frequencies. However, it will present a much lower resistance at the higher diathermy frequencies and so other stray circuits can occur, which may result in burns. For example, if there is a poor connection between the passive electrode and the patient, and a sharp earthed object comes into contact with the patient (Figure 21.3), then an alternate current path could be formed for the return current. The current in this case flows from **a** (Figure 21.3) to the active electrode, into the patient, then some of current leaves by the earthed object and flows to earth. To complete the circuit back to the connection **b**, the current passes from earth to the passive electrode via the stray capacitance. Burns can arise at the connection between the sharp object and the patient, but can also arise at parts of higher current density where only certain parts of the passive electrode are touching the patient such as an improperly applied diathermy plate. Burns can be minimised by ensuring

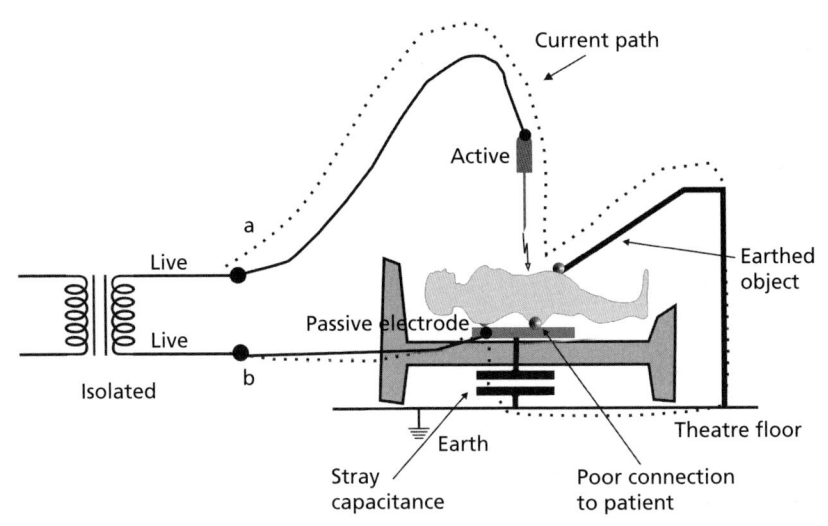

Fig. 21.3 The diathermy circuit demonstrating stray capacitance. **a** and **b** are outputs from the diathermy machine, and the patient has a poor contact with the passive electrode. An earthed object, such as a drip stand, is touching the patient. Stray capacitance is shown between the passive electrode and the theatre floor. [Figure adapted from Tooley [2004] with permission.]

that the passive electrode is securely and maximally attached to the patient (so that the correct circuit is the best route for the current to flow). Also it must be ensured that no other objects are touching the patient when the diathermy is on. Alarms can be installed in the equipment (Figure 22.1) to warn users that the device is active, of faulty electrode connections, and of high stray leakage currents. The stray currents gets worse with increasing frequency, so this is why modern isolated systems tend to be designed to operate at the lower frequency range of diathermy machines (500 kHz) to minimise this effect. Older machines used to operate at frequencies up to 5 MHz.

Interference with other equipment

Surgical diathermy uses radio waves and the equipment can therefore act as a short range transmitter, and sensitive monitoring equipment, such as ECG monitors can be a receiver for these signals. Ideally the ECG electrodes and leads should be placed as far from the diathermy electrodes as possible, and care with interpretation must be made when diathermy is in use. Pacemakers can be similarly interfered with, and sometimes the interference can result in the pacemaker wrongly detecting a tachycardia. In these cases the pacemaker should be put into a fixed rate mode, either automatically or by using a manual method, such as by applying a magnet to the pacemaker.

References

Tooley, M. (2004) Surgical diathermy, *Anaesthesia and Intensive Care Medicine*, **5**(11), pp. 369–71.

Chapter 22

Gas supply and the anaesthetic machine

This chapter contains: supply of anaesthetic gases; pressure regulators, gas flow control and anaesthetic machine safety features; the anaesthetic machine and equipment checklist.
The chapter links with: Chapters 7, 9, 11, 12, 14, 15, 16, 24, 25, 26 and 28.

Supply of anaesthetic gases

In Europe and other advanced medical communities, medical gases are generally supplied by pipeline, with cylinders available as back up.

Oxygen

Large hospitals usually have oxygen supplied and stored in liquid form, since one volume of it provides 840 volumes of gaseous oxygen at 15°C. It is stored in a secure *Vacuum Insulated Evaporator* (VIE) on the hospital site. The arrangement is shown in Figure 22.1. The VIE consists of an insulated container, the inner layer of which is made of stainless steel, the outer of which is made of carbon steel. The liquid oxygen is stored in the inner container at about −160°C (lower than the critical temperature of −118°C) at a pressure of between 700 and 1200 kPa. There is a vapour withdrawal line at the top of the VIE, from which oxygen vapour can go via a restrictor to a superheater, where the gas is heated towards ambient temperature. Where demand exceeds supply from this route, there is also a liquid withdrawal line from the bottom of the VIE, from which liquid oxygen can be withdrawn; the liquid can be made to join the vapour line downstream of the restrictor and pass either through the superheater or back to the top of the VIE. The liquid can also be made to pass through an evaporator before joining the vapour line. After passing through the superheater, the oxygen vapour is passed through a series of pressure regulators to drop the pressure down to the distribution pipeline pressure of 410 kPa. It should be remembered that no insulation is perfect and there is a pressure relief valve on top of the VIE in case lack of demand and gradual temperature rise results in a pressure build up in the container. There is a filling port and there is usually considerable wastage in filling the VIE; the delivery hose needs

288 | GAS SUPPLY AND THE ANAESTHETIC MACHINE

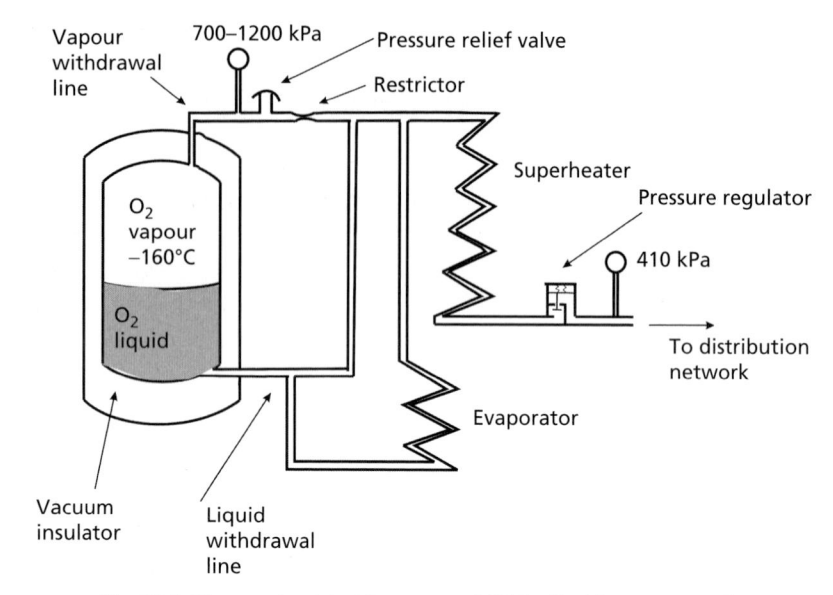

Fig. 22.1 Vacuum Insulated Evaporator (VIE) for liquid oxygen supply.

to be cooled to below the critical temperature, using the tanker liquid oxygen itself to cool the delivery pipe. The whole VIE device is mounted on a hinged weighing scale and is situated outside the hospital building, protected by a caged enclosure, which also houses two banks of reserve cylinders. These take over automatically if the VIE output falls.

In a smaller hospital, in a country in which delivery of cylinders of oxygen is not a problem, banks of large cylinders can be used to deliver piped oxygen. Appropriate valves, monitoring and alarms need to be in place to ensure that the supply automatically switches to a full cylinder.

Gas pressure in a full cylinder is of the order of 135 atm (13 700 kPa); therefore this represents the number of times the volume of the cylinder is multiplied to calculate the deliverable gas volume, remembering to use cylinder (gauge) pressure plus one atmosphere pressure to do this calculation, an application of Boyle's law.

In some countries, neither pipelines nor cylinders of oxygen are readily available and a device called an *oxygen concentrator* is useful. This takes in and compresses ambient air, which is then passed through a *zeolite* structure [Li et al. 1998], which adsorbs nitrogen, leaving an oxygen rich atmosphere at the outlet. Since only nitrogen is adsorbed, leaving argon as well as oxygen in the gas, the output gas never quite reaches 100% oxygen. A zeolite is often made of an aluminium hydroxide lattice, which allows nitrogen molecules to be adsorbed and trapped. Adsorption requires the device to be pressurised, and on depressurisation the nitrogen is desorbed and released. Hence the system also includes a compressor and, if a pair of zeolites that can be alternately compressed and decompressed is used, a relatively constant output of oxygen enriched gas can be produced, though the devices available have their limitations [Easy et al. 1988, Harris et al. 1985]. These include an efficiency limited by the pressure change available and the need to absorb water vapour from the gas first. Other methods of oxygen production, using chemical means, are available, but are intended for use in military and field anaesthesia [Bhisman et al. 1996, Hall et al. 1986].

Nitrous oxide

Nitrous oxide has a critical temperature above room temperature in temperate climates and is therefore generally stored as a liquid in pressurised cylinders, with N_2O vapour present in the space above the liquid. The actual pressure of a full cylinder is, of course, the SVP at room temperature, but generally lies between 4400 and 5000 kPa. As demand is placed on the supply, more liquid is vaporised. This vaporisation requires energy from the local environment, the latent heat of vaporisation (see Chapter 9). If the demand from the system is high, this can result in a significant fall in temperature within the cylinder's pressure regulator, thus freezing any water vapour present and causing possible obstruction of the regulator outlet. Thermostatically controlled regulators are available to prevent this.

Banks of nitrous oxide cylinders work in the same way as oxygen cylinders, with one bank of perhaps three large cylinders being the 'running' bank, the other bank being the reserve.

On site manufacture of nitrous oxide is not yet developed for use, but could be achieved by heating ammonium nitrate. Impurities in this process include nitrogen dioxide, which must be washed out with water and sodium hydroxide, before removing the water itself with aluminium silicate.

Entonox

This is 50% nitrous oxide, 50% oxygen. It is used as a safe and effective analgesic gas mixture, usually in the Obstetric Delivery Suite or in the Accident Department. The gas mixture is stored in cylinders or cylinder banks and delivered to the patient using a two stage pressure regulator, the second incorporating a demand valve, shown in Figure 22.2. If the cylinder temperature falls below −6°C, the *pseudocritical* temperature, the oxygen and

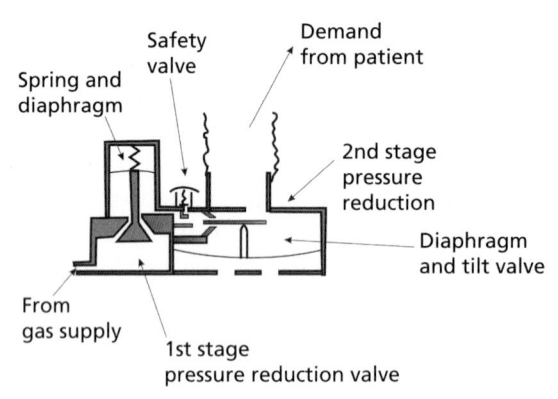

Fig. 22.2 Two stage 'Entonox' pressure regulator and demand valve.

nitrous oxide separate into layers, known as *lamination* or the *Poynting effect*. The effects of lamination may be minimised by two means: storing the cylinders horizontally at a temperature of 5°C or more for 24 hours allows the layers to mix. There is also a long tube from the valve housing at the top of the cylinder to a point a small distance from the bottom that prevents the withdrawal of pure N_2O from the cylinder in the event of lamination. Further measures to prevent excessive cooling of the cylinder include using a number of cylinders open simultaneously, so that no single cylinder has excessive demand placed on it.

Medical compressed air

This is more widely used in anaesthesia and intensive care than previously, as the range of intravenous anaesthetic techniques expands, and concerns about the side effects and environmental effects of nitrous oxide become apparent. Much greater purity is required from medical compressed air than from industrial compressed air. In particular, oil mist, which is often present in industrial compressed air to lubricate power tools, must be removed from medical air. This is partly because of the risk of inhalation of oil mist and partly because of the risk of explosion in an environment of raised oxygen partial pressure. Medical air for breathing devices is supplied from the pipeline at 410 kPa; when supplied at 700 kPa, it is also used for powering tools in the operating theatre, when the oil lubricant has to be added again. It is clearly important not to confuse the two sources, since ICU gas mixers are calibrated to a 410 kPa supply.

Compressed air can be supplied either by banks of cylinders or by using an air compressor. The air is drawn into such a duty compressor, with a standby compressor in reserve. The air must also be filtered and dried and regulated down to 700 kPa and 400 kPa outlet pressures. It is possible to use a small compressor for an operating or other department, to use the same compressor to run an oxygen concentrator as well, or to use a larger compressor for the whole hospital.

Piped medical vacuum

Anybody who has used portable suction devices will know how much more efficient and readily available wall suction is. The outlet on the wall is one of many on a *ring main*, from where the pipe work goes via drainage, filtering and valve mechanisms towards a vacuum reservoir. Major drainage of fluids that have been suctioned takes place and thence through bacterial filters towards a vacuum pump; flexible hoses are used for this transmission in order to reduce noise. Once again duty and standby pumps are in the system. The pump output is passed through a silencer.

Gas cylinders

Where anaesthetic gases are not supplied from one of the sources already described, they are generally supplied by cylinders. These are made from an alloy of molybdenum and steel that is resistant to corrosion.

The difference between a gas and a vapour has already been discussed (Chapter 9). Oxygen, nitrogen, air and helium are stored in cylinders as gases. Nitrous oxide, carbon dioxide and cyclopropane (which is no longer available in UK for anaesthetic use) are stored as liquids in equilibrium with vapour at the pressures indicated on the cylinder gauges at ambient temperature. When a nitrous oxide cylinder is full, up to 0.8 of its volume contains liquid N_2O. The ratio of the weight of nitrous oxide the cylinder could hold to the weight of water it could hold is the *filling ratio* and this is usually 0.75 in temperate climates, 0.67 in tropical climates. The mounting of such liquid containing cylinders must be vertical. Carbon dioxide is included here, although it is no longer supplied for use in anaesthesia in many countries.

Table 22.1 and 22.2 [Medishield 1979] shows cylinder sizes, dimensions and capacities, as well as the physical properties of the relevant gases or vapours contained within them. Clearly the contents of a gas-containing cylinder are usefully indicated by a pressure gauge, whereas this is a less useful indicator for a liquid–vapour containing cylinder. Table 22.2 also shows the International Standards Organisation (ISO) colour coding (and that of the

GAS SUPPLY AND THE ANAESTHETIC MACHINE

Table 22.1 Medical gas cylinder nomenclature, dimensions and gas capacities

Cylinder size	C	D	E	F	G	J
Dimensions, mm	356 × 89	457 × 102	788 × 102	865 × 140	1248 × 40	1450 × 29
Capacities, l						
Oxygen	170	340	680	1360	3400	6800
Nitrous oxide	450	900	1800	3600	9000	–
Entonox	–	500	–	2000	5000	–
Air	–	–	–	–	3200	6400
Carbon dioxide	450	900	1800	–	–	–

Table 22.2 Physical properties and colour coding of medical gases in cylinders

	Oxygen	Nitrous oxide	Entonox	Air	Carbon dioxide
Physical state in cylinder	Gas	Liquid	Gas	Gas	Liquid
Pressure at 15°C, kPa	13 700	4 400	13 700	13 700	5000
Critical Temperature, °C	−118.4	36.5	Separation at −7	−140.7	31
Critical Pressure, kPa	5079	7260		3773	7380
Boiling point at 1 atm, °C	−183	−89	–	−194.4	Sublimation −78.5
Flammability	Supports combustion	Supports combustion	Supports combustion	Supports combustion	No
ISO colour coding: body	black	blue	blue	grey	grey
:shoulder1	white	blue	white	white	grey
:shoulder2	white	blue	blue	black	grey
US colour coding	green	blue	——	yellow	grey

USA to demonstrate that even International Standards are not universal) for gases. The mechanical integrity of cylinders is tested every 5 years.

The ISO *pin index system* is a method of preventing inappropriate attachment of the wrong cylinder to the wrong yoke on the anaesthetic machine. It allows keying in of the cylinder to its own yoke. The valve block at the top of the cylinder has the pin index holes and the appropriate yoke has the

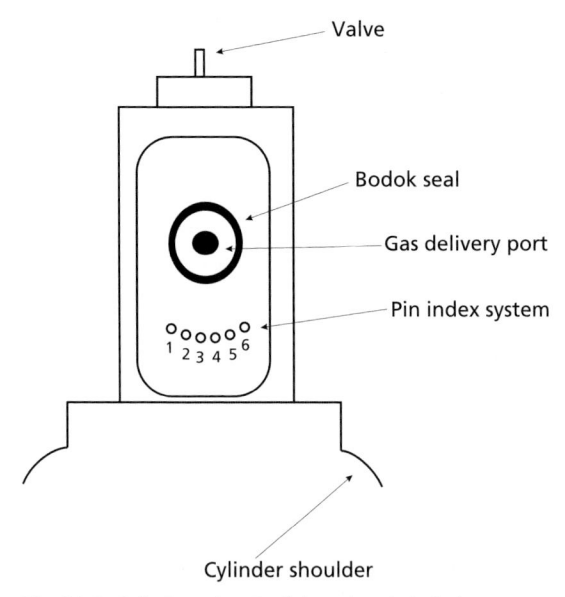

Fig. 22.3 Cylinder valve, Bodok seal and pin index system to ensure non-interchangeability of cylinders and yokes: Pins 2 and 5 – Oxygen; Pins 3 and 5 – Nitrous oxide; Pins 1 and 5 – Air; Pins 1 and 6 – Carbon dioxide.

mating pins. Figure 22.3 shows the pin index configuration on valve blocks to ensure unique fitting for each gas. When attaching a cylinder to an anaesthetic machine yoke, the presence of a sealing washer, the *Bodok* seal, should be noted at the point where the cylinder gas outlet attaches. If the Bodok seal is absent or in poor condition, there will not be a gas-tight fit. The valve block is fused to the cylinder by a material that melts at a relatively low temperature so that, in the event of fire, the cylinder contents can escape, so reducing the chance of an explosion.

Pressure regulators, gas flow control and anaesthetic machine safety features

Gas is delivered to the anaesthetic machine from cylinders or from the pipeline at different pressures, as indicated previously. There are one or two stages of pressure reduction within the anaesthetic machine, designed to minimise the danger to patients. Between an oxygen cylinder and the anaesthetic machine, there is a reduction from 13 700 kPa to 420 kPa and on some makes (Ohmeda), a further pressure reduction to 140 kPa. Figure 22.4(a) shows a pressure reduction valve (pressure regulator) in its simplest form.

High pressure gas, at pressure P_H, acts across the small cross sectional area a of the outlet nozzle from the high pressure stage. In the low pressure stage, the gas is at a reduced pressure P_R, acting across the diaphragm of larger cross sectional area A. The high pressure nozzle outlet and the diaphragm are mechanically connected by the J piece. As P_R reduces, forces on the J piece open the nozzle, admitting more gas to the low pressure side. Equilibrium of vertical forces on the J piece (using gauge pressures) gives:

$$P_H a = P_R A.$$

Hence the pressure reduction ratio is inversely proportional to cross sectional area ratios, and P_R reduces as P_H reduces, although the constant of proportionality can be reduced to a minimum as a/A is reduced. Figures 22.4(b) and 22.4(c) show variations on this model, with the force from a spring or the pantograph biasing the equation above and with a slightly different relationship between the forces in each case. For the example in Figure 22.4(b), the force F_S from the spring, which is designed to depress the diaphragm from the outside, means the equation for equilibrium of vertical forces on the J piece now becomes

$$P_H.a + F_S = P_R.A$$

i.e.

$$P_R = P_H \frac{a}{A} + F_S \frac{a}{A}.$$

By using a very small nozzle on the high pressure side, and a very large diaphragm on the low pressure side, a/A can be minimised, and by making the spring strong enough to make F_S high, P_R may be kept relatively constant, even in the face of changing P_H.

Correct functioning of pressure regulators is essential for safe anaesthesia [Puri et al. 1987].

A final stage of pressure reduction is provided by a flow restrictor at the needle valve, which controls flow into the Rotameter. The performance of the needle valve is such that the mass flow through it is largely independent of downstream pressure, even if the back pressure changes, referred to in Chapter 14, cause a change in the position of the Rotameter bobbin [Hutton et al. 1986]. In modern anaesthetic machines Rotameters themselves no longer exist as such, but there is an electronic icon of 'Rotameters' on the monitoring screen. The mechanism for flow control is via a series of electronically controlled solenoid valves.

PRESSURE REGULATORS, GAS FLOW CONTROL AND SAFETY FEATURES | 295

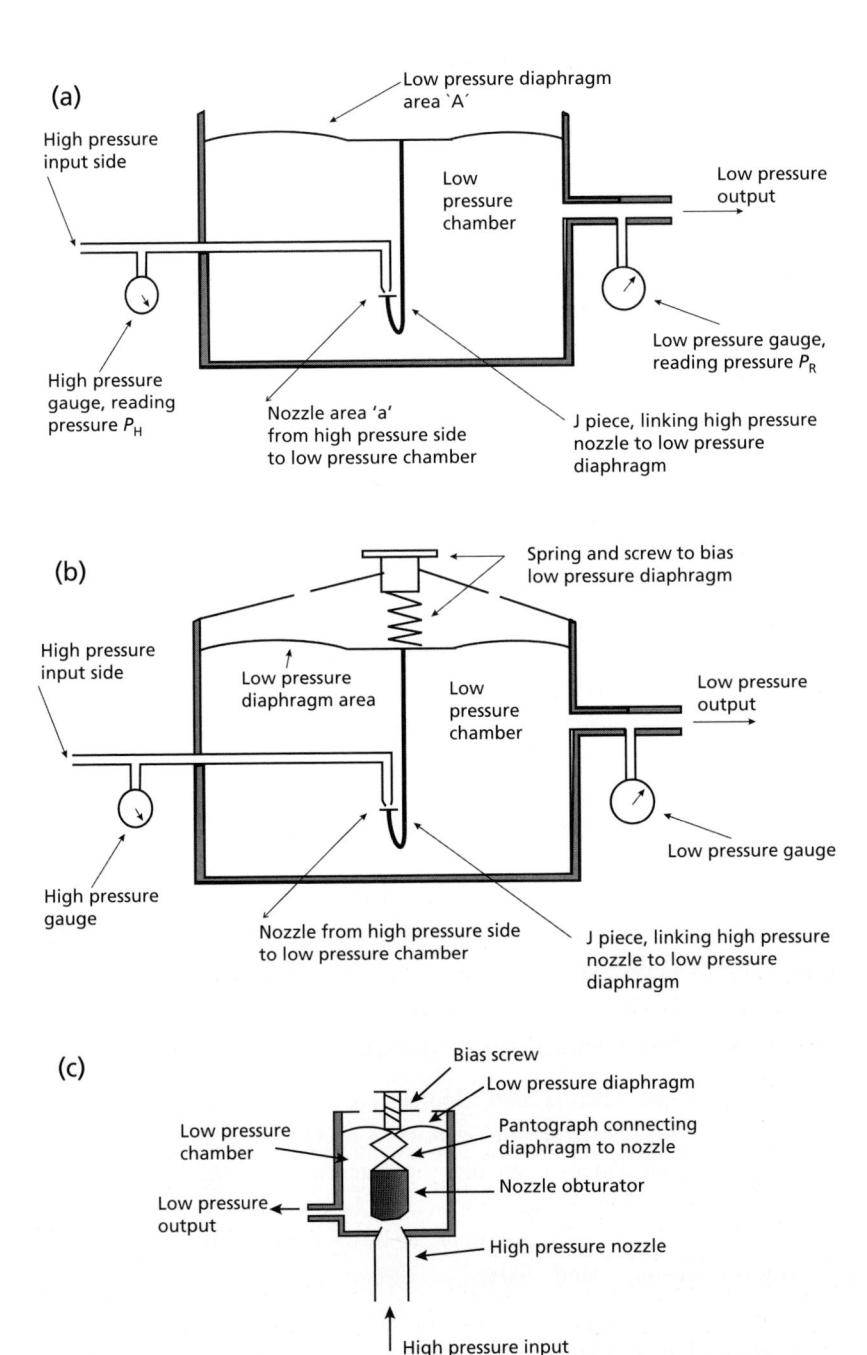

Fig. 22.4 (a) A pressure reduction valve in its simplest form. (b) A pressure reduction valve. (c) A further example of a pressure reduction valve.

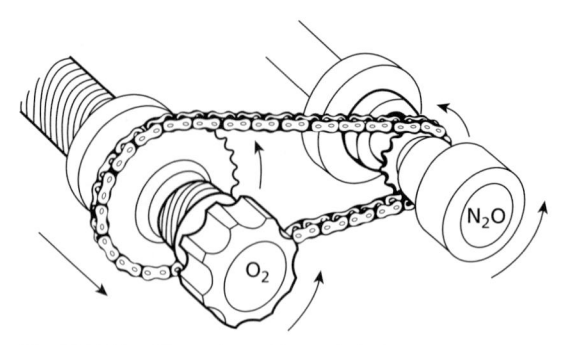

Fig. 22.5 The Ohmeda anti-hypoxic linkage mechanism.

The pressure beyond the rotameters in the anaesthetic machine backbar ranges from 1 to 8 kPa, depending on what the gas flow rate is and whether the vaporiser, situated on the backbar, is switched on. The backbar has a pressure relief valve that activates at about 40 kPa.

Oxygen failure warning devices are activated when oxygen delivery pressure falls below 200 kPa. There are some additional safety features to minimise the possibility of delivery of a hypoxic gas mixture to the patient, apart from the rotameter bank design referred to in Chapter 14 and apart from the monitoring devices discussed elsewhere. There are means of limiting nitrous oxide flow if oxygen delivery pressure falls or fails. One is the Ohmeda chain link between the oxygen and nitrous oxide flow controls [Cicman *et al.* 1993], shown in Figure 22.5. The other is the Draeger solution, shown in Figure 22.6. The oxygen flush button is connected directly to the high pressure oxygen source. Modern safety dictates that the button should not be lockable in the depressed position.

The anaesthetic machine and equipment checklist

[Association of Anaesthetists 2004, Klopfenstein *et al.* 1994, Jackson *et al.* 1993, Barthram *et al.* 1992, Weightmann *et al.* 1992, McQuillan *et al.* 1987].

It is mandatory for patient safety that the anaesthetist checks the anaesthetic machine and all equipment he or she is using before staring an operating list. This is based on the checklist recommended by the Association of Anaesthetists of Great Britain and Ireland [2004].

Anaesthetic machine

Check that the anaesthetic machine and associated equipment is connected to the electrical supply.

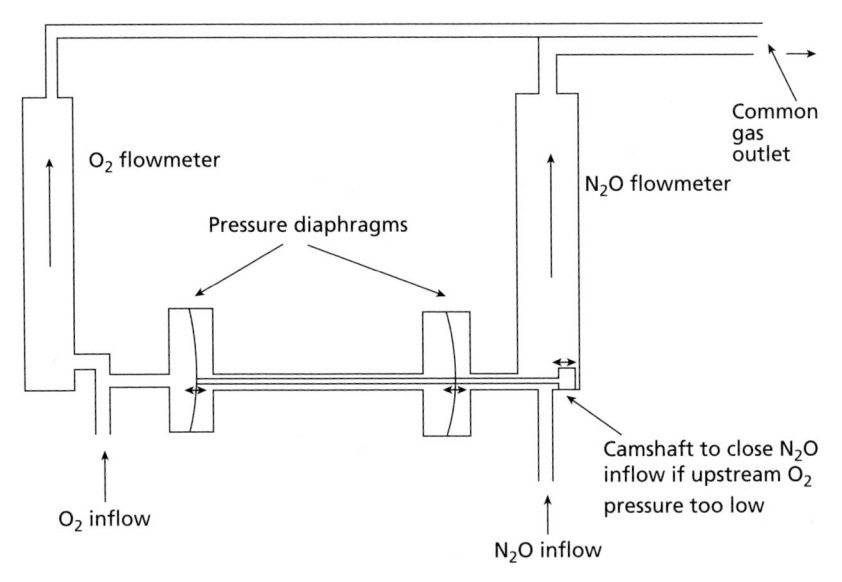

Fig. 22.6 The Draeger anti-hypoxic mechanism.

Oxygen analyser

The oxygen analyser nowadays is usually integrated into the anaesthetic machine, but in any event should be placed where it can monitor gas composition at the common gas outlet. The analyser should be switched, checked and calibrated.

Medical gas supplies

Identify the gases being supplied by the pipeline, checking with a tug test that each pipeline is correctly inserted into the appropriate gas supply terminal. Check that the anaesthetic machine is connected to a supply of oxygen and that an adequate reserve is available from a spare cylinder. Check that adequate supplies of other gases are available and appropriately connected. Cylinders should be securely attached and switched off. Carbon dioxide cylinders should not normally be present [Nunn 1990] on the anaesthetic machine and a blanking plug should be fitted to any empty cylinder yoke. All pressure gauges for connected pipelines should read 400 kPa. Check the smooth operation of flowmeters, ensuring that flow control valves operate smoothly and that the bobbin moves freely. With only the oxygen flow control valve open and a flow of 5 l min^{-1}, check that the oxygen analyser display approaches 100%. Turn off all flow control valves. Operate the emergency oxygen bypass control and ensure that flow occurs without significant fall in the pipeline supply pressure. Confirm that the oxygen analyser display approaches 100%. Ensure that the emergency oxygen bypass control ceases to operate when released.

Vaporisers

Check that the vaporisers for the required volatile agents are fitted correctly to the anaesthetic machine, that any backbar locking mechanism is fully engaged and that the control knobs rotate fully through the full range. Ensure the vaporiser is not tilted. Switch off vaporisers. Check the vaporisers are adequately filled and that the filling port is closed. Set a flow of 5 l min^{-1} of oxygen and, with the vaporiser turned off, temporarily occlude the common gas outlet. There should be no leaks and the flowmeter bobbin should dip. Repeat this test for each vaporiser present. Turn off all vaporisers and bobbins. If vaporisers are changed at any point this test must be repeated.

Breathing systems

[Adams 1994]
Check all breathing systems to be used. Visual inspection is required to ensure correct assembly and that all connections, including the ones to the anaesthetic machine should be secured by a 'push and twist'. Ensure that there are no leaks or obstructions in the reservoir bag or breathing system. A pressure leak test should be performed by occluding the patient end of the breathing system and compressing the reservoir bag. Each system should be checked according to its structure; in particular the inner tube of a Bain's system should be occluded to ensure correct attachment. Check that the adjustable pressure limiting valve can be fully opened and closed. The correct operation of the unidirectional valves of a circle system should be carefully checked. If it is intended to use low fresh gas flows, there must be a means of analysing oxygen concentration in the inspiratory limb, and end tidal carbon dioxide must also be monitored in this situation.

Ventilators

Check that the ventilator is correctly configured for use and that tubing is securely attached. Set the controls for use and ensure that adequate pressure is generated in the inspiratory phase. Check that the disconnection alarm and the relief valve functions correctly. Ensure that there is an alternative means to ventilate the patient's lungs.

Scavenging

Ensure that the scavenging system is switched on and functioning and is attached to the appropriate expiratory port of the breathing system or ventilator.

Ancillary equipment

Ensure all the ancillary equipment that is needed is present, including intubation aids, facemasks of correct size, airway management devices. Check that laryngoscopes are working. Check that suction apparatus is working and that the bed or trolley can be rapidly tilted head down.

Monitoring

Ensure that appropriate monitoring equipment is present, switched on and calibrated and that alarm limits are set appropriately.

References

Adams, A. P. (1994) Symposium on mishap or negligence. Breathing system disconnections, *Br. J. Anaesth.*, **73**, pp. 46–54.

Association of Anaesthetists of Great Britain and Ireland (2004) *Checklist for Anaesthetic Apparatus*, Association of Anaesthetists, London.

Barthram, C. and McClymont, W. (1992) The use of a checklist for anaesthetic machines, *Anaesthesia*, **47**, pp. 1066–69.

Bhisman, N. R., Rout, C. C. and Murray, W. B. (1996) Laboratory Assessment of oxygen delivery from a portable chemical generator, *Anaesthesia*, **51**, pp. 1127–28.

Cicman, J. et al. (1993) Oxygen failure protection device. In *Operating Principles of Narkomed Anesthesia Systems*, North American Draeger, pp. 2–17.

Easy, W. R., Douglas, G. A. and Merrifield, A. J. (1988) A combined oxygen concentrator and compressed air unit with assessment of a prototype and discussion of potential applications, *Anaesthesia*, **43**, pp. 37–41.

Hall, L. W., Kellegher, K. E. B. and Fleet, K. J. (1986) A portable oxygen generator, Anaesthesia, **41**, 516–518.

Harris, C. E. and Simpson, P. J. (1985) The 'Mini-O_2' and 'Healthdyne' oxygen concentrators. Performance and potential application, *Anaesthesia*, **40**, pp. 1206–1209.

Hutton, P. and Boaden, R. W. (1986) Performance of needle valves, *Br. J. Anaes.*, **58**, pp. 919–24.

Jackson, I. J. B. and Wilson, R. J. T. (1993) Association of Anaesthetists checklist for anaesthetic machines; problem with detection of significant leaks, *Anaesthesia*, **48**, pp. 152–53.

Klopfenstein, C. E., Bernstein, M., van Gessel, E. and Forster, A. (1994) Preop checking of the anaesthetic machine. A survey in a University department of anaesthesiology, *Br. J. Anaesth.*, **72**(suppl 1), p. 11.

Li, Y. Y., Perera, S. P. and Crittenden, B. D. (1998) Zeolite monoliths for air separation, *Trans. I. Chem. E,* **76**(A), pp. 921–30.

McQuillan, P. J. and Jackson, I. J. B. (1987) Potential leaks from anaesthetic machines. Potential leaks through open Rotameter valves and empty cylinder yokes, *Anaesthesia,* **42**, pp. 1365–2044.

Medishield (1979) *General Information; Medical Gases and Cylinders,* Medishield.

Nunn, J. F. (1990) Carbon dioxide cylinders on anaesthetic apparatus, *Br. J. Anaesth.,* **65**, pp. 155–56.

Puri, G. D. *et al.* (1987) Awareness under anaesthesia due to defective gas-loaded regulator, *Anaesthesia,* **42**, pp. 539–40.

Weightmann, W. M. *et al.* (1992) Functionally crossed pipelines. An intermittent condition caused by a faulty ventilator, *Anaesthesia,* **47**, pp. 500–502.

Chapter 23

Airway management devices

> This chapter contains: the artificial airway; the facemask; the laryngeal mask airway; the I-gel; the cuffed oropharyngeal airway; the endotracheal tube. The chapter links with: Chapters 7, 22, 24, 25 and 26.

Introduction

The most important interface between the breathing system and the patient's lungs is an *airway management device* (AMD). Post-operatively it can be considered to be a means of delivering oxygen enriched air to the patient. Intraoperatively it is intended to secure the patient's airway, which might otherwise obstruct due to deep anaesthesia, to provide a reasonably gas tight seal to ensure accurate delivery of anaesthetic gases and, if necessary, to protect the lungs against aspiration of gastric contents. Postoperatively, the AMD can be nasal prongs or a variable performance mask, whose efficiencies may not be predictable [Wagstaff *et al.* 2007]. Intraoperatively it might be an artificial airway with a facemask, a supraglottic airway of one of the many types now available or an endotracheal tube (ETT). A supraglottic airway is one that sits in the pharynx or larynx above the vocal cords and these days is usually a laryngeal mask airway (LMA) of the numerous types now available, a cuffed oropharyngeal airway (COPA), or a Combitube. The LMA types available consist of: the classical LMA; the flexible (reinforced) LMA with a flexible tube to the breathing system; the 'Proseal', which has a gastric drainage tube as well as a gas transport tube; the intubating LMA, a device with a rigid right angled tube that acts as a ventilation conduit in the usual way, but through which an endotracheal tube may also be blindly introduced into the trachea; the 'I-gel' which has a gastric and a respiratory port as does the Proseal, but is less bulky, and whose bowl does not require inflation with air, but is filled with a gel that expands with body heat to form a seal. These days, almost all devices are made of material that excludes latex, but care should be taken to ensure this is indeed the case when there is a latex sensitive patient. Depending on the exact surgical and anaesthetic circumstances, the anaesthetist's experience and equipment availability, a choice is made between these devices to secure the airway for

Fig. 23.1 The Guedel oropharyngeal airway; sizes 0, 1 and 4 are shown.

a given operation. Additionally, there are other devices available to assist in securing the airway, such as the laryngoscope, the fibre optic bronchoscope and the cricothyrotomy tube.

The artificial airway

This is often an oropharyngeal device, shown in Figure 23.1, designed to prevent the tongue from obstructing the pharynx, to provide an elliptically shaped patent airway for gases to flow and to provide a route for suction of the pharynx. It is important that a suitable size is chosen for efficacy, from size 000 (smallest for neonates), through sizes 00, 0, 1, 2, 3, to size 4 (biggest for adult males). Its use does not preclude the need for effective manual airway management. It is usually made of a polyvinyl chloride (PVC) and problems with the Guedel airway include damage to bridged or capped teeth and intolerance in a patient with inadequately attenuated upper airway reflexes. Some oropharyngeal airways are designed to allow the introduction of a fibre optic bronchoscope into the mouth (see Chapter 24); they are also constructed as two longitudinally separable halves, so that an endotracheal tube can be subsequently introduced over the fibre optic bronchoscope into the trachea.

Alternatively a nasopharyngeal airway can be inserted through the nose. It is of circular cross-section, made of soft polyurethane, and its tip reaches the pharynx (Figure 23.2). Usually one size down from the equivalent endotracheal tube size is used, e.g. size 6 or 7. Epistaxis is a potential problem. Artificial airways are usually used in conjunction with a facemask.

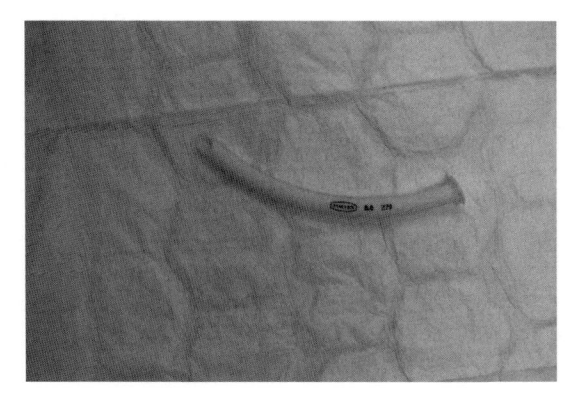

Fig. 23.2 A nasopharyngeal airway.

The facemask

The facemask is designed to fit snugly over the patient's face, and the freedom from gas leak is maintained by an adjustable air-filled cushion around its edge. The mask body is made of a rigid PVC, often transparent to allow visualisation of the exhaled breath, secretions or vomit. The apex of the mask contains the means of connecting the catheter mount or angle piece for connecting the mask to the breathing system. Some facemasks have a separate or combined port to allow introduction of a fibre optic laryngoscope during anaesthesia. In an anaesthetic setting, the mask is usually applied gently but efficaciously to the patient's face but, in an ICU setting, where continuous positive airway pressure is required in a spontaneously breathing patient, the mask is tightly applied to the face with straps round the back of the head. In neonatal facemask anaesthesia, it is necessary to minimise the dead space in a mask, and the *Rendell-Baker mask* achieves this.

As indicated earlier, masks for use in the recovery ward or on other wards do not attempt to have a gas tight seal, rather the reverse. A 3 l min^{-1} supply of oxygen can be delivered to an *MC mask* (Mary Catterall) so that the mask acts as an oxygen reservoir, but it also has some holes in the side that act as orifices to entrain air during peak inspiration and to allow excretion of alveolar gas during exhalation. This is a variable performance device, since the precise inhaled concentration of oxygen depends on the patient's peak inspiratory flow. A slightly more accurate variant on this is the so-called *Venturi mask*, which allows a precise amount of air to be entrained through a manually adjustable orifice, which will yield a more accurately metered level of inspired oxygen. The performance of such devices can be analysed [Dorrington 1989], and unless the mask contains a diffuser or some other device resembling a Venturi tube, then it acts by entrainment rather than by the Venturi principle. There is a choice of devices for managing the airway in the immediate postoperative phase, in order to optimise oxygenation. These include the laryngeal mask airway (LMA), which can be left *in situ* post-operatively until the patient wakes up, the MC facemask, and nasal prongs, all of which are capable of providing adequate oxygenation [Stausholm *et al.* 1995, Peyton *et al.* 2000, Wilkes *et al.* 1999, Martins *et al.* 1997, McBrien *et al.* 1995]. Extreme care must be taken when connecting a high pressure oxygen source to any device attached to the patient's airway to ensure there is a low impedance conduit for exhalation [Newton 1991].

The laryngeal mask airway

The laryngeal mask airway (LMA), developed by Dr Archie Brain in the 1980s and in use in the UK since then, provides a means of securing the patient's airway, leaving the anaesthetist's hands free from holding a facemask, without the need for endotracheal intubation. The functionally important part of the device is an oval bowl shaped structure with an inflatable cuff around the rim, which is introduced into the pharynx. The cuff ensures that the pharyngeal structures remain open and that the airway is surrounded by a relatively gas-tight seal, although that is not guaranteed. The cuff at the tip of the LMA sits over the oesophagus, partially occluding it. The fenestrated lumen in the centre of the bowl, which sits over the glottis, is continuous with a tube connecting the bowl to the outside of the mouth; the tube is the means of introducing the LMA through the mouth and of connecting the LMA to the breathing system. LMAs come in sizes 1 (neonates to 6.5 kg), 1½ (infants to 10 kg), 2 (children to 20 kg), 2½ (to 30 kg), and 3, 4 and 5 (small adults to large adults). Each size requires a maximum volume of air to be inflated into the cuff of $10(n-1)$ ml, where n is the LMA size; the exception to this is size 1, which requires 4 ml of air. Inserting the LMA is achieved by sliding the device through the mouth with the cuff deflated, maintaining bowl contact with the palate while doing so. Sometimes the tip of the cuff gets caught at the right angled turn of the pharynx, or the insertion process allows the epiglottis to become folded down to obstruct the glottis. Otherwise there is no positive endpoint to indicate successful insertion, other than watching the quality of the ventilation. If there is any doubt about the success of the insertion there should be a low clinical threshold for removing the LMA and trying again [Broderick *et al.* 1989]. The LMA lends itself to the introduction of a fibre optic laryngoscope into the tube to just beyond the bowl in order to inspect the glottis. An example of the LMA is shown in Figures 23.3(a) and (b).

THE LARYNGEAL MASK AIRWAY | 305

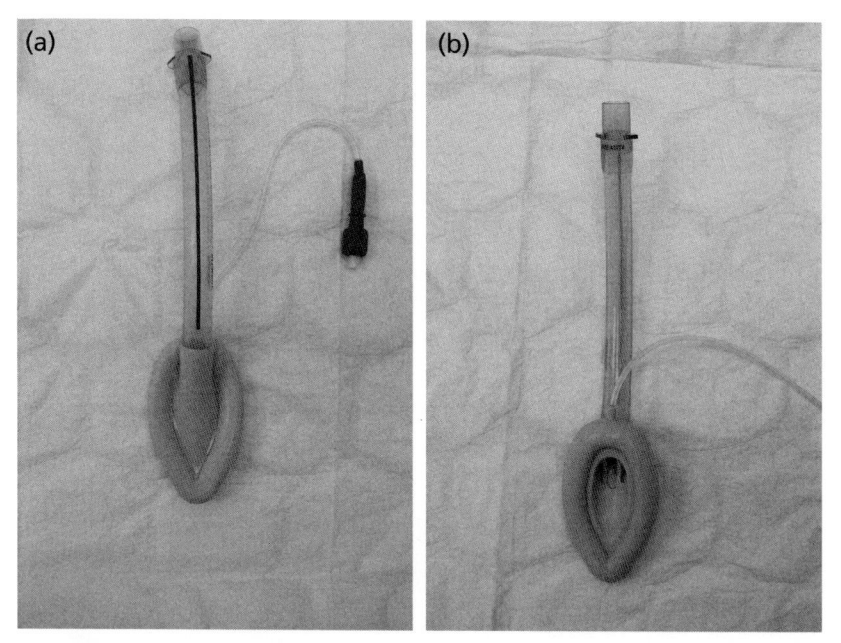

Fig. 23.3 The classical laryngeal mask airway (LMA): (a) dorsal view. (b) ventral view.

There are variations on the standard LMA. The *reinforced LMA* (flexible LMA) has a wire reinforced tube, which is rather narrower than the tube on the classical LMA, and which provides greater flexibility while preventing the tube from kinking. This is useful when the operative site is in or near the mouth itself. However these LMAs may require temporary reinforcement of the tube to aid LMA insertion. Such a tube can be used for airway laser surgery [Lockey *et al.* 1997].

The *intubating LMA* (ILMA) [Baskett *et al.* 1998, Brain *et al.* 1997] is a device with a rigid right angled tube attached to the bowl, shown in Figure 23.4. The tube is made of low friction metal, so that an endotracheal tube can be introduced inside the tube and inserted through the glottis [Asai *et al.* 1993].

The *Proseal LMA* [Brain *et al.* 1995] essentially has two tubes in the same mould connected to the bowl; one has a fenestrated lumen in the centre of the bowl that sits opposite the glottis for ventilation as normal; the other tube has a lumen that goes to a point on the tip of the cuff, to sit opposite the oesophagus and allow gastric drainage. This makes the Proseal a rather bulkier device to insert than the classic LMA, and it usually needs a metal scoop to aid insertion. The device is shown in Figures 23.5(a) and (b).

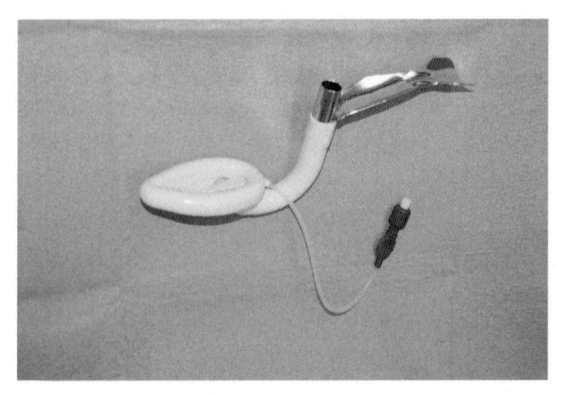

Fig. 23.4 The intubating LMA (ILMA), lateral view.

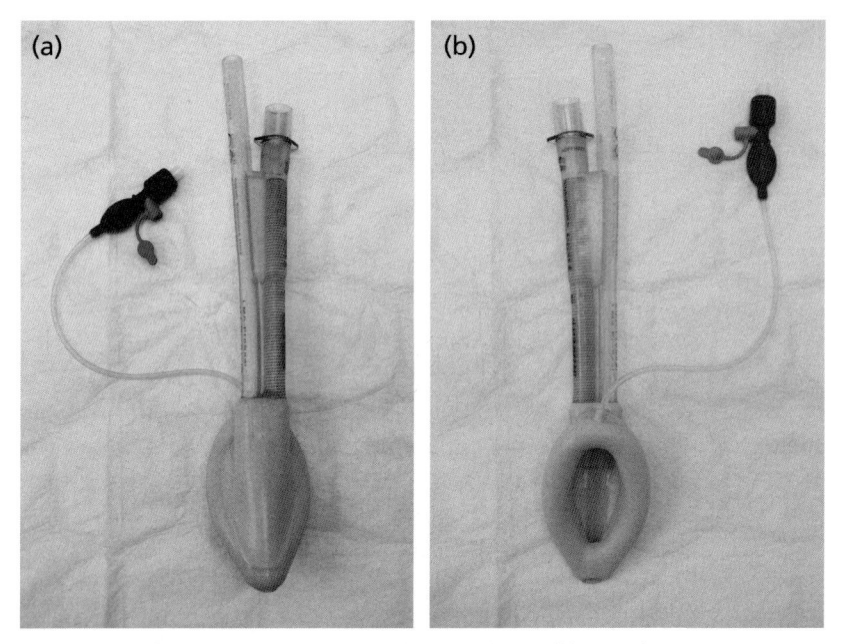

Fig. 23.5 The Proseal LMA: (a) dorsal view, (b) ventral view.

The LMA is often used for patients breathing spontaneously, although there is some increase in airway resistance [Righini et al. 1995, Bhatt et al. 1992] compared with a facemask. However it can also be used for artificial ventilation, although it should not be assumed to provide the same protection against aspiration of gastric contents as an endotracheal tube, even if a Proseal is used. Furthermore, the seal provided by the bowl cuff may not be adequate to ventilate the patient if airway pressures are high, such as during laparoscopic surgery. XIIth cranial nerve damage from the cuff has been reported [King *et al.* 1994].

The I-gel

The I-gel is a device that is functionally like a Proseal in that it acts like an LMA, sitting in the oropharynx, and also has a gastric drainage port, shown in Figure 23.6(a) and (b). It is of a more compact and rigid construction than a Proseal, and can be introduced without the aid of a scoop. In particular, its sealing cuff does not require inflation, but contains a gel within it that expands on exposure to body heat, resulting in an effective seal for positive pressure ventilation where moderate pressures are involved [Uppal *et al.* 2009].

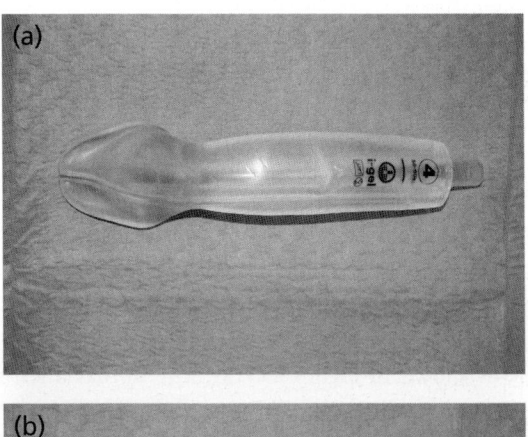

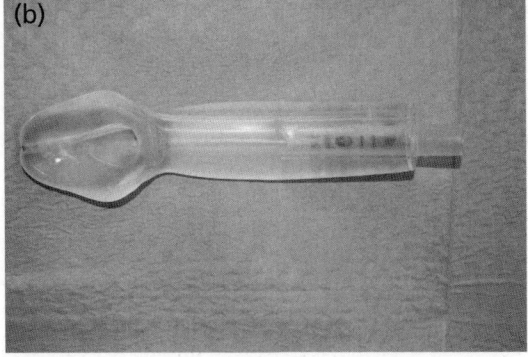

Fig. 23.6 The I-gel airway: (a) dorsal view, (b) ventral view.

The cuffed oropharyngeal airway

These are devices with some functional similarities to the classic LMA and Proseal LMA. The cuffed oropharyngeal airway (COPA) [Rees et al. 1999] consists of a ventilating tube, introduced into the pharynx through the mouth; it has a single circumferential cuff to keep it centrally placed over the glottis.

The *Combitube* [Hartmann et al. 2000] is a double lumen, double cuff device, designed with one lumen at its tip and the other lumen being circumferentially placed around the tube, near, but not at its tip; the two lumina are functionally separated on the outside of the device by the distal cuff. The proximal cuff is to keep the tube approximately in the middle of the pharynx. The tube is inserted and if the distal lumen goes into the glottis, the circumferential lumen can be used for gastric aspiration. If the distal lumen ends up in the oesophagus, the distal cuff functionally separates the oesophagus from the glottis and the proximal lumen can be used to ventilate the patient; a gastight seal for this purpose is assured by the proximal cuff. Some device for detecting which tube was fulfilling which function would be required, such as a Wee oesophageal detector (see later section).

The endotracheal tube

When absolute security from gastric aspiration [Oikkonen et al. 1997], or muscle relaxation and artificial ventilation are required, endotracheal intubation is still the anaesthetists' choice. Endotracheal tubes (ETTs) are introduced through the mouth or the nose, usually using a laryngoscope. The orotracheal tube is usually about 30 cm long if left uncut; these days it is disposable and made from clear PVC rather than red rubber, with a curvature down its length that corresponds to the anatomical curvature from the mouth to the trachea, as shown in Figure 23.7(a). The tip has a left handed bevel to allow easier visual insertion and there may be an additional lumen on the opposite side, a *Murphy's eye*, to ensure that gas can pass through the tube if the bevelled end is obstructed. The size by which the tubes are described is the internal diameter in cm and ranges from 2.5 cm for a neonatal patient to 9.0 cm for a large male patient. The size necessary for a given patient is dictated by the need to pass the tube easily through the narrowest part of the airway, which is the glottis in an adult and the cricoid cartilage in a child. The larger the tube used, the lower will be the resistance to gas flow through the tube, but damage to the vocal cords and the tracheal mucosa must be avoided. In general, cuffed tubes are used in adults to ensure a leak free seal to the airway, and uncuffed tubes are used in neonates and infants to avoid tracheal mucosal damage (Figure 23.8).

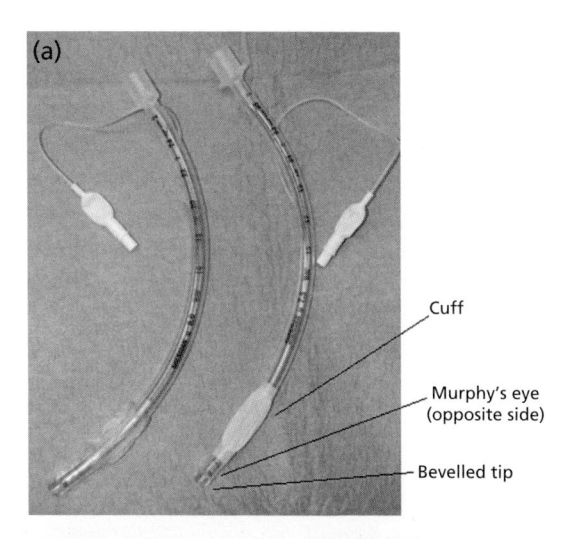

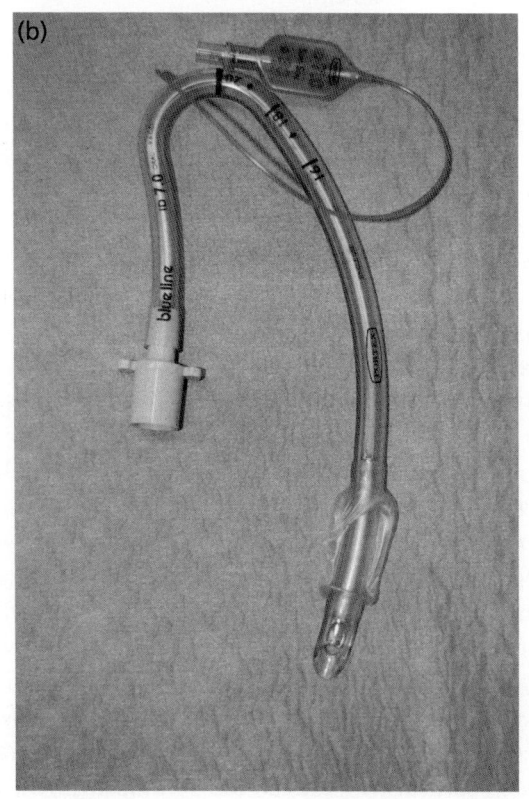

Fig. 23.7 (a) The standard cuffed, adult endotracheal tube; (b) the preformed, downward facing Rae tube.

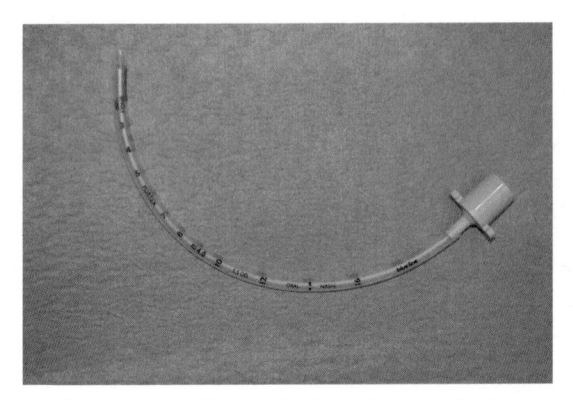

Fig. 23.8 Uncuffed paediatric endotracheal tubes.

In children the tube size is given by

$$\frac{age \text{ yr}}{4} + 4.5 \text{ cm},$$

Although $\{(age/4) + 4.0\}$ cm often works better (personal experience of clinical author). In neonatal anaesthesia, or where there is a need to use a particularly narrow ETT in an adult, such as a micro laryngeal tube in pharyngoscopy, thought must given to the fact that the airway resistance down such a tube is extremely high, being inversely proportional to the fourth power of the radius (see Chapter 7). In such a situation, tube design is very important [Bleu et al. 1985] and adequate time must be given to allow for full expiration of the tidal volume; furthermore, it should be recognised that high peak measured airway pressure does not necessarily mean high alveolar pressure. Similarly, if any sort of device is introduced down the length of an ETT, such as a fibre optic bronchoscope or a suction catheter, there will be a dramatic increase in airway resistance [Magee 2008].

The wish to avoid causing ischaemic damage to the tracheal mucosa is the reason that a leak is tolerated around the tubes used in neonates and children. However in recent years there has been an increasing use of cuffed endotracheal tubes in children. A reliable seal with lack of stridor due to tracheal mucosal oedema after extubation has been demonstrated [Weiss et al. 2009]. In adults, an inflatable cuff around the distal end of the ETT is always used to ensure freedom from gas leak and absolute security from gastric contents entering the trachea. Old red rubber tubes had low volume high pressure cuffs, which had a small area of contact with the trachea and might therefore have caused high contact pressures, particularly if the cuff was over inflated. When autoclaved multiple times, the old cuffs also had a tendency to herniate and obstruct the end of the ETT. Modern ETTs have large volume low pressure cuffs, designed to have large contact areas and low contact pressures with the mucosa, although these too can cause damage if over inflated [Messahel *et al.* 1994]. The pressure that the cuff exerts on the tracheal mucosa is a function of the cuff material elasticity as well as the cuff pressure [Magee 1991]. Ideally the cuff pressure should be measured, monitored and maintained at no more than 2.5 kPa (25 cmH_2O) [Miller 1992, Morris *et al.* 1985]. Automatic cuff pressure controllers that make rapid changes to cuff pressure have been found to reduce the integrity of the seal [Weiss *et al.* 2008]. The cuffs are normally filled with air through a narrow cuff inflation tube that also has a pilot balloon, but will enlarge if nitrous oxide is absorbed into the cuff volume, unless it is prefilled with nitrous oxide instead [Karasawa *et al.* 2001]. Another problem with air filled ETT cuffs is the change in their volume according to Boyle's law with change in altitude when an intubated patient is being transported by air. A solution to both these problems is to fill the cuff with saline [Smith *et al.* 2002] or to use a foam-filled cuff [Power 1990]. This is also a solution to prevent an airway fire during airway laser surgery. There are ETTs whose pilot balloons have a secondary latex pilot balloon within them; these provide an extra reservoir for additional gas being absorbed within the cuff, thereby preventing the ETT cuff from enlarging excessively.

The length of an ETT is dictated by the need for the tip to reach into the trachea, but to avoid being so long as to enter a main bronchus, usually the right, and also to avoid adding unnecessarily to airway resistance. The correct length of tube is judged visually on laryngoscopy. In adults the correct length of the tube in the trachea is usually between 20 and 25 cm. As a guide, the length of an ETT in children is given by

$$\frac{age \text{ yr}}{2} + 11.0 \text{ cm}.$$

However, the adequacy of ventilation in both lungs should be checked visually and by auscultation.

Nasotracheal tubes are naturally longer than orotracheal tubes and the tubes are often made of soft polyurethane or silicone rubber in order to cause less trauma to the nasal mucosa. It is usually necessary to use a nasal tube one size smaller than the equivalent oral tube.

Any PVC ETT can kink as it heats up to body temperature. Care must be taken in securing the tube to a catheter mount to avoid this and this connection should be visible where possible. There are preformed, specially shaped ETTs to try to minimise the chance of kinking, such as RAE tube (see Figure 23.7(b)). Alternatively, the ETT can be reinforced with a wire coil down its length.

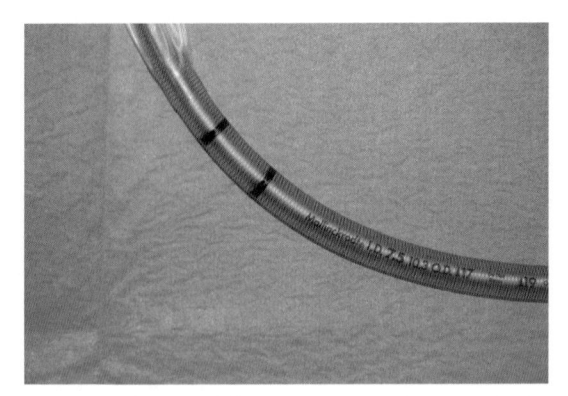

Fig. 23.9 A flexo-metallic endotracheal tube for use in the presence of lasers.

Methods of confirming correct placement of an ETT include visual and auscultatory; the CO_2 analyser, pulse oximetry, pH indicator [Gedeon et al. 1995] and the Wee detector, which consists of a tube attached at one end to a large rubber bulb. The bulb is initially squeezed to evacuate it and when the other end is attached to the visible end of the ETT, if the bulb reinflates, the ETT is in the trachea; if it remains deflated, it is in the oesophagus. Another detection device is the SCOTI device (*Sonomatic Confirmation of Tracheal Intubation*), which differentiates the acoustic properties of the trachea and the oesophagus [Nandwani et al. 1996, Murray et al. 1995]. A comparative study showed the Wee device to be more reliable than the SCOTI [Lockey et al. 1997].

Other ETTs for special circumstances include tubes for use in airway laser surgery, which have a metallic surface incorporated in the wall in order to reflect the laser beam away from the tube and prevent the tube material from igniting in the oxygen or nitrous oxide environment [Hunton et al. 1988], see Figure 23.9; there are often two cuffs in case one is damaged and rendered useless by the laser.

Tracheostomy tubes are much shorter than ETTs and have a non-bevelled end, with the cuff closer to the end, and a right-angled bend. It is possible to use a percutaneous dilatational technique to insert a tracheostomy tube on ICU [Oldroyd et al. 1995]. A variant, used for emergency access to the airway through the tracheal wall, is the cricothyrotomy tube [Patel et al. 2008]. However, care must be taken in ventilating patients through high pressure oxygen sources, that surgical emphysema is not inadvertently caused by malposition of the cricothyrotomy needle. Care must also be taken to ensure that there is an open pathway for expiration if pulmonary barotrauma is to be avoided.

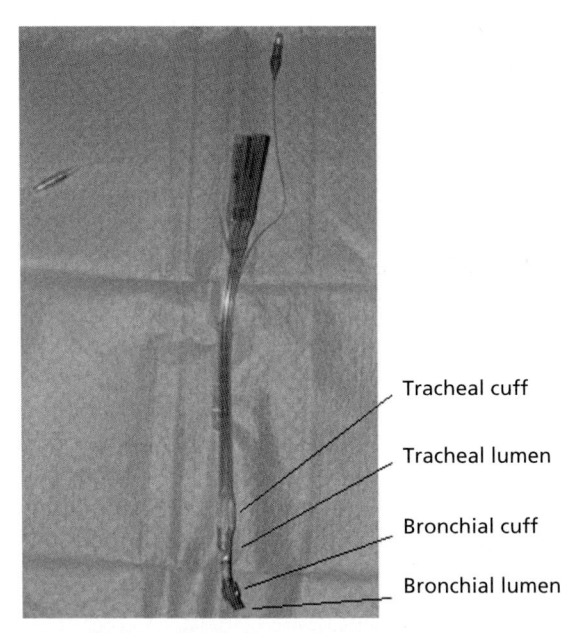

Fig. 23.10 A typical left sided double lumen tube, ventral view.

Double lumen tubes, for use in thoracic surgery where the two lungs need to be functionally separated, are made to be right or left sided, have a tracheal lumen and an endobronchial lumen and a carinal hook to aid correct placement. The right sided tubes have an additional bronchial opening to ensure adequate ventilation of the right upper lobe of the lung. Because the two lumina are constructed back to back, they are D-shaped and the sizes are described in French gauge, usually available from FG 35 to 41 (see Figure 23.10) Each lumen has different flow characteristics [Chiaranda *et al.* 1989]. When inserting a double lumen tube, the tube must be inserted with the endobronchial end facing forward initially, then rotated appropriately. Then the cuffs must be inflated separately and sequentially and finally together to confirm each lumen is where it is intended to be. Fibre optic bronchoscopy is the best method to confirm appropriate placement of a double lumen tube. Excessive inflation of the bronchial cuff can be a cause for concern.

References

Asai, T., Latto, I. P. and Vaughan, R. S. (1993) Distance between the grill of the LMA and the vocal cords, *Anaesthesia*, **48**, pp. 667–69.

Baskett, P. J. F., Parr, M. J. A. and Nolan, J. P. (1998) The intubating laryngeal mask. Results of a multicentre trial of experience of 500 cases, *Anaesthesia*, **53**, pp. 1174–79.

Bhatt, S. B., Kendall, A. P., Lin, E. S. and Oh, T. E. (1992) Resistance and additional inspiratory imposed by LMA, *Anaesthesia*, **47**, pp. 343–47.

Bleu, H., Rytlander, M. and Wisborg, T. (1985) Resistance of tracheal tubes 3.0 and 3.5 mm ID, Comparison of four commonly used types, *Anaesthesia*, **40**, pp. 885–88.

Brain, A. I. J. *et al.* (1997) The ILMA: development of a new device for intubation of the trachea, *Br. J. Anaesth.*, **79**, pp. 699–703.

Brain, A. I. J., Verghese, C. *et al.* (1995) A new laryngeal mask prototype; a preliminary evaluation of seal pressures and glottic isolation, *Anaesthesia*, **50**, pp. 42–48.

Broderick, P. M. *et al.* (1989) The LMA. One hundred patients spontaneously breathing, *Anaesthesia*, **44**, pp. 238–41.

Chiaranda, M. *et al.* (1989) Measurement of flow resistive properties of double lumen bronchial tubes in vitro, *Anaesthesia*, **44**, pp. 335–40.

Dorrington, K. L. (1989) Medical breathing systems. Chapter 2, in *Anaesthetic and Extracorporeal Gas Transfer*, Oxford Medical Engineering Series 9, Oxford Science Publications, Oxford, pp. 31–86.

Gedeon, A., Knill, P. and Mobius, C. (1994) A new colorimetric breath indicator. A comparison of the performance of two CO_2 indicators, *Anaesthesia*, **49**, pp. 798–803.

Hartmann, T. *et al.* (2000) The oesophageal-tracheal Combitube small adult. An alternative airway for ventilatory support during gynaecological laparoscopy, *Anaesthesia*, **55**, pp. 670–75.

Hunton, J. and Oswal, V. H. (1988) Anaesthesia for CO_2 laser laryngeal surgery in infants, *Anaesthesia*, **43**, pp. 394–96.

Karasawa, F. *et al.* (2001) An assessment of a method of inflating cuffs with a nitrous oxide gas mixture to prevent n increase in intracuff pressure in five different tracheal tube designs, *Anaesthesia*, **56**, pp. 155–59.

King, C. and Street, M. K. (1994) Twelfth cranial paralysis following use of the laryngeal mask airway, *Anaesthesia*, **49**, 786–87.

Lockey D. J. and Woodward, W. (1997) SCOTI versus Wee. An assessment of two oesophageal intubation detection devices, *Anaesthesia*, **52**, pp. 242–43.

Magee, P. T. (1991) Endobronchial cuff pressures of double lumen tubes, *Anesth. Analg,.* **72**, pp. 265–66.

Magee, P. T. (2008) A study of gas flow in coaxial tubes, *Anaesthesia*, **63**, pp. 905–906.

Martins, J. and Brimacombe, J. R. (1997) Oxygen enrichment during emergence with the LMA-T-Bag vs. T piece, *Anaesthesia*, **52**, 1195–98.

McBrien, M. E. and Sellars, W. F. S. (1995) A comparison of three variable performance devices for postoperative oxygen treatment, *Anaesthesia*, **50**, pp. 136–38.

Messahel, B. F. *et al.* (1994) Total tracheal obliteration after intubation with low pressure cuffed endotracheal tube, *Br. J. Anaesth.*, **73**, pp. 697–99.

Miller, D. M. (1992) A pressure regulator for the cuff of a tracheal tube, *Anaesthesia,* **47**, pp. 594–96.

Morris, J. V. and Latto, I. P. (1985) An electropneumatic instrument for measuring and controlling the pressures in the cuffs of tracheal tubes: 'the Cardiff cuff controller', *J. Med. Eng. Tech.*, **9**, pp. 229–30.

Murray, D., Ward, M. E. and Sear, J. W. (1995) SCOTI – a new device for identification of tracheal intubation. *Anaesthesia*, **50**, pp. 1062–64.

Nandwani, N. *et al.* (1996) Configuration of the SCOTI device with different tracheal tubes, *Anaesthesia*, **51**, pp. 932–34.

Newton, N. I. (1991) Supplementary oxygen – a potential for disaster (editorial), *Anaesthesia*, **46**, pp. 905–906.

Oikkonen, M. and Aromaa, U. (1997) Leakage of fluid around low-pressure tracheal tube cuffs, *Anaesthesia*, **52**, pp. 567–69.

Oldroyd, G. J. and Bodenham, A. R. (1995) An evaluation of a new percutaneous tracheostomy kit, *Anaesthesia*, **50**, pp. 49–51.

Pandit, L. J., Chambers, P. and O'Malley, S. (1997) KTP laser resistant properties of the reinforced LMA, *Br. J. Anaesth.*, **78**, pp. 594–600.

Patel, B. and Frerk, C. (2008) Large bore cricothyrotomy devices, *Br. J. Anaesth.* Continuing Education Series, **8**(5), pp. 157–60.

Peyton, P., Cowie, D. and Howard, W. (2000) Supplementary oxygenation with the LMA: a comparison of four devices, *Anaesthesia*, **55**, pp. 992–99.

Power, K. J. (1990) Foam cuffed tracheal tubes. Clinical and laboratory assessment, *Br. J. Anaesth.*, **65**, pp. 433–37.

Rees, S. G. O. and Gabbott, D. A. (1999) Use of the oropharyngeal airway for manual ventilation by non-anaesthetists, *Anaesthesia*, **54**, pp. 1089–93.

Righini, E. R. *et al.* (1997) Additional inspiratory resistance imposed by the LMA: in vitro vs. in vivo comparison, *Anaesthesia*, **52**, pp. 872–78.

Smith, R. P. R. and McArdle, B. H. (2002) Pressure in cuffs of tracheal tubes at altitude, *Anaesthesia*, **57**, pp. 374–78.

Stausholm, K. *et al.* (1995) Comparison of three devices for oxygen administration in the postoperative period, *Br. J. Anaesth.*, **74**, 607–609.

Upal, V., Fletcher, G. and Kinsella, J. (2009) Comparison of I-gel with a cuffed endotracheal tube during pressure controlled ventilation, *Br. J. Anaesth.*, **102**, pp. 264–68.

Wagstaff, T. and Soni, N. (2007) Performance of six types of oxygen delivery devices at varying respiratory rates, *Anaesthesia*, **62**, pp. 492–503.

Wee, M. Y. K. (1988) The oesophageal detector device. Assessment of a new method to distinguish oesophageal from tracheal intubation, *Anaesthesia*, **43**, pp. 27–29.

Weiss, M., Dullenkopf, A., Fischer, J. *et al.* (2009) Prospective randomised controlled multi-centre trial of cuffed or uncuffed endotracheal tubes in small children, *Br. J. Anaesth*,. **103**, pp. 867–73.

Wilkes, A. R. and Vaughan, R. S. (1999) The use of breathing system filters as oxygen delivery devices, *Anaesthesia*, **54**, pp. 552–58.

Weiss, M., Doell, C., Koepfer, N. *et al.* (2009) Rapid pressure compensation by automated cuff pressure controllers worsens sealing in tracheal tubes, *Br. J. Anaesth.*, **102**, pp. 273–78.

Chapter 24

Aids to intubation

> This chapter contains: the laryngoscope; fibre optics; bougies and catheters.
> The chapter links with: Chapter 23.

The laryngoscope

This device was invented by Sir Ivan Magill and Sir Robert Macintosh to visualise the vocal cords to aid intubation. The curved blade of the Macintosh laryngoscope is still popular as the standard and its design has been reshaped in recent years to reduce the biomechanical forces on the teeth [Bucx *et al.* 1997, Bucx *et al.* 1994]. The straight blade of the Magill laryngoscope can make the view easier under some circumstances, as do modern variants [Henderson 1997]. The handle has a battery and a light source in it, and the light is transmitted by a fibre optic cable to the tip of the blade, which is usually at right angles to the handle. There are many different shaped blades to suit different circumstances, particularly to visualise the anatomically different airway of the neonate. Other variants to aid difficult intubations include the Polio laryngoscope, in which the angle between the handle and the blade is an obtuse one and the McCoy laryngoscope, which has a lever to manipulate the tip of the laryngoscope blade to improve the view of the vocal cords [McCoy *et al.* 1993]; some of these devices are shown in Figure 24.1.

In recent years a new range of laryngoscopes has been introduced with fibre optic systems that allow indirect visualisation of the vocal cords via an eyepiece or a small video screen. These devices clearly improve access to the airway where there would otherwise be a laryngoscopic view with a high Cormack-Lehane score. They currently include the McGrath [Ray *et al.* 2009], the Glidescope, the Airtraq [Lange *et al.* 2009] (shown in Figures 24.2(a) and (b)), the C-MAC [McElwain *et al.* 2010 and others] and testing to date has been on manikins. As with all new devices, their efficacy depends on the skill of the user; one study has demonstrated greater skill with the familiar Macintosh laryngoscope than with the newer ones [Powell *et al.* 2009].

With the emergence of Jakob–Creutzfeldt disease and the recognition of the infectious risk of prions from tonsillar tissue on laryngoscope blades, a range

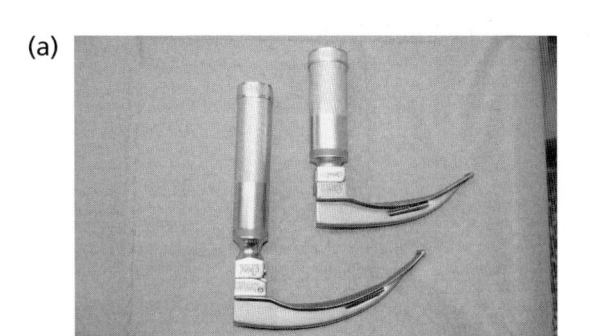

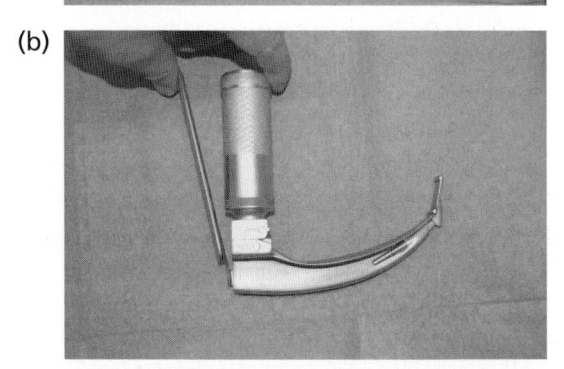

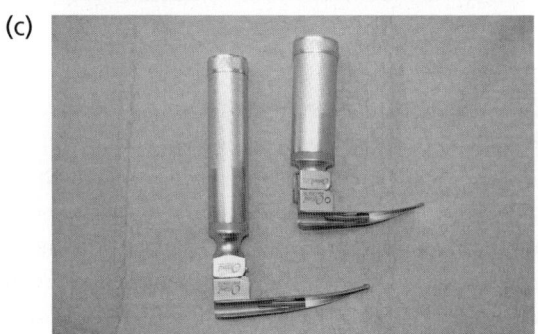

Fig. 24.1 Some different laryngoscopes and blades to aid endotracheal intubation; (a) Standard MacIntosh laryngoscopes, medium and long blades; (b) McCoy laryngoscope with adjustable blade tip; (c) two variants of blade for paediatric laryngoscopy.

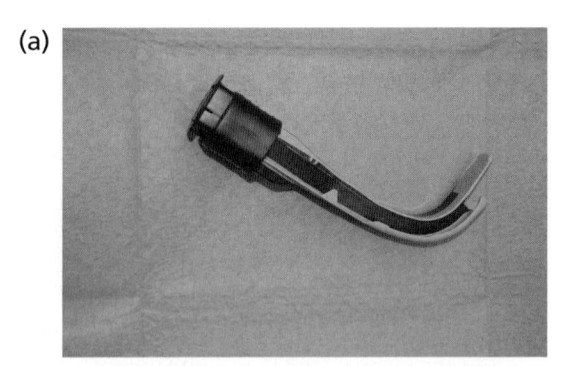

Fig. 24.2 Airtraq optical laryngoscope: (a) lateral view showing gutter for endotracheal tube placement; (b) proximal end showing optical eyepiece.

of disposable paediatric blades has been produced, with variable quality of mechanical flexibility and light quality [Goodwin *et al.* 2006].

Where the surgical procedure itself is laryngoscopy and the airway must be ventilated by a jet attached to the side of the laryngoscope, the laryngoscope is designed to allow a jet of high pressure oxygen at its proximal end to entrain a much larger volume of air into the lungs, which may or may not be by the Venturi effect, depending on the precise design of the attachment (see Chapter 7). This is usually done manually by the anaesthetist.

Fibre optics

Optical fibres can be used with lasers (see Chapter 29) to direct the beam, but they are also used in the design of endoscopes and bronchoscopes to permit the clinician to see around corners. Another medical use is for electrical isolation of the patient (as discussed in Chapter 6) and interference free transmission

FIBRE OPTICS | 319

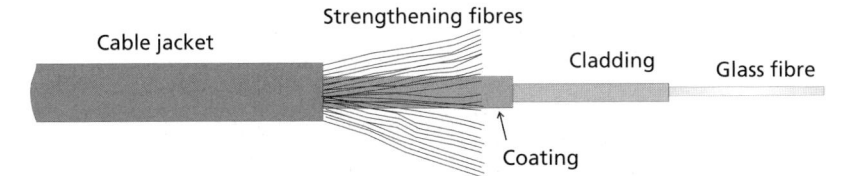

Fig. 24.3 The fibreoptical cable showing the coating, cladding and core.

of signals from an EEG head box to the main amplifier via an optical fibre link. In these cases the electrical signals are converted to modulated light via a photodiode, sent down the fibre and then converted back via a detector.

The fibre is shown in Figure 24.3. The light propagates along the fibre by the process of total internal reflection.

Reflection and refraction of light

When light travelling in a transparent material meets the surface of another transparent material, two things happen: some of the light is reflected, and some of the light is transmitted into the second transparent material.

The light which is transmitted usually changes direction when it enters the second material (e.g. the glass cladding of the optical fibre). This bending is called refraction and it depends upon the fact that light travels at one speed in one material and at a different speed in another. Therefore each material (glass core, cladding) has its own refractive index. In this case the light goes from a higher index (glass core) to a lower one (cladding). In Figure 24.4 θ_2 is always

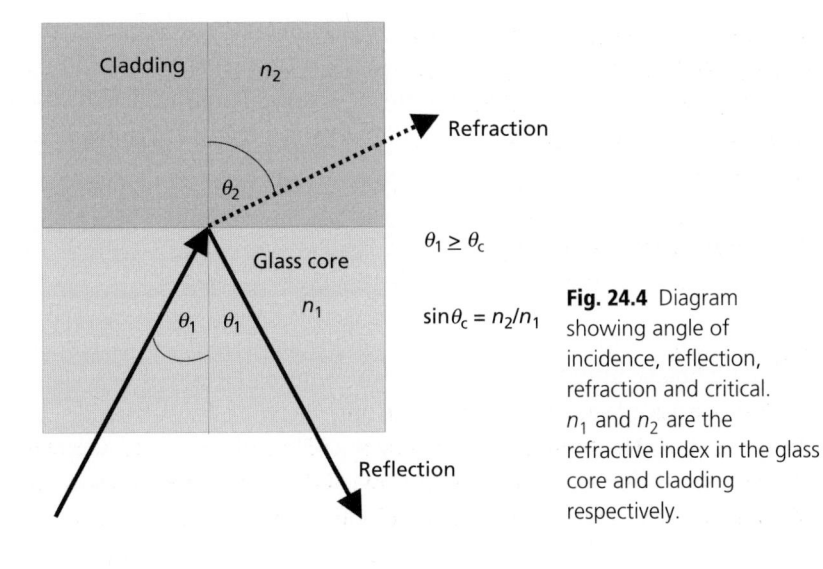

Fig. 24.4 Diagram showing angle of incidence, reflection, refraction and critical. n_1 and n_2 are the refractive index in the glass core and cladding respectively.

$\theta_1 \geq \theta_c$

$\sin\theta_c = n_2/n_1$

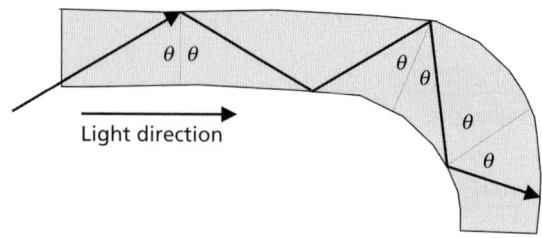

Fig. 24.5 The bending of optical fibres.

greater than θ_1, and at the critical angle θ_c, all the light will be reflected. As long as the angle of incidence is greater than the critical angle (which is set by the smallness of the glass core), then total internal reflection will occur. Even if the optical fibre is bent (to certain limits), then light will be transmitted along the tube as demonstrated in Figure 24.5.

Fibre optic intubation

The use of a fibre optic scope to aid awake intubation is well established. Where there is some doubt about the ability to visualise the vocal cords with orthodox laryngoscopy, or to maintain the integrity of the airway following intravenous induction of anaesthesia, awake intubation using an endotracheal tube railroaded on to the fibre optic laryngo- or broncho-scope is an extremely valuable technique. Fibre optic visualisation of the trachea and bronchi is also helpful following insertion of a double lumen tube to ensure correct positioning. It can also be used through an intubating or classical LMA or an I-gel to visualise the state of the glottis prior to intubation in an anaesthetised patient, and subsequently to access the trachea with a railroaded endotracheal tube [de Lloyd *et al.* 2010].

Bougies and catheters

These simple devices as aids to intubation should not be forgotten in the quest for technology. Frequently, when the view of the glottis is impaired, it is easy to insert a bougie into the trachea, over which the endotracheal tube can be railroaded, when there may be some difficulty inserting a tube directly. The bougie must be relatively new, with the right level of stiffness along its length, with an adjustable tip. Care must be taken not to damage mucosa as it is inserted. A Cook catheter is like a hollow bougie, which is easy to insert and through which the patient may be ventilated while awaiting the insertion of the railroaded endotracheal tube.

References

Bucx, M. J. L. *et al.* (1997) Reshaping the Macintosh blade using biomechanical remodelling. A prospective comparative study in patients, *Anaesthesia*, **52**, pp. 662–67.

Bucx, M. J. L., Sniders, C., van Geel, R. T. M. *et al.* (1994) Forces acting on the maxillary incisor teeth during laryngoscopy using the McIntosh laryngoscope, *Anaesthesia*, **49**, pp. 1064–70.

de Lloyd, L., Hodzovic, I., Voisey, S. *et al.* (2010) Comparison of fibrescope guided intubation via the classic laryngeal mask airway and I-gel in a manikin, *Anaesthesia*, **65**, pp. 36–43.

Goodwin, N., Wilkes, A. and Hall, J. (2006) Flexibility and light emission of disposable paediatric Miller 1 laryngoscopic blades, *Anaesthesia*, **61**, pp. 792–99.

Henderson, J. J. (1997) The use of the paraglossal straight blade laryngoscopy in difficult tracheal intubation, *Anaesthesia*, **52**, pp. 552–60.

Lange, M., Frommer, M., Redel, A. *et al.* (2009) Comparison of Glidescope and Airtraq optical laryngoscopes in patients undergoing direct microlaryngoscopy, *Anaesthesia*, **64**, pp. 323–28.

McCoy, E. P. and Mirakhur, R. K. (1993) The levering laryngoscope, *Anaesthesia*, **48**, 1993, pp. 516–519.

McElwain, J., Malik, M., Harte, B. *et al.* (2010) Comparison of the C-MAC videolaryngoscope with the MacIntosh, Glidescope and Airtraq laryngoscopes in easy and difficult laryngoscopy scenarios in manikins, *Anaesthesia*, **65**, pp. 483–89.

Powell, L., Andrzejowski, J., Taylor, R. and Turnbull, D. (2009) Comparison of four laryngoscopes in a high fidelity simulator using normal and difficult airway, *Br. J. Anaesth.*, **103**, pp. 755–60.

Ray, D. C., Billington, C., Kearns, P. K. *et al.* (2009) A comparison of McGrath and Macintosh laryngoscopes in novice users: a manikin study, *Anaesthesia*, **64**, pp. 1207–1210.

Chapter 25

Breathing systems

This chapter contains: the Mapleson classification of semi-closed rebreathing systems; the Humphrey ADE system; Venturi systems; the circle system. The chapter links with: Chapters 7, 11, 14, 15, 16 and 26.

Introduction

An anaesthetic breathing system is a means of transferring the breathing gas mixture from the anaesthetic machine common gas outlet to the patient. It is also the means of transferring the exhaled gas from the patient to the outside world, usually via a scavenging system. Alternatively, after the carbon dioxide is absorbed from the exhaled gas, the unused fresh gas components of the exhaled gas are recirculated back to the patient. In general, a breathing system consists of a fresh gas limb, an inspiratory and expiratory limb, an expiratory valve, a reservoir bag and it may also consist of one or more unidirectional valves and a CO_2 absorber. The simpler devices have fewer components and usually involve some rebreathing of expiratory gas, depending on the level of fresh gas flow. The ability to minimise rebreathing at as economical a fresh gas flow as possible is a measure of the breathing system's efficiency. Depending on the precise design of the breathing system, such efficiency will vary depending on whether the patient is breathing spontaneously or is undergoing controlled artificial ventilation (see Chapter 26). The more complex systems ensure minimum rebreathing by the use of unidirectional valves and CO_2 absorption systems; in this way, the additional complexity allows more economical use of fresh gas and volatile agent.

The systems that use higher fresh gas flows (FGF) and involve some rebreathing were classified in 1954 by Professor Mapleson, according to their behaviour in terms of the FGF requirement to prevent CO_2 rebreathing [Mapleson 1954]. At the time and for three decades beyond, they were the most popular breathing systems in UK anaesthetic practice. The *Mapleson Classification* of rebreathing systems is shown in Figure 25.1. Their design lends their structure and function to mathematical analysis [Dorrington 1989].

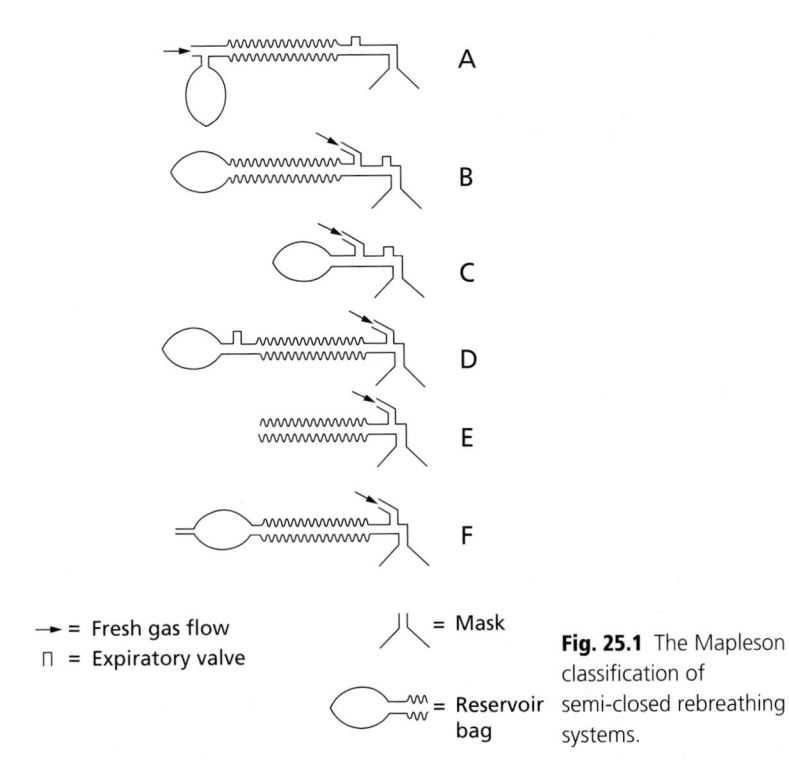

Fig. 25.1 The Mapleson classification of semi-closed rebreathing systems.

The Mapleson classification of semi-closed rebreathing systems

Mapleson A systems

The *Magill* breathing system was invented by Sir Ivan Whiteside Magill in the early twentieth century. As shown in Figure 25.1A and Figure 25.2, the system is characterised by having the expiratory valve close to the patient and the fresh gas inflow at a distance from the patient, but close to the reservoir bag. Because of this particular configuration, the system is very economical in spontaneous breathing. On inspiration fresh gas is stored in the reservoir bag and travels down the tubing towards the patient. The expiratory valve is closed at this point of the respiratory cycle and the deadspace of the system contains a small amount of alveolar (CO_2 containing) gas; the system deadspace lies between the patient's mouth and the expiratory valve. This is the first gas to be inhaled on the next inspiration. The last gas to be inhaled on inspiration is fresh gas, which therefore resides in the patient's own deadspace and becomes the first gas to be exhaled on expiration. On early expiration the pressure in the

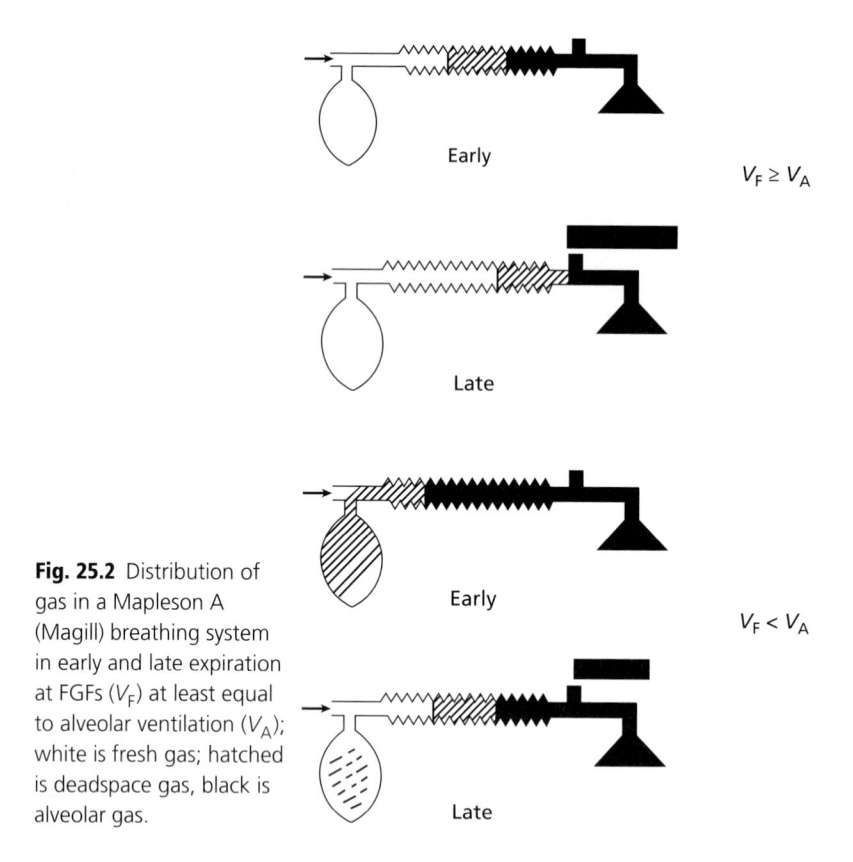

Fig. 25.2 Distribution of gas in a Mapleson A (Magill) breathing system in early and late expiration at FGFs (V_F) at least equal to alveolar ventilation (V_A); white is fresh gas; hatched is deadspace gas, black is alveolar gas.

breathing system has not risen adequately for the expiratory valve to open and the fresh gas from the patient's deadspace is stored in the tubing. As expiration proceeds, the expiratory valve opens and alveolar gas is preferentially exhaled. In the meantime, fresh gas from the gas supply is being stored in the tube and reservoir bag. This means that relatively low fresh gas flows can be used while keeping rebreathing to a minimum, which also reduces the likelihood of spilling fresh gas from the valve. In fact, no significant rebreathing occurs when the FGF falls to as low as 70% of the patient's minute ventilation, and this efficiency is unrelated to the respiratory pattern [Cook 1996a, b]. If the FGF falls any lower, then more of the expired alveolar gas is stored in the tubing for subsequent rebreathing. Figure 25.2 shows the distribution of fresh gas, deadspace gas and alveolar gas in early and late expiration, at low and high FGFs. The system also imposes minimum respiratory work on the patient [Ooi et al. 1993b].

A modern variant of the Magill system is shown in Figure 25.3; it is known as the *Lack* breathing system. It consists of two concentric tubes, inspiration occurring down the outer, expiration down the inner. The expiratory inner tube must therefore be of sufficient diameter in order not to create significant resistance to expiration and it must be accommodated within the inspiratory outer tube. Therefore the concentric Lack is 30 mm in diameter, wider than the standard 22 mm tubing [Barnes *et al.* 1980]. If the diagram is studied carefully and compared with the Magill system, it can be seen that the Lack system is functionally identical to the Magill, and is probably as efficient as the MaGill [Nott *et al.* 1977, Miller *et al.* 1989]. The expiratory outlet from the common part of the system is close to the patient, although the valve itself is at some distance and the fresh gas inlet and reservoir bag are distant from the patient. There is also a version called the *parallel Lack* system, where the tubes lie side by side rather than concentrically and may superficially resemble a circle system [Ooi *et al.* 1993a].

If such Mapleson A systems are used for controlled ventilation, the efficiency of the gas usage is relatively poor. The design, which preferentially excretes alveolar gas and stores fresh gas in spontaneous breathing, does the opposite when positive pressure breathing is applied. To facilitate inspiration, if the reservoir bag is squeezed, or if a bellows is applied to the reservoir limb, the pressure in the system increases, the partially closed expiratory valve near the patient opens and the fresh gas stored in the tube is preferentially excreted, although some will go to the patient. On expiration, the system preferentially fills with alveolar gas, partly because of the partially closed valve, and it is ready to be rebreathed on the next inspiration [Waters *et al.* 1961]. Mapleson A systems are the least appropriate systems for controlled ventilation and a FGF of four times minute ventilation, with an expiratory time of four times inspiratory time, is required to prevent rebreathing [Tyler *et al.* 1989].

The efficiency of the Mapleson A system in controlled ventilation could be improved if the escape of fresh gas during inspiration could be prevented. Figure 25.4 shows such a modification, known as the *Miller modification*

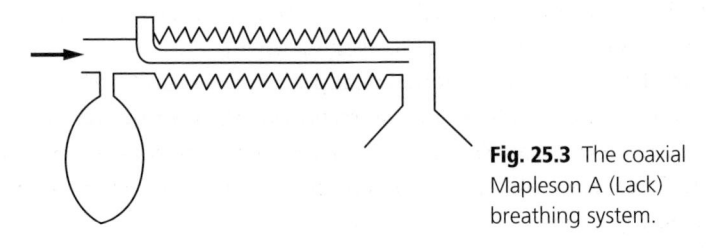

Fig. 25.3 The coaxial Mapleson A (Lack) breathing system.

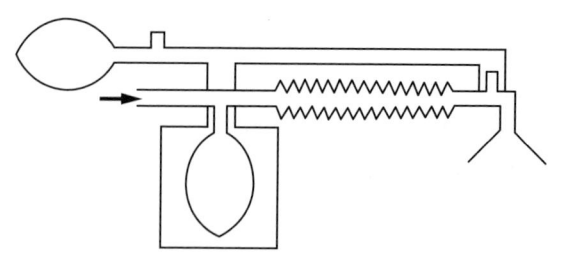

Fig. 25.4 The Miller modification (Enclosed MaGill) breathing system.

[Miller et al. 1988, Bruce et al. 1989]. One study has demonstrated a greater efficiency of the enclosed Mapleson A in controlled ventilation than the Mapleson D [Tham et al. 1993]. The generic term for the Miller modification is the *enclosed afferent reservoir system* (EAR) and has been shown to minimise rebreathing at a FGF of at least 80 ml kg^{-1} min^{-1} [Jennings et al. 1992].

Mapleson B and C systems

These are shown in Figure 25.1B and C for historical interest only, since they were once commonly used in UK anaesthetic rooms.

Mapleson D, E and F systems – T pieces

These are shown in Figure 25.1 D, E and F and are otherwise known as *T pieces*, because the fresh gas is delivered to the system at a T junction close to the patient. This feature is common to all T pieces, whether the rest of the system has merely tubing or an open tailed reservoir bag and an expiratory valve. Functionally T pieces have a common mode of performance, due to the close proximity of the fresh gas delivery close to the patient end of the system, which ensures preferential delivery of fresh gas. Figure 25.5 shows a sinusoidal inspiratory waveform and a constant FGF. In the early part of inspiration, the FGF exceeds the inspiratory requirement. Area 'A' represents fresh gas that could be stored in the system for later use; area 'B' represents a time when the inspiratory requirement exceeds the FGF; if B ≤ A, then the gas rebreathed from the system consists solely of stored fresh gas, providing it has remained unmixed with alveolar gas. T pieces need a higher FGF than Mapleson A systems to minimise rebreathing in spontaneous rebreathing, but are more efficient than Mapleson A in controlled ventilation. A FGF of about twice the minute ventilation is needed, but the exact FGF required depends on the inspiratory flow waveform, V_D to V_T and I to E ratios [Mapleson 1954,

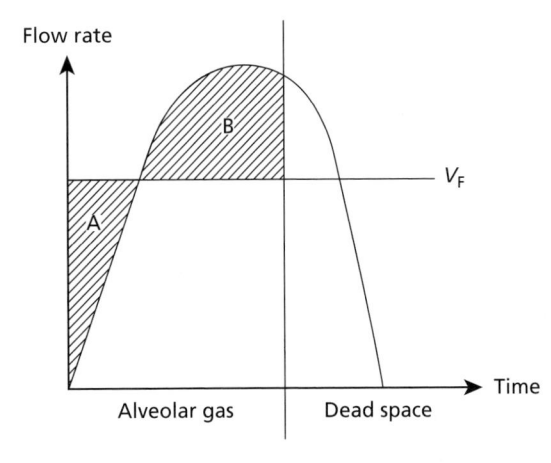

Fig. 25.5 The functional gas storage property of T pieces; by storing the fresh gas and deadspace gas in the reservoir of a T piece, rebreathing can be avoided at FGFs less than peak inspiratory flow rate.

Dorrington 1989, Mapleson 1958, Willis *et al.* 1975]. The performance of these systems is highly dependent on a long expiratory pause [Cook 1996a,b].

Mapleson D

This system has a reservoir bag and expiratory valve distant from the patient. In spontaneous breathing, the patient inspires fresh gas from the T junction and some alveolar gas from the tubing and reservoir bag. On expiration, the anatomical deadspace fresh gas is lost to the system, alveolar gas follows and it is eventually excreted from the distant expiratory valve. As fresh gas continues to be supplied during expiration, it too is stored in the tubing. The higher the fresh gas flow, the greater will be the ratio of fresh gas to alveolar gas stored in the tubing for the next inspiration and the more quickly will alveolar gas be excreted. In controlled ventilation, squeezing the reservoir bag in inspiration uses the column of stored gas (alveolar and fresh) in the tubing to push a bolus of relatively pure fresh gas into the patient. Expiration follows the same pattern as in spontaneous breathing. It can be seen from Figures 25.1D and 25.5, that factors that exacerbate rebreathing in this system are a low FGF and an inadequate length of tubing: the tubing should generally be about 1 m long.

Figure 25.6 shows a modern variant of the Mapleson D, known as the *Bain system*, designed by Bain and Spoerel in 1972 [Bain *et al.* 1972]. It is a system of concentric tubes and superficially resembles the Lack. However, the FGF is delivered down the relatively narrow bore inner tube, which delivers the gas close to the patient (as a T piece). Expiration occurs down the standard diameter outer tube towards a distant reservoir bag and expiratory valve. Care must be taken with this, as with all coaxial systems, that the inner coaxial tube

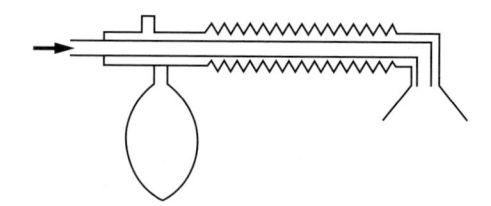

Fig. 25.6 The coaxial Mapleson D (Bain) breathing system.

does not become detached at either end or the system's deadspace become much larger [Heath *et al.* 1991]. There are two tests for the integrity of such a system before use: in one an obturator is applied to the inner tube to ensure its integrity from leakage; in the other, on passing a high flow of oxygen through the inner tube, collapse of the reservoir bag must be observed to demonstrate its continuity from one end to the other. Coaxial systems should not be used with intermittent flow anaesthetic machines, since the narrow inner tube does not allow the patient's inspiratory effort to be transmitted to the demand valve of the machine.

In spontaneous breathing, a FGF of between 2.5 and 3 times the patient's minute ventilation [Conway *et al.* 1977] or 200 ml kg^{-1} min^{-1} [Jonsson *et al.* 1986] is required to prevent rebreathing. Any reduction in this value leads to a sudden increase in rebreathing or ventilation or both. The coaxial T piece may not be as efficient as the orthodox Mapleson D because the coaxial arrangement at the patient end of the system may encourage gas mixing.

Figure 25.7 shows graphs of FGF delivery plotted against patient minute volume required for controlled ventilation from T pieces in adults for different values of P_{ECO_2}. Generally a FGF of between 70 and 100 ml kg^{-1} min^{-1} is

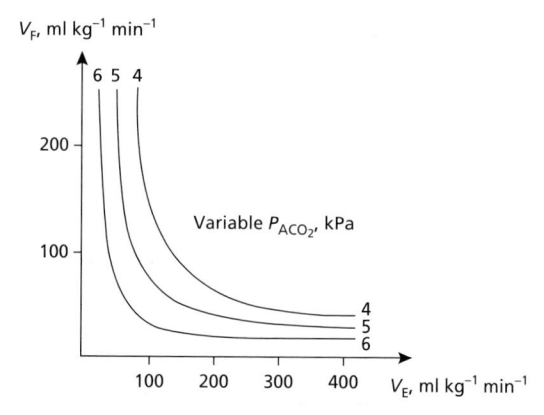

Fig. 25.7 A graphical aid to estimating alveolar P_{CO_2} (P_{ACO_2}) in kPa, from a given combination of FGF (V_F) and minute ventilation (V_E).

sufficient for normocapnia, providing minute ventilation is kept sufficiently high at between 120 and 150 ml kg^{-1} min^{-1} [Henville *et al.* 1976, Bain *et al.* 1973].

With the advent of the need to ventilate patients from a distance in an MRI scanner, the long Bain breathing system has been introduced. The additional length increases the compliance of the system such that smaller tidal volumes and higher end expiratory pressures result [Sweeting *et al.* 2002].

Mapleson E and F

These systems are otherwise known respectively as the *Ayre's T piece* and with its *Jackson Rees modification*, shown in Figure 25.1E and F. The Ayre's T piece was invented by Phillip Ayre in 1937 for use in cleft lip and palate surgery in children. In deciding on the length of the reservoir tube, too long a tube would add unnecessary deadspace, which might permit rebreathing; too short a tube would allow dilution of fresh gas with air. A tube of a volume one third that of the patient's tidal volume is the optimal length [Ayre 1956]. The valveless system is simple and provides low resistance to spontaneous breathing for small children [Hatch 1987], although there are now some variants with an overflow valve. Jackson Rees' modification is the addition of an open tailed reservoir bag to allow visible monitoring of breathing, as well as a means of manual controlled ventilation. The FGF required to avoid rebreathing is twice the child's minute ventilation, V_{min}:

$$V_{min} = V_T \times f,$$

where V_T is tidal volume and f is respiratory rate = 7 ml kg^{-1} body weight × (15–30 breaths min^{-1}).

Table 25.1 shows some suggestions for the FGF for this particular T piece for controlled ventilation in children and these techniques have been confirmed in clinical use [Park *et al.* 1998].

Humphrey ADE system

This system was designed in 1983 by Humphrey in South Africa, so that, at turn of a lever, a system with Mapleson A characteristics could be used for spontaneous ventilation, and a system with D and E characteristics was available for controlled ventilation [Humphrey *et al.* 1986] and can be used for paediatric patients as well [Orikowski *et al.* 1991].

Table 25.1 Fresh gas flow requirements for children per minute using a T piece under controlled ventilation

	Patient weight, kg	$\dfrac{V_E}{V_F}$		
Rosea and Froese			Predicted PaCO$_2$ kPa	
			4.9	4
	10.0–30.0	2	1000 ml + 100 ml kg^{-1}	1600 ml + 100 ml kg^{-1}
	> 30	2	2000 ml + 50 ml kg^{-1}	3200 ml + 50 ml kg^{-1}
Hatch			Measured PECO$_2$ = 4.5 kPa, PiCO$_2$ < 0.5 kPa	
	3.0–20.0	0.67	1000 ml + 200 ml kg^{-1}	

Venturi systems

A Venturi can be incorporated into a Bain system to deliver fresh gas and entrain expiratory gas that has had its CO$_2$ absorbed by a soda lime canister, to reduce the FGF requirements to 30 ml kg^{-1} min^{-1} [Jörgensen et al. 1985]. However, care must be taken in its design if adequate inspired oxygen concentrations are to be preserved [Campbell et al. 1996]. A similar device has been incorporated into a Magill system to reduce FGF requirements to 40 ml kg^{-1} min^{-1} [Jakubazsko et al. 1989].

The circle system

This is altogether a more complex system in which exhaled gas is recirculated. It incorporates unidirectional valves and a means of absorbing carbon dioxide. Its advantage is the ability to use low, very low or minimal FGF [Virtue 1974], which is appropriate with the use of expensive modern volatile agents. Its disadvantage is its complexity; there are many components and connections and the inspired gas concentrations are not closely related to the concentrations dialled up on rotameters or the vaporiser, particularly at low FGFs. Therefore, in the modern era, sophisticated gas monitors are needed and have become standard.

In deciding where to place the various components of a circle system, shown in Figure 25.8, Eger [Eger 1974] suggested three rules:

- there must be a unidirectional valve between the reservoir bag and the patient, on both the inspiratory and expiratory sides;
- fresh gas must not enter the system between the patient and the expiratory valve;

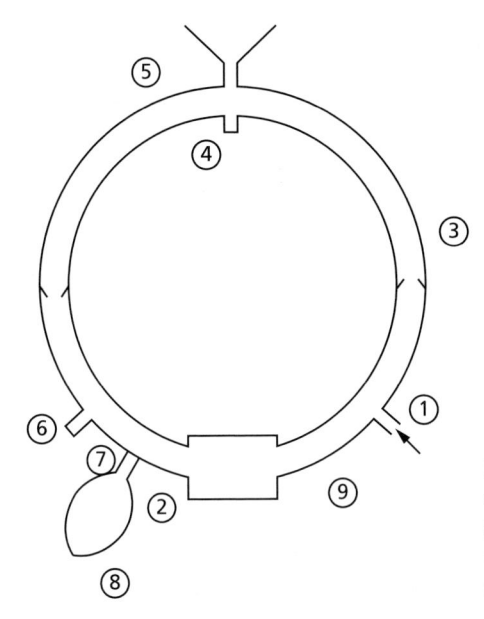

Fig. 25.8 Arrangement of components of a circle system. See text for referral to numbered points.

- the overflow valve must not be placed between the patient and the inspiratory unidirectional valve.

Different configurations of circle systems have different efficiencies, as defined by fresh gas utilisation [[Zbinden 1991]. Figure 25.9 shows three configurations of the system, A, C and H, that obey these rules and that otherwise have slightly different characteristics [Eger *et al.* 1968]. In evolving circle system designs, it was found that the greatest efficiency in spontaneous breathing occurred where the overflow valve is close to the patient (cf. the Magill system); this arrangement is least efficient in controlled ventilation, except when both unidirectional valves are also close to the patient (system H). The differences between systems are not great, but become accentuated as FGF is reduced.

Referring again to Figure 25.8, in deciding where to put the FGF inlet, the best compromise is achieved in position 1, just downstream of the CO_2 absorber. Fresh gas can collect retrogradely in the absorber and reservoir bag, pushing expired gas back towards the overflow valve. At low flows, the vented alveolar gas will not yet have passed through the absorber, thus economising on its use. As the FGF is increased, so does the likelihood of venting expired gas that has already passed through the absorber, as well as venting fresh gas itself. Placing the FGF inlet at position 2, upstream of the absorber, could help

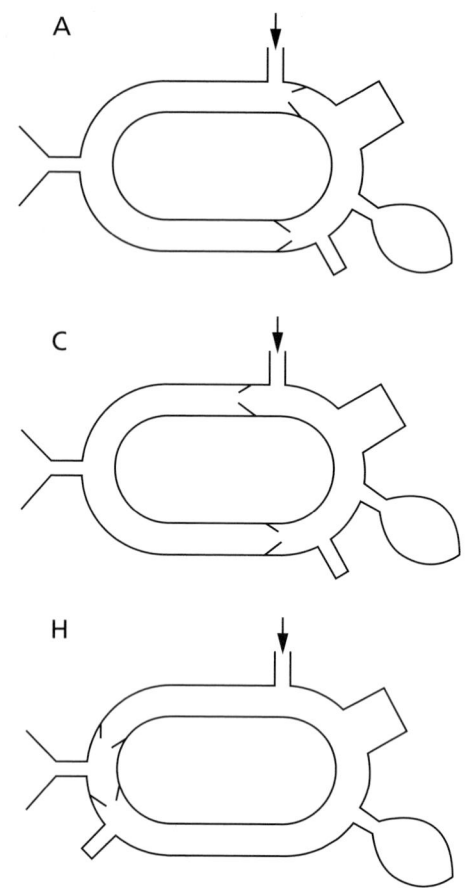

Fig. 25.9 Three arrangements of the circle system with different behavioural characteristics.

humidify the fresh gas, but may also cause unnecessary venting of fresh gas and venting of absorbent dust into the inspiratory limb, as well as preventing retrograde flow of recently exhaled alveolar gas.

During controlled ventilation, gas spills out of the overflow valve during inspiration. In order to minimise the chance of this being fresh gas, the best position for the overflow valve is at 6 or 7 as shown in Figure 25.8, which will preferentially allow for venting of alveolar gas before it has reached the absorber.

Equal efficiency can be obtained during spontaneous and controlled ventilation with the reservoir bag in positions 8 or 9. If the bag is downstream of the absorber at position 9, then in controlled ventilation a squeeze of the bag forces retrograde flow of gases already cleared of CO_2 back through the absorber. Placing the bag upstream of the absorber in position 8 allows retrograde filling of the bag with fresh gas and this is the most common position of the reservoir bag. A modern development of circle system structure is characterised by a coaxial inspiratory limb [Cook *et al.* 1996]; such a modification improves the response time at low FGFs to step changes in fresh gas concentrations.

A most critical part of the circle system is the CO_2 absorber, which has to be designed to allow low resistance to flow and to avoid *channelling*, where gas flows through the absorbent crystals along paths of least resistance, thus avoiding contact with the absorbent in other parts of the absorber. Although absorbent canisters have traditionally been large and arranged in two halves to allow partial change of absorbent, modern canisters are smaller and anticipate more frequent exchange of crystals [Shaw *et al.* 1998].

Two absorbents in common use are soda lime and Baralyme. Soda lime consists of 4% sodium hydroxide, 1% potassium hydroxide, 14–19% water and the rest as calcium hydroxide [Dorsch *et al.* 1984]. Additionally there are small amounts of silica for drying, kieselguhr for hardening and dye indicators to show when the crystals are spent. The water is essential for the reaction, which occurs as:

$$CO_2 + H_2O \Leftrightarrow H_2CO_3$$

$$2NaOH + 2H_2CO_3 + Ca(OH)_2 \Rightarrow CaCO_3 + Na_2CO_3 + 4H_2O.$$

These reactions occur with the liberation of heat and water [Lumley *et al.* 1976], both of which are useful by-products in the context of heat and humidity preservation of the patient.

Baralyme is a mixture of 20% barium hydroxide, 80% calcium hydroxide and some potassium hydroxide. The water for the reaction is present as the octahydrate of barium hydroxide.

In all cases the monovalent hydroxides are more reactive than the divalent $Ca(OH)_2$. Where crystals have been allowed to dry out, such as leaving the gas flowing through the canister overnight, these highly reactive monovalent hydroxides can also produce significant amounts of carbon monoxide and even formaldehyde [Low Flow Anaesthesia Society Proceedings 1999]. An absorbent consisting of only $Ca(OH)_2$ (Amsorb, Belfast) has been produced to counteract this [Bedi 2001].

As well as absorbing CO_2, the absorbents also tend to absorb volatile agents [Grodin *et al.* 1982]; this is generally acceptable, though wasteful, with a major exception: when the volatile agent *trichloroethylene* reacts with soda lime, it produces dichloroacetylene gas, otherwise known as phosgene, a neurotoxic agent used in the trenches of World War I. When Sevoflurane was being developed, it was thought to react with soda lime to produce a renally toxic agent, *Compound A*; this was not found to be of clinical significance [Baxter et al. 1998] although in the USA the Federal Drug Administration recommends keeping the FGF at more than 3 L min^{-1} under these conditions. Soda lime that is allowed to dry out has been found to encourage carbon monoxide formation [Harrison et al. 1996]. Indicators added to the absorbent show when it needs replacing. Ethyl violet is a commonly used example, although its usefulness can be impaired by photo deactivation by fluorescent lighting [Andrews et al. 1990].

The purpose of using a circle system is to use low FGF in order to economise on the consumption of increasingly expensive volatile agents and minimise pollution; an agent concentration on the vaporiser of 1% of a low FGF uses much less agent than 1% at a high FGF. At the start of an anaesthetic, none of the body tissues contains anaesthetic agent. To enhance anaesthetic uptake and excretion of nitrogen, high FGF is generally used at the start [Mapleson 1998]. When clinically desirable levels are reached in relevant tissues, these FGF's can be reduced, theoretically to metabolic O_2 consumption rate, say 250 ml min^{-1}, assuming no gas leakage and no further uptake by tissues or tubing. In reality both exist and some nitrous oxide and volatile agent is also switched on. The response time of the system is proportional to the difference between gas inflow and tissue uptake and inversely proportional to system volume [Conway 1986]. At high FGF, gas concentrations delivered to the circle system from the anaesthetic machine are nearly equal to those received by the patient. As FGF becomes a significantly smaller part of what the patient receives, the majority being recirculated gas, the gas concentrations received by the patient bear significantly less resemblance to that delivered by the anaesthetic machine [Mapleson 1960]. At the beginning of an anaesthetic, when anaesthetic gas uptake by the patient from the alveoli to the tissues is high (oxygen uptake being relatively constant), alveolar O_2 concentrations appear higher than expected and this effect becomes very marked as FGF is reduced. As the anaesthetic proceeds and tissues become saturated, less anaesthetic gas is absorbed from the system, alveolar O_2 concentrations are lower than expected and this effect also becomes more marked as FGF is reduced [Conway 1981]. This degree of uncertainty means that O_2, volatile agent and N_2O monitoring, with gases sampled from the patient end of the system, are highly desirable, possibly mandatory. The capital cost of such monitoring is offset within a short time against the revenue savings on volatile agent. Furthermore, the oxygen consumption of patients can vary, as can the rate of uptake of anaesthetic agents.

Circle systems can be used for paediatric anaesthesia, using a 1 L reservoir bag and smaller bore tubing. Despite the presence of valves, the work of breathing has been found to be acceptable [Conterato *et al.* 1989].

References

Andrews, J. J., Johnston, R. V., Bee, D. E. and Arens, J. F. (1990) Photodeactivation of ethyl violet: a potential hazard of Sodasorb, *Anesthesiology*, **72**, pp. 59–64.

Ayre, P. (1956) The T-piece technique, *Br. J. Anaesth.*, **28**, pp. 520–23.

Bain, J. A. and Spoerel, W. E. (1972) A streamlined anaesthetic system, *Canad. Anaesth. Soc. J.*, **19**, pp. 426–35.

Bain, J. A. and Spoerel, W. E. (1973) Flow requirements for a modified Mapleson D system during controlled ventilation, *Canad. Anaesth. Soc. J.*, **20**, pp. 629–36.

Barnes, P. K., Conway, C. M. and Purcell, G. R. G. (1980) The Lack anaesthetic system, Anaesthesia **35**, pp. 393.

Baxter, A. *et al.* (1998) Formation of carbon monoxide from difluoro-methyl ether anaesthetics, *Anesthesiology*, **89**, pp. 937–40.

Bedi, A. (2001) The in vitro performance of carbon dioxide absorbent with and without strong alkali, *Anaesthesia*, **56**, pp. 546–50.

Bruce, W. E. and Soni N. C. (1989) Preliminary evaluation of the enclosed MaGill breathing system, *Br. J. Anaesth.* **62**, pp. 144–49.

Campbell, D. J. and Fairfield, M. C. (1996) The delivery of oxygen by a venturi T-piece, *Anaesthesia*, **51**, pp. 558–60.

Conterato, J., Lindahl, S. G., Meyer, D. M. and Bires J. A. (1989) Assessment of spontaneous ventilation in anesthetised children with use of a pediatric circle or a Jackson-Rees system, *Anesth. Analg.*, **69**, pp. 484–90.

Conway, C. M. *et al.* (1976) An experimental study of gaseous homeostasis and the MaGill circuit using low fresh gas flows, *Br. J. Anaesth.*, **48**, pp. 447–55.

Conway, C. M., Seeley, H. F., Barnes, P.K. (1977) Spontaneous ventilation with the Bain anaesthetic system, *Br. J. Anaesth.*, **49**, pp. 1245–49.

Conway, C. M. (1981) Alveolar gas relationships during use of the circle system with carbon dioxide absorption, *Br. J. Anaesth.*, **53**, pp. 1135–46.

Conway, C. M. (1986) Gaseous homeostasis and the circle system; factors influencing anaesthetic gas exchange, *Br. J. Anaesth.*, **58**, pp. 1167–80.

Cook, L. B. (1996a) Respiratory pattern and rebreathing in the Mapleson A, C and D systems with spontaneous ventilation. A theory, *Anaesthesia*, **51**, pp. 371–85.

Cook, L. B. (1996b) The importance of the expiratory pause. Comparison of the Mapleson A, C and D rebreathing systems using a lung model, *Anaesthesia*, **51**, pp. 453–60.

Cook, L. B. and Chakrabarti M. K. (1996) Circle systems with a coaxial inspiratory limb, *Anaesthesia*, **51**, pp. 247–54.

Dorrington, K. L. (1989) Medical breathing systems. Chapter 1 in *Anaesthetic and Extracorporeal Gas Transfer*, Oxford Medical Engineering Series, Oxford University Press, Oxford, pp. 31–87 and 150–172.

Dorsch, J. and Dorsch S. E. (1984) Breathing systems I–IV. In *Understanding Anesthesia Equipment*, Williams & Wilkins, pp. 140–220.

Eger, E. I. II and Ethans, C.T. (1968) The effects of inflow, overflow and valve placement on the economy of the circle system, *Anesthesiology*, **29**, pp. 93–100.

Eger, E. I. II (1974) Anesthetic systems: construction and function. Chapter 13 in *Anesthetic Uptake and Action*, Williams & Wilkins, pp. 206–27.

Grodin, W. K. and Epstein, R. A. (1982) Halothane absorption complicating the use of soda lime to humidify anesthetic gases, *Br. J. Anaesth.*, **54**, pp. 555–59.

Harrison, N., Knowles, A. C. and Welchew, E. A. (1996) Carbon monoxide within circle systems, *Anaesthesia*, **51**, pp. 1037–40.

Hatch, D. J. (1987) Paediatric anaesthetic equipment, *Br. J. Anaesth.*, **57**, 672–84.

Heath, P. J. and Marks, L. F. (1991) Modified occlusion tests for the Bain breathing system, *Anaesthesia*, **46**, pp. 213–216.

Henville, J. D. and Adams, A. P. (1976) The Bain anaesthetic system. An assessment during controlled ventilation, *Anaesthesia*, **31**, pp. 247–56.

Humphrey, D., Bock-Utne, J. G. and Downing, J. W. (1986) Single lever Humphrey ADE low flow universal breathing system: Part I: comparison with dual lever ADE, MaGill and Bain systems in spontaneously breathing adults. Part II: Comparison with Bain system in anaesthetised adults during controlled ventilation, *Canad. Anaesth. Soc. J.*, **33**, pp. 698–718.

Jakubazsko, J., Christensen, P. and Jorgensen, S. (1989) The MaGill–Venturi attachment: studies in volunteers and patients, *Acta Anaesthesiol. Scand.*, **33**, pp. 422–25.

Jennings, A. D. *et al.* (1992) Physical characteristics of the EAR breathing system, *Br. J. Anaesth.*, **68**, pp. 625–29.

Jonsson, L. O. and Zetterstrom, H. (1986) Fresh gas flow in coaxial Mapleson A and D circuits during spontaneous breathing, *Acta Anaesthesiol. Scand.*, **30**, pp. 588–93.

Jorgensen, S. and Hansen, L. K. (1985) The Venturi anaesthetic circuit. 1. An all purpose breathing system for anaesthesia, *Acta Anaesthesiol. Scand.*, **29**, pp. 269–72.

Low Flow Anaesthesia Society Proceedings (1999) York, UK.

Lumley, J. and Morgan, M. (1976) The temperature inside carbon dioxide absorbers, *Anaesthesia*, **31**, p. 63.

Mapleson, W. W. (1960) The concentration of anaesthetics in closed circuits, with special reference to halothane. I. A theoretical study, *Br. J. Anaesth.*, **32**, pp. 298–309.

Mapleson, W.W. (1954) The elimination of rebreathing in various semi-closed anaesthetic systems, *Br. J. Anaesth.*, **26**, pp. 323–32.

Mapleson, W. W. (1998) The theoretical ideal FGF sequence at the start of low flow anaesthesia, *Anaesthesia*, **53**, pp. 264–72.

Mapleson, W. W. (1958) Theoretical considerations of the effects of rebreathing in two semi closed anaesthetic systems, *Br. Med. Bull.*, **14**, pp. 64–68.

Miller, D. M. and Miller, J. C. (1988) Enclosed afferent reservoir breathing systems, *Br. J. Anaesth.*, **60**, pp. 469–75.

Miller, S. W., Barnes, P. K., Soni, N. and Tennant, R. (1989) Comparison of the MaGill and Lack anaesthetic breathing systems in anaesthetised patients, *Br. J. Anaesth.*, **62**, pp. 153–58.

Nott, M. R., Walters, F. and Norman, J. (1977) A comparison of the Lack and Bain semi closed circuits in spontaneous respiration, *Br. J. Anaesth.*, **49**, pp. 512.

Ooi, R., Lack, J. A., Soni, N., Whittle, J. and Pattison, J. (1993a) The parallel anaesthetic Lack breathing system, *Anaesthesia*, **48**, 409–414.

Ooi, R., Pattison, J. and Soni, N. (1993b) Additional work of breathing imposed by Mapleson A systems, *Anaesthesia*, **48**, pp. 599–603.

Orlikowski, C. E., Ewart, M. C. and Bingham, R. M. (1991) The Humphrey ADE system: evaluation in paediatric use, *Br. J. Anaesth.*, **66**, pp. 253–57.

Park, J. W. *et al.* (1998) Predictable normocapnoea in controlled ventilation of infants with the Jackson-Rees or the Bain system, *Anaesthesia*, **53**, pp. 1180–84.

Shaw, M., Scott and D. H. T. (1998) Performance characteristics of 'to and fro' disposable soda lime canister, *Anaesthesia*, **53**, pp. 454–60.

Sweeting, C. J., Thomas, P. W. and Sanders, D. J. (2002) The long Bain system: an investigation into the implications of remote ventilation, *Anaesthesia*, **57**, pp. 1183–86.

Tham, E. J. *et al.* (1993) Efficiency of breathing systems A and D in the Carden Ventmasta ventilator, *Br. J. Anaesth.*, **71**, pp. 741–46.

Tyler, C. K. G., Barnes, P. K. and Rafferty, M. P. (1989) Controlled ventilation with a Mapleson A (MaGill) breathing system: reassessment using a lung model, *Br. J. Anaesth.*, **62**, pp. 462–66.

Virtue, R. W. (1974) Minimal flow nitrous oxide anesthesia, *Anesthesiology*, **40**, pp. 196–98.

Waters, D. J. and Mapleson, W. W. (1961) Rebreathing during controlled ventilation with various semi-closed anaesthetic systems, *Br. J. Anaesth.*, **33**, pp. 374–81.

Willis, B. A., Pender, J. W. and Mapleson, W. W. (1975) Rebreathing of a T-piece: volunteer and theoretical studies of the Jackson-Rees modification of Ayre's T-piece during spontaneous respiration, *Br. J. Anaesth.*, **47**, pp. 1239–46.

Zbinden, A. M., Feigenwinter, P. and Hutmacher, M. (1991) Fresh gas utilization of eight circle systems, *Br. J. Anaesth.*, **67**, pp. 492–99.

Chapter 26

Artificial ventilators

This chapter contains: respiratory mechanics during ventilation; ventilator mechanics; some ventilator types; intensive care ventilators; manual resuscitators; portable ventilators; other methods of gas exchange.
The chapter links with: Chapters 3, 4, 7, 11, 14, 15, 16, 22, 23, 24 and 25.

Respiratory mechanics during ventilation

When pressure is applied by the ventilator to drive gas into the lungs, energy is expended to overcome airway resistance R to gas flow in the airways, in order to store gas in the alveoli, whose readiness to having their volume increased is represented by the concept of compliance, C. The storage of gas within individual compliances represents potential energy storage (see Chapter 3).

The acceleration of gas and anatomical components within the system represent kinetic energy change, resisted by the inertance, I, of the system. At conventional ventilation frequencies, these kinetic energy changes are negligible compared with the other energy changes taking place. Inertance can be ignored and the system behaves like a flow resistor in series with a compliance. These variables determine the pressure and volume changes that take place within the lung. As ventilation frequency increases into the high range, inertance becomes significant and the frequency response of anatomical structures becomes important, with phase differences between pressure and volume signals occurring [Lin et al. 1989]. Mechanical resistance, R, in the system is largely due to resistance to gas flow down airways and is defined as pressure change per unit flow $\Delta P/Q$, typically 4 cm H_2O l^{-1} s. at 0.5 l s^{-1}. However there is a contribution from viscous resistive forces in the lung and chest wall tissues. High resistance may require long inspiratory times, while expiratory times that are too short may lead to gas trapping in alveoli. Excessive resistance may mean that the power required to ventilate the patient may exceed that available to the ventilator.

Compliance, C, is a measure of the capacitative properties of the alveoli and is defined as volume change per unit pressure change $\Delta V/\Delta P$. It is not uniform throughout the respiratory cycle and has values in the range 0.05–0.10 L $(cmH_2O)^{-1}$. Dynamic compliance is the value given to this variable throughout the inspiratory period to the end of inspiration, when airway pressure is highest. During the inspiratory pause, airway pressure falls to a plateau during which the static compliance can be measured, which is greater than the dynamic compliance. Compliance differs between these two parts of inspiration, because of the redistribution of gas between the lung units during the pause and because of a degree of stress relaxation in the elastic tissues of the lung. Such a phenomenon leads to hysteresis in the pressure–volume curve, shown in Figure 3.3. If compliance is low, short inspiratory times may result in inadequate lung filling; if compliance is high, normal driving pressures may be high enough to cause barotrauma.

Ventilator mechanics

Ventilator classification

Numerous methods of classifying artificial ventilators have evolved [Smallwood 1986, Loh et al. 1987, Dorrington 1989], but Mapleson's classification [Mushin et al. 1980] remains the most useful, defining ventilators as pressure generators or flow generators. Modern ventilators can function as both types of generator under different conditions. This is shown in Table 26.1, which also shows that the ventilator has to carry out an inspiratory phase, a change from inspiration to expiration, an expiratory phase and a change from expiration to inspiration. A ventilator is classified as a *pressure generator* or a *flow generator* during the inspiratory and possibly the expiratory phase; it may be a constant or non-constant pressure or flow generator. Cycling is according to one of the variables shown in Table 26.1, namely, time, volume, pressure, flow or patient triggering.

An ideal pressure generator maintains a preset pressure, P_{aw}, at its output, which is then delivered to the patient. The flow, Q, generated in the airways depends on the resistance and compliance of the respiratory system, which in turn determines the tidal volume, V_T, delivered to the lungs. Figure 26.1 shows the curves for P_{aw}, Q and V_T under these conditions. It also shows that, in the presence of reduced compliance or increased airway resistance, the rate of fall of Q varies.

Conceptually, a constant pressure generator consists of a weighted bellows; if these bellows are spring loaded, the generator is a decreasing pressure generator, which can be thought of as a discharging compliance. A flow generator (see below) with a leak takes on the characteristics of a pressure generator in series with a resistance. These hydraulic definitions have parallels in electrical analogies (see Chapter 4).

Table 26.1 Ventilator classification and cycling modes.

Pressure generator Constant Non-constant	Inspiratory phase	Flow generator Constant Non-constant
Cycling	Time Pressure Volume Flow	
Pressure generator Constant Non-constant	Expiratory phase	Flow generator Constant Non-constant
Cycling	Time Pressure Volume Flow Patient	

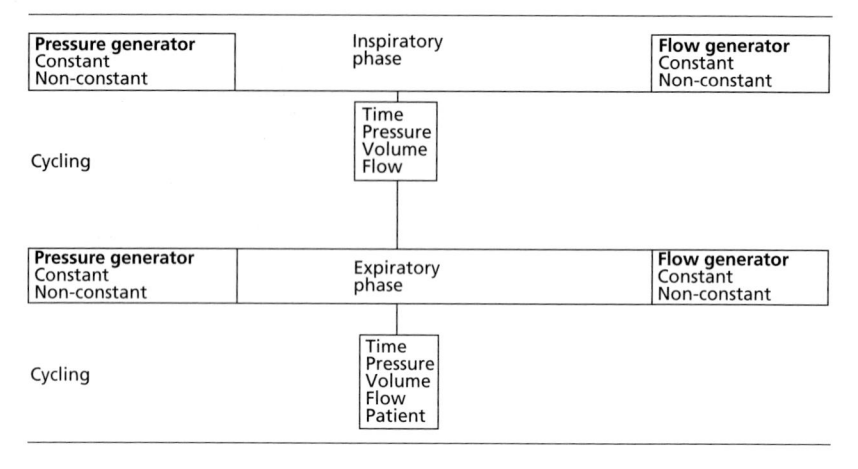

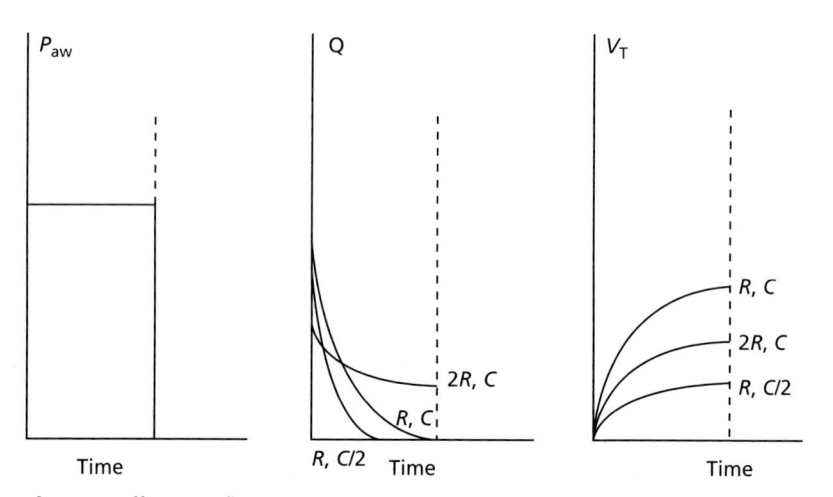

Fig. 26.1 Effects on flow and tidal volume of altered airway resistance and lung compliance when using a constant pressure generator.

A pressure generator delivers gas through its own internal resistance and through the resistance of the tubing. It can be thought of as generating a pressure, which is dropped across this *source resistance* R_S, and the series airway resistance R_L, shown in Figure 26.2. The presence of R_S reduces the pressure available to R_L, which may reduce V_T. Therefore an ideal pressure generator should keep internal and connecting tubing resistance as low as possible. The electrical analogy is of a voltage source in series with resistances R_S and R_L.

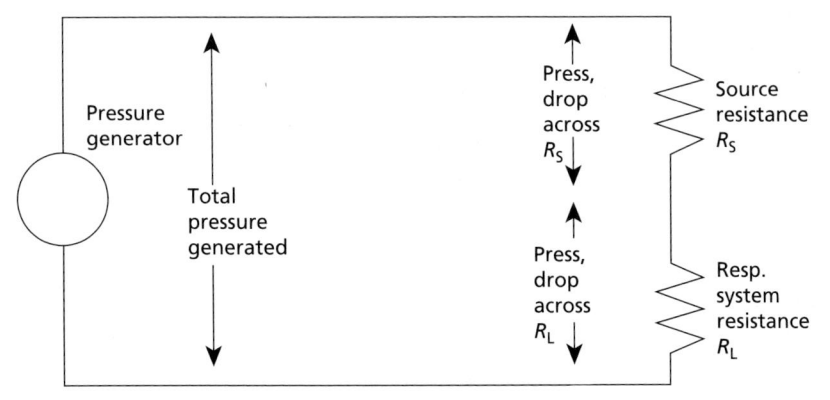

Fig. 26.2 Electrical analogue of a constant pressure generator.

Although mechanically simpler, pressure generators may not be able to deliver adequate V_T if lung R and C are unfavourable.

Ideal flow generators determine the preset flow pattern (constant, ramp, decelerating) that is delivered to the patient. This preset flow is maintained whatever the patient's airway resistance or lung compliance may be. Changes in patient R and C do not affect Q, but P_{aw} may change under these circumstances and under unfavourable circumstances P_{aw} may rise high enough to cause barotrauma. Figure 26.3 shows graphs of Q, P_{aw} and V_T under conditions of constant Q and the way in which low lung compliance and high R alters P_{aw} is also shown.

Conceptually, constant flow generation is achieved by delivering a high generated pressure through a nozzle or high source resistance, which is how a constant current generator works in the electrical analogy. Thus a continuum of behaviour of ventilators has now been considered, despite the initial discrete separation of flow and pressure generators. Indeed many practical ventilators have characteristics of both. Sinusoidal or other positive displacement pumps generate variable but predictable flow. Gas injectors do not necessarily produce constant flow but, as indicated, have characteristics of constant pressure generators in series with high source resistance.

The loss of gas volume V_S, from a flow generator to a patient's respiratory system can be thought of as an internal or source compliance C_S, in parallel with the lung compliance C_L, as in Figure 26.4. Thus delivered tidal volume V_T is the sum of V_S and patient tidal volume V_L. The electrical analogy is of a current source connected to two parallel capacitors, C_S and C_L. Parallel compliance becomes significant at high ventilation frequencies and in patients with low lung compliance, such as infants. Source compliance should be minimised,

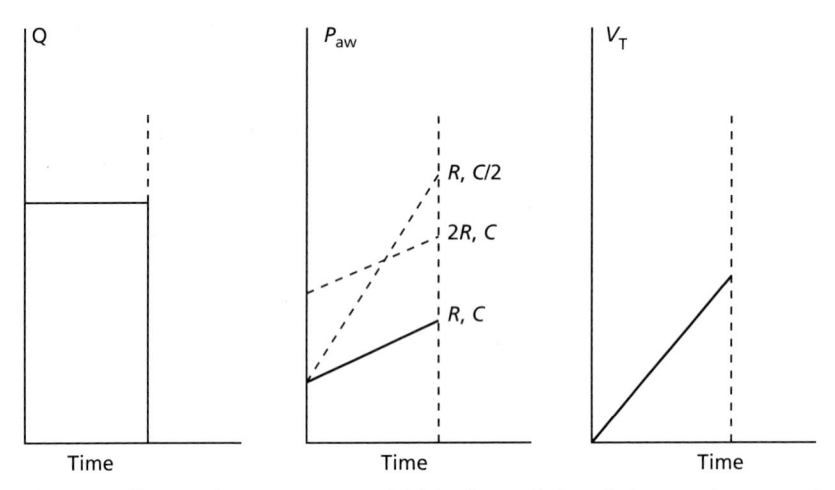

Fig. 26.3 Effect on airway pressure and tidal volume of altered airway resistance and lung compliance when using a constant flow generator.

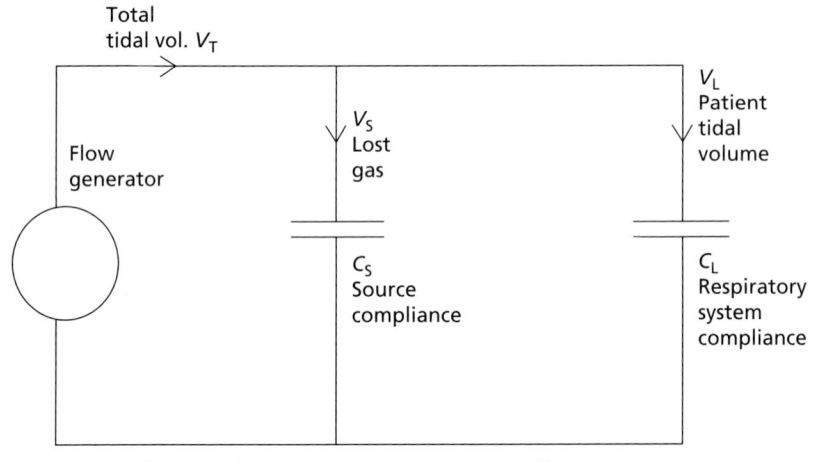

Fig. 26.4 Electrical analogue of a constant flow generator.

but this implies narrow rigid tubing, which is not ideal if the patient is to breathe through the same tubing, unless adequate bias flow is also present [Lin et al. 1989].

Power

This is required to drive the ventilator during the active part of the inspiratory phase and, in selecting a ventilator, thought should be given to its power requirement and whether this will be adequate for the intended use. Required power can be estimated from the product of airway pressure and flowrate and a typical value is 20 W. Peak power production is likely to be the limiting variable in a ventilator and must be greater than the peak airway pressure times peak flowrate plus the energy losses described above. A high proportion of the required power is in overcoming the internal resistance of the ventilator; this can be as much as 200 W if the device is a flow generator, where internal resistance is high. Although more flexible, this aspect makes flow generators mechanically less efficient than pressure generators. The power provided may be from the fresh gas flow itself in *minute volume dividers* (see the section on the *Manley* ventilator), electrical power or compressed gas, usually oxygen.

Cycling

Cycling is according to one of the variables shown in Table 26.1; the ventilator incorporates a device to cycle at a predetermined value of those variables. Many ventilators require careful inspection to determine which variable is being used at a given moment.

In cycling from inspiration to expiration, time cycling implies that the inspiratory phase is determined solely by the ventilator and not by the characteristics of the lungs. V_T, P_{aw} and Q are all free to vary and the value at the moment of cycling depends on the effect on the lungs of the generator of flow or pressure. The mechanism is usually electronic or fluidic gating, an electromechanical device such as a solenoid, or a pneumatic spool valve [*Moyle et al.* 1998]. In volume cycling P_{aw} and inspiratory time vary depending on lung characteristics, but the variation is small in a constant flow (high pressure) generator. In a constant (low) pressure generator, if lung compliance was low enough, V_T might never be delivered and the ventilator might never volume cycle. The volume cycling device is usually a concertina bellows, whose volume can be mechanically or electrically preset. In pressure cycling, the inspiratory time alters when R and C vary and usually takes the form of a diaphragm toggle. Usually a ventilator that can be triggered by a patient breath cycles using this mechanism.

Table 26.2 summarises the effects on V_T of the different forms of cycling [Barnes 1982].

Expiration

This phase is generally passive and is usually a matter of discharging the gas stored in the lungs under compliance C, through airways and tubing resistance R to atmosphere. There are times when it is clinically useful to raise the expiratory pressure above atmospheric pressure in order to increase FRC and

Table 26.2 Effect on tidal volume of altered airway resistance and lung compliance for different cycling modes when using ventilators with different generation characteristics.

Cycling mode	Time			Pressure			Volume		
Lung state	Normal	2R	C/2	Normal	2R	C/2	Normal	2R	C/2
Pressure gen.	V_T-	$V_T\downarrow$	$V_T\downarrow$	Not		Possible	V_T-	\multicolumn{2}{l	}{Cycling volume may not be reached}
Flow gen.	V_T-	$V_T-\!*T$	$V_T-\!*T$	V_T-	$V_T\downarrow$	$V_T\downarrow$	V_T-	$V_T-\!*$	$V_T-\!*$

improve oxygenation; this is known as *PEEP, positive end expiratory pressure*. This can be done either by retarding expiratory flowrate via an expiratory limb resistor, or by using a threshold resistor that closes below a preset pressure in the expiratory limb. The methods include a flow restrictor, an underwater blow-off or a pressure sensitive solenoid valve. *Negative end expiratory pressure (NEEP)*, which was introduced to minimise adverse cardiovascular effects, is no longer used because it may induce premature airway closure.

Safety features

In inexperienced hands a ventilator can be a dangerous device. All users should have a low threshold of suspicion for fault finding and be familiar with its use. Numerous variables are monitored on a ventilator (see below) and there should be appropriate alarms on these monitors so that inadequate or excessive volume or airway pressure can be detected before the patient is harmed. Safe ventilators have evolved by design and experience, but there are still potential and real problems.

For example, some ventilators used to be of the descending bellows type [Gattinoni 1988]; in the case of a patient becoming disconnected, gravity would continue to expand the bellows, drawing in room air; subsequent compression of the bellows in the inspiratory phase generates adequate pressure to fail to trigger a disconnection alarm. In a modern ventilator of the ascending bellows type, a patient disconnection would cause the bellows to collapse and pressure and volume alarms would annunciate. Pressure sensitive alarms that are inappropriately used may fail when used with ventilators, which can be thought of as 'discharging compliances' such as the Manley MP3 ventilator [Campbell *et al.* 1996]. Other mechanical causes of hypoventilation or excessive airway pressure due to bad design have been reported [Schreiber 1985, Somner, 1988, Roth *et al.* 1986, Sara *et al.* 1986, Pryn *et al.* 1989] and include blocked expiratory valves and misconnected tubing.

Testing

Before being put into clinical use, all ventilators are rigorously tested on a lung model containing a variable R and a variable C. Testing according to an *ISO (International Standards Organisation) standard* includes endurance, waveform and volume performance tests, internal compliance testing and many other aspects.

Special features of paediatric ventilators

Ventilators used for babies and children need far more flexibility in some respects than adult ventilators. Respiratory rate needs to be in the range 15 to 40 breaths minute^{-1}, tidal volume may need to vary from 16 to 500 ml. Lung compliance of babies can vary from 2.5 to 30 ml (cm H_2O)$^{-1}$ and airway resistance from 2 to 30 cm $H_2O\, l^{-1}$. For neonates the respiratory time may need to be as short as 0.5 seconds, inspiratory flow as low as $2\,l\,min^{-1}$ and peak airway pressure limited to 6–7 cm H_2O. As mentioned earlier, source compliance C_S can become significant under these conditions and it is preferable that neonatal and infant ventilators are valveless [Chakrabarti 1990], so that the infant can breathe without doing too much respiratory work. Ventilators such as the *Bear Cub* and *Sechrist* occlude gas escape to atmosphere from a T piece during the inspiratory phase. However, it has been shown that some infant ventilators do not successfully limit peak inspiratory pressure reliably, resulting in possible barotrauma [Kirplani *et al.* 1988, Synnott *et al.* 1987]. The combination of short inspiratory times and small diameter endotracheal tubes may mean that there is a significant difference between indicated airway pressure and alveolar pressure, while short expiratory times may lead to incomplete expiration [Tipping *et al.* 1991]. Appropriate alarms should be available and used to respond to the tight constraints of excessive or inadequate measured values of preset variables. Adequate humidification is mandatory.

Some ventilator types

A few examples of mechanisms of ventilators will be given here with some examples, but the list is by no means complete. Although they are not necessarily the most modern, they are still widely used and the principles of their structure and function are relevant to all ventilators.

The Bag-in-bottle

Most modern anaesthetic machines have an integrated ventilator that appears as a bellows inside a Perspex cylinder. It is clear from observation that the bellows is squeezed flat from a pressurised driving gas source inside the cylinder (but outside the bellows) to facilitate inspiration of the patient's lungs, and is allowed to expand by being passively filled with breathing gas, while passive expiration occurs. The generic name for this type of ventilator is the 'bag-in-bottle', and the bellows is usually of the ascending type, as alluded to earlier, in order that a leak in the breathing system would immediately become visible by the collapse of the bellows under its own weight. This configuration provides a small amount of inherent PEEP, although it may also add some expiratory resistance. A descending bellows arrangement is seldom seen these days because, if a leak develops in the breathing system, the bellows continues to passively expand by falling, and to be actively compressed; not only would this fail to give a clear visual indication of failure to ventilate the patient but, depending on the magnitude of the leak, the pressures generated in the system may not trigger a low pressure alarm. Note that in both cases, the driving gas, usually pressurised oxygen, is separated from the breathing gas by the bellows; if the bellows itself develops a leak, the breathing gas inside it is merely contaminated by oxygen. The mechanism to intermittently power the driving gas to compress the bellows can either be by a piston, an electric motor driving a cam, or by an electronically controlled flow valve.

Datex-Ohmeda ADU ventilator

This is an example of a 'bag-in-bottle'. It is a volume preset, time cycled ventilator with pneumatic power and an ascending bellows. It acts as a pressure generator in the inspiratory phase. Respiratory frequency and I:E ratio (inspiratory to expiratory time ratio) are preset; the ventilator can be set to be volume controlled (when it acts primarily as a flow generator) when the volume is preset; or it can be set to be pressure controlled (when it acts as a pressure generator) and the peak pressure is preset. This ventilator comes as part of the Datex-Ohmeda AS3 and AS5 anaesthetic machines and the associated monitoring integrated into the anaesthetic machine itself makes graphical and digital outputs of numerous ventilation parameters available. A smaller bellows can be inserted if tidal volumes of smaller than 200 ml are required. A diagram of the device is shown in Figure 26.5. During inspiration (Figure 26.5(a)) the driving gas passes into the pressure chamber and compresses the bellows, which thus delivers breathing gas to the patient's breathing system via the CO_2 absorber. It can be seen that the driving gas pressure also keeps the overflow valve of the breathing system closed. In expiration (Figure 26.5(b)), the driving gas pressure reduces to a minimum, the bellows expands, admitting a mixture of fresh breathing gas and mixed expired gas, and the overflow valve is allowed to open so that any excess gas may be excreted to atmosphere. The driving gas flowrate adjusts itself to deliver the preset variables according to the

SOME VENTILATOR TYPES | **347**

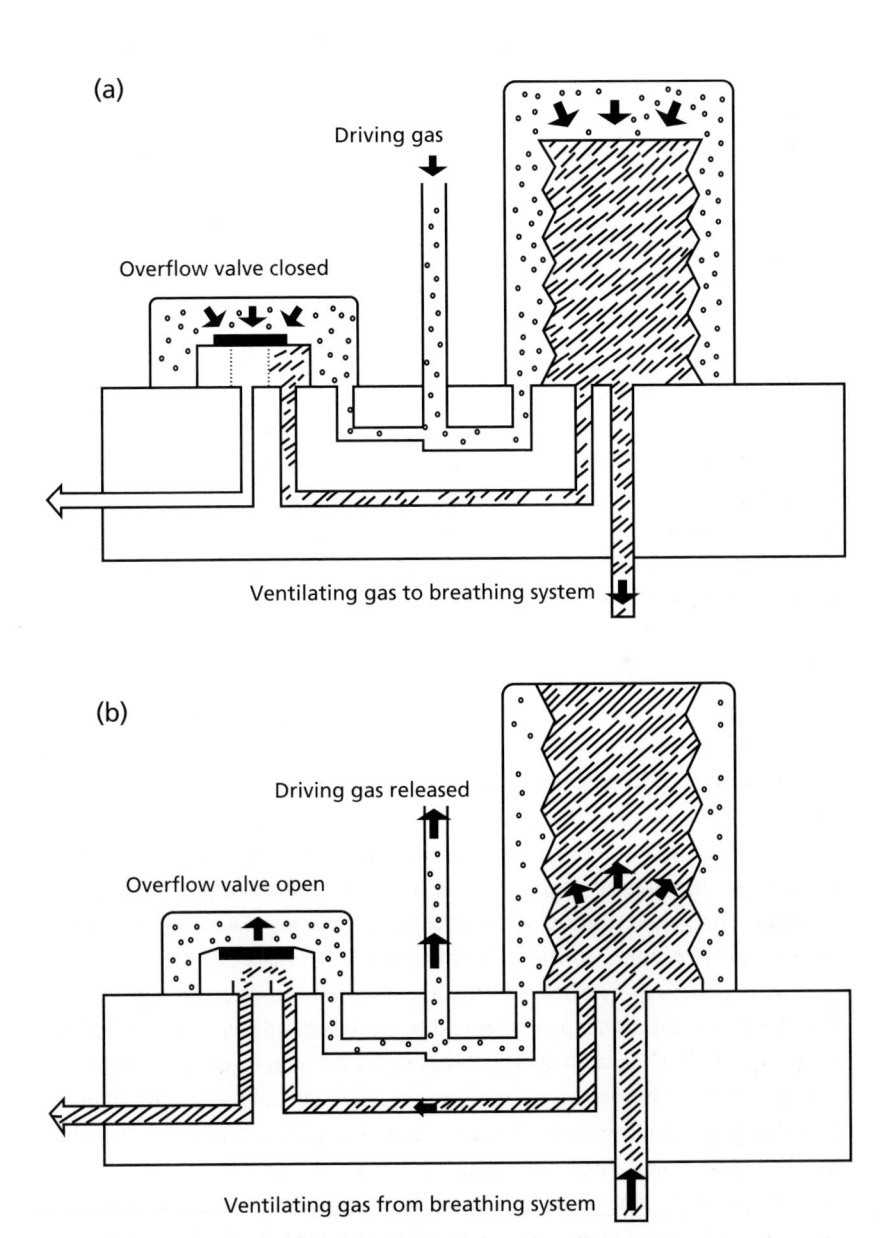

Fig. 26.5 Datex Ohmeda ADU anaesthetic ventilator: (a) inspiratory phase; (b) expiratory phase.

ventilator's setting as a pressure or volume controlled device. This ventilator also compensates the tidal volume setting for different fresh gas flows and for the system's compressible volume.

Minute volume dividers

These devices do not have separate driving gas and breathing gas supplies. The only gas supply to the ventilator is from the common gas outlet of the anaesthetic machine, and it both drives the mechanism and supplies the patient with breathing gas.

The Manley MP3 ventilator

This is the most recently developed model in the Manley series [Classic Paper, *Br. J. Anaesth.* 1995] of ventilators. Figure 26.6 shows that the essential mechanism consists of two sets of bellows and three unidirectional valves. It is a time cycled pressure generator.

During inspiration, the smaller bellows B_1 receives fresh gas and stores it, while the main bellows B_2 delivers its gas tidal volume to the patient. For this to be possible, valve V_1, between bellows B_1 and B_2 is closed, inspiratory valve V_2 between B_2 and the patient is open and expiratory valve V_3 is closed. The valves are pneumatically connected to allow this to occur. The inspiratory time control determines the extent to which B_1 fills in the inspiratory phase. Meanwhile B_2 is delivering the selected tidal volume due to the pressure on the bellows from the adjustable weight W.

At the end of inspiration, the expanded bellows B_1 activates a trip lever, which triggers the start of expiration. Under the influence of a powerful spring, B_1 delivers its stored volume as well as the continuing FGF to bellows B_2, through the now open valve V_1. In this period V_2 is closed and V_3 is open to allow the patient to exhale the breath. B_2 fills until another trip lever is activated by a catch on an arm attached to B_2, which triggers the start of the next inspiration. This catch can be adjusted on the arm to give a preset tidal volume. Taps in the primary circuit and expiratory unit allow the manual and spontaneous ventilation modes to occur, with an *APL valve* (adjustable pressure limiting) and reservoir bag in the circuit.

The main advantages of this anaesthetic ventilator are its mechanical simplicity and that it does not need electrical power. If the FGF from the common gas outlet of the anaesthetic machine is disconnected, the ventilator stops functioning. Since it comes without its own integral alarms, modern anaesthetic practice requires supplementary alarms to be provided. As the ventilator is a pressure generator, the preset tidal volume may not be completely

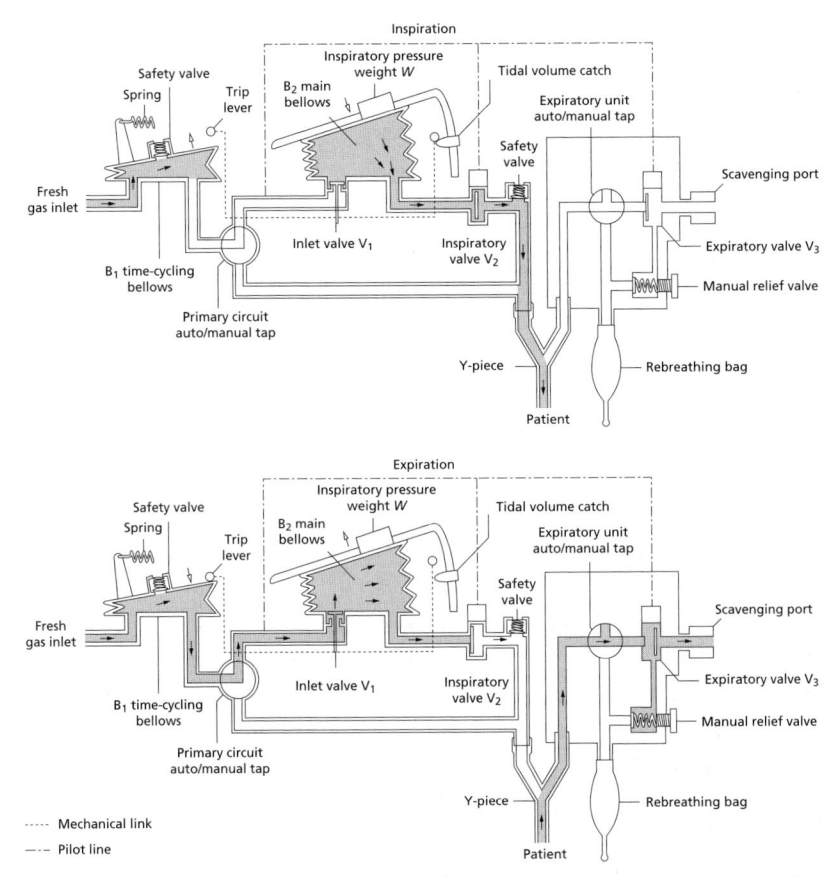

Fig. 26.6 Manley MP3 Minute Volume Divider Ventilator. Functional parts in inspiration and expiration.

delivered to patients with poor lung compliance. This would be clear from the end expiratory position of B_2. An analogue dial connected to an aneroid barometer shows the pressure in the system.

Intermittent blowers

These devices are driven by a gas source, usually air or oxygen pressurised to between 250 and 400 kPa. Prior to the advent of microprocessor control of ventilators the most important component of this type of ventilator for anaesthetic use was the control module, described in the example below and shown in Figure 26.7, which consists of a pneumatically controlled spool valve and an inspiratory and expiratory timer; such a device does not require electrical power and divides the breathing gas into tidal volumes controlled

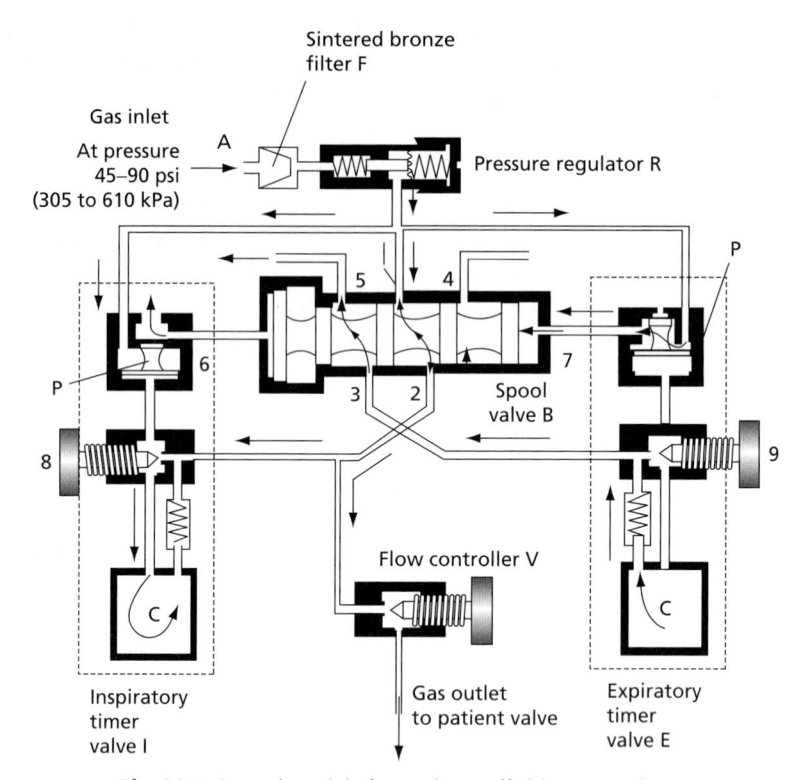

Fig. 26.7 Control module for Penlon Nuffield 200 ventilator.

by adjusting the timers. To separate inspiratory and expiratory phases, more sophisticated intensive care ventilators in this class use an electrically switched magnetic field in a coil to displace an iron core to open a valve that is otherwise held closed by a spring. It may be designed to open and close the valve, or to allow a variable flow through the valve, depending on the current flowing through the coil (solenoid) when it is a *proportional flow valve*. This system works well with a short response time in a high pressure, low compliance system using a small valve aperture. The same mechanism operates to close a PEEP valve when used in a low pressure system with a large aperture valve, otherwise kept open by a light spring. On modern ventilators microprocessors control the electrical signal to a proportional flow valve to provide a range of ventilator modes and variables.

Penlon Nuffield 200 ventilator

This is a relatively compact and versatile ventilator in this class, and can be thought of as a time cycled flow generator. As such, it is versatile enough to be used with patients of a wide range of size, airway resistance and lung compliance. When used for adults, it is volume preset and when used for babies, it is pressure preset. It can be used with different breathing systems, ones which lend themselves to controlled ventilation. The driving gas and the breathing gas are separated pneumatically, hence tubing connecting the breathing system to the ventilator must be of sufficient volume (a tidal volume at least) and length (1 m) to ensure continuous separation of these two gas flows, otherwise dilution of the breathing gas can occur. The valve block on the ventilator consists of a port connecting tubing to the breathing system reservoir bag mount, an exhaust port to the scavenging system and a 60 cm H_2O pressure relief valve. The valve block can be exchanged for a *Newton valve* for paediatric use, which opens for expiration at a lower pressure than the standard valve for adult use.

The control module, accessible to the user, has an airway pressure gauge, an on–off switch, a flow control knob and controls for inspiratory and expiratory times. Inside the control module there is, shown in Figure 26.7, a spool valve B and a pair of timers controlling inspiratory time, I and the expiratory time, E. The spool valve has five ports for controlling the direction of gas flow, as well as a pneumatic port at each end that actuates the spool valve itself. Each of the timers I and E has a piston P, which is driven by gas that is metered by a variable flow restrictor and stored in a capacitor, C. When the pressure within C is high enough, driving gas is delivered to that end of the spool valve, driving the spool to the opposite end of the valve, thus switching gas flows to allow the other half of the ventilatory cycle to occur.

The driving gas, which can be oxygen, O_2-N_2O, or air at regular 400 kPa pipeline pressure, enters the control module at A, thence through a filter F and a pressure regulator R, thence to port 1 of spool valve B. When the spool valve is to the left of the valve housing, the gas passes from port 1 to port 2, thence to the patient valve via a variable flow controller V; this is the inspiratory phase. As indicated earlier, part of this flow also goes to the inspiratory timer valve I to determine the inspiratory time; since the gas flow is constant, this determines tidal volume delivered. At the end of inspiration, the spool is moved to the right hand side via port 6 and the expiratory phase begins; driving gas entering port 1 is now diverted via port 3 to the expiratory timer E, containing a variable flow restrictor, which determines the expiratory time, a volume capacitor C and a piston P, identical to those in the inspiratory timer. At the end of expiration, the spool is moved to the left via port 7 and inspiration begins again. Note that the inspiratory and expiratory flow restrictors are independently variable.

The control module for the portable paediatric ventilator, the Pneupac, is almost identical to that of the Penlon Nuffield. The additional piece of

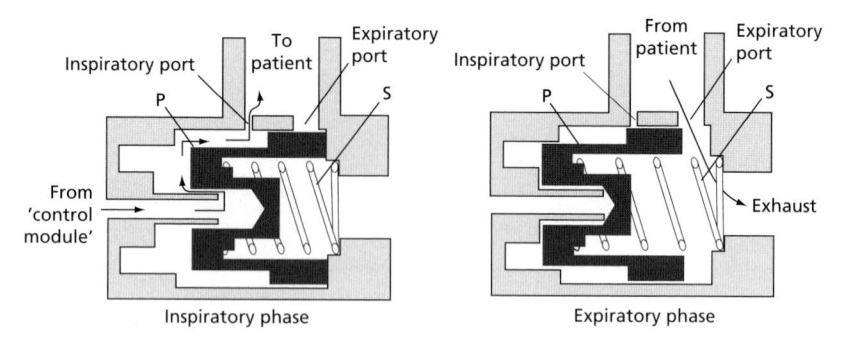

Fig. 26.8 Pneupac portable ventilator patient valve.

technology that the Pneupac has is its patient valve, shown in Figure 26.8; it could usefully be compared to those in manual resuscitators (see below). The Pneupac Ventipac ventilator is a useful device for ventilating paediatric patients during transfer, but ideally it needs an additional means of confirming delivered tidal volume [McCluskey *et al.* 1995]. Both these ventilators have been modified for use in a hyperbaric chamber, when such a facility is needed to care for a critically ill diver, for example [Lewis 1991, Spittal *et al.* 1991].

Intensive care ventilators

Ventilators for use in intensive care have a more complex task than those for use in anaesthesia. They are required to ventilate lungs with a wider range of compliance, often through airways of increased resistance. They provide a wide range of respiratory support to accommodate the patient's changing dependence on such support, from intermittent positive pressure ventilation with positive end expiratory pressure, volume or pressure assisted ventilation, and other modalities. The subtle control provided by microprocessors affords the patient with more complexity and more flexibility than the anaesthesia ventilators described hitherto. Two examples are described as representatives of number of ICU ventilators on the market that will have similar components and structure.

Veolar ICU ventilator

This intensive care ventilator consists of a pneumatic flow system and an electronic control system, shown in Figure 26.9. The pneumatic flow system consists of an air–oxygen pressure regulator and gas mixer, a gas reservoir, a servo controlled flow valve, a patient overpressure valve, an ambient air

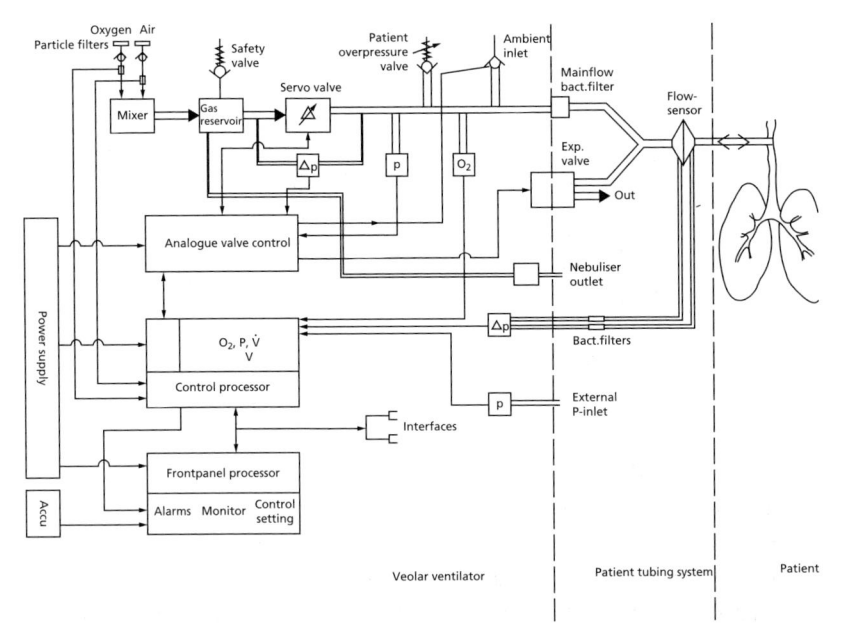

Fig. 26.9 Functional diagram of the Veolar intensive care ventilator.

inlet, a flow sensor, an expiratory valve and a pressure source for a nebuliser. The electronic control system is used to check the pneumatic flow system and consists of an analogue valve control, pressure sensors, a front panel microprocessor and a control microprocessor. The function of the whole system is continually checked by the two microprocessors.

Pneumatic flow system: Figure 26.10 shows the *air–oxygen pressure regulator and mixer*. Air and oxygen, at pressures of between 2 and 8 bar are delivered to the ventilator; the gases then pass through 5 µm filters and one way valves to the gas pressure regulators, which drop gas pressures to 1.5 bar. The gas mixer allows manual adjustment of the oxygen concentration, which alters the position of a piston inside a cylinder with a number of radial holes. Each gas enters the cylinder from its own pressure regulator in an amount that depends on the position of the piston in the cylinder. The gas mixture enters the gas reservoir at a maximum flow rate of 90 l min^{-1}, through another pressure regulator, which keeps the reservoir pressure at 350 cmH$_2$O, with an accuracy of ±5%.

The gas reservoir: This tank holds a gas volume of 8 l, compressed to 350 cmH$_2$O pressure, which means that the gas delivery to the patient can

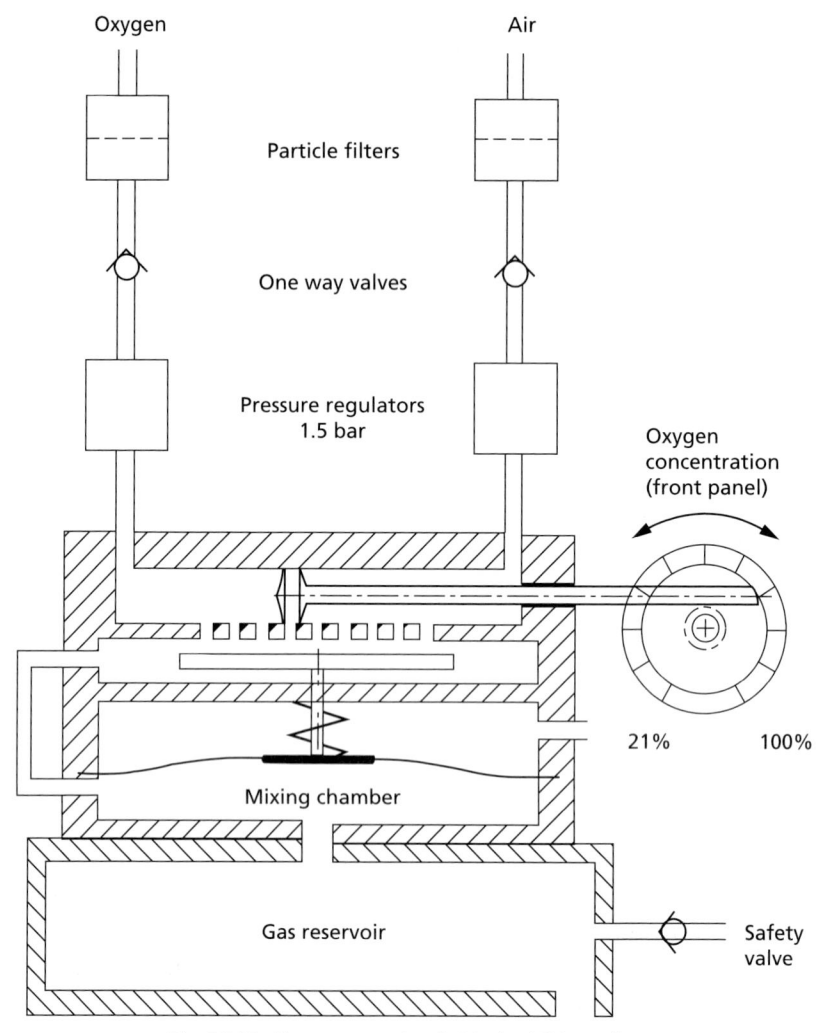

Fig. 26.10 Air–oxygen mixer in Veolar ICU ventilator.

be largely independent of supply flow rate. The reservoir has an added safety valve to prevent overpressure.

The servo-controlled flow valve: Figure 26.11 shows this high speed device, a series of solenoid valves that are able to deliver flows in the range 20 to 300 ml s^{-1} with precision. The valve uses an electrodynamic motor, an electromagnetic device similar to a solenoid, such as is found in a loudspeaker, to move a vertical piston. The piston is connected, by a rod, to a small cylindrical shaped plunger that has a triangular shaped orifice on two sides. The rod

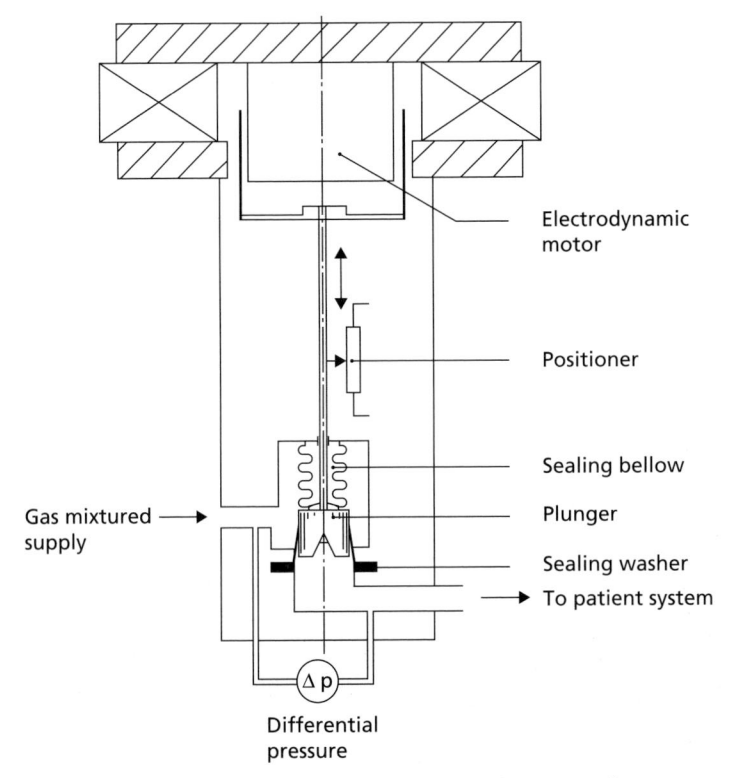

Fig. 26.11 Servo controlled flow valve in Veolar ICU ventilator.

is also connected to a high precision linear potentiometer, as in a control system. The motor raises the plunger and allows gas flow through the valve to the patient. The triangular shaped orifice in the plunger outlet allows a precise relationship between the triangular height, the pressure differential across the plunger orifice and the flowrate; an electronic servo control loop is driven by this relationship. Depending on the ventilator control settings, the control microprocessor determines the correct signal to send to the motor (solenoid device) to raise the plunger. The analogue valve controller compares the perceived flow, determined by the actual height of the plunger from the information provided by the potentiometer, to the pressure drop across the plunger. The plunger in this case acts as a flowmeter, the actual flow rate being proportional to the measured pressure drop. The analogue valve controller then readjusts the current to the motor to deliver the flow needed.

The response time of this system is about 100 ms, due to the precision of the control loop and the small system deadspace. It allows precise control of tidal volume, with any desired flow pattern and at any pressure, with an accuracy of 5%, regardless of back pressure. Because of multiple checking of measured flow against plunger height and motor current, the microprocessor assures accurate function under all conditions. Numerous alarm systems back up this function. In spontaneous modes of ventilation, the valve acts as a demand valve, matching desired flow and pressure.

The flow sensor: This is incorporated within the patient breathing system, using the principle of measurement of pressure differential across a variable orifice in the form of a fine Mylar membrane. The pressure transducer thus formed is an accurate, differential electronic device. This arrangement gives a linear relationship between Q and ΔP over a range of 200 to 3000 ml s^{-1} (see Chapter 7). However, further electronic linearisation can allow flow measurement down to 30 ml s^{-1}, maintaining a $\pm 5\%$ accuracy, which is only marginally affected by humidity or nebulised drugs. The flow sensor is used to provide information on flow, volume and timing of the respiratory cycle to the alarm systems, but is not used to control inspiratory and expiratory valves.

The expiratory valve: This is shown in Figure 26.12 and consists of a large surface area silicone rubber membrane inside a plastic housing. Occlusion of the membrane against the patient expiratory tubing during the inspiratory phase, or partial closure during PEEP or CPAP, occurs by the action of a solenoid, which closes the valve. During expiration, the current to the motor is off and a spring opens a valve to allow expiration to occur. Although the valve housing and silicone rubber membrane are labelled, it is unfortunately still possible to place the membrane back into the housing back-to-front after cleaning.

The electronic control system

This is divided into two independent systems, controlled by two separate microprocessors, operating in parallel and simultaneously; in addition, each microprocessor checks the other.

The front panel processor: This manages control settings and alarms, determining the setting of every control knob and sending the relevant information to the control processor. All monitored data are displayed, alarm conditions are determined and identified and internal circuits and voltage levels are all checked by this processor.

The control processor: This controls inspiratory and expiratory valves via the analogue valve control. It also contains *A to D* (ADC) and *D to A* (DAC) converters for flow and pressure control. It integrates flows to give tidal volume, calculates all pressure values and determines a number of alarm conditions.

INTENSIVE CARE VENTILATORS | 357

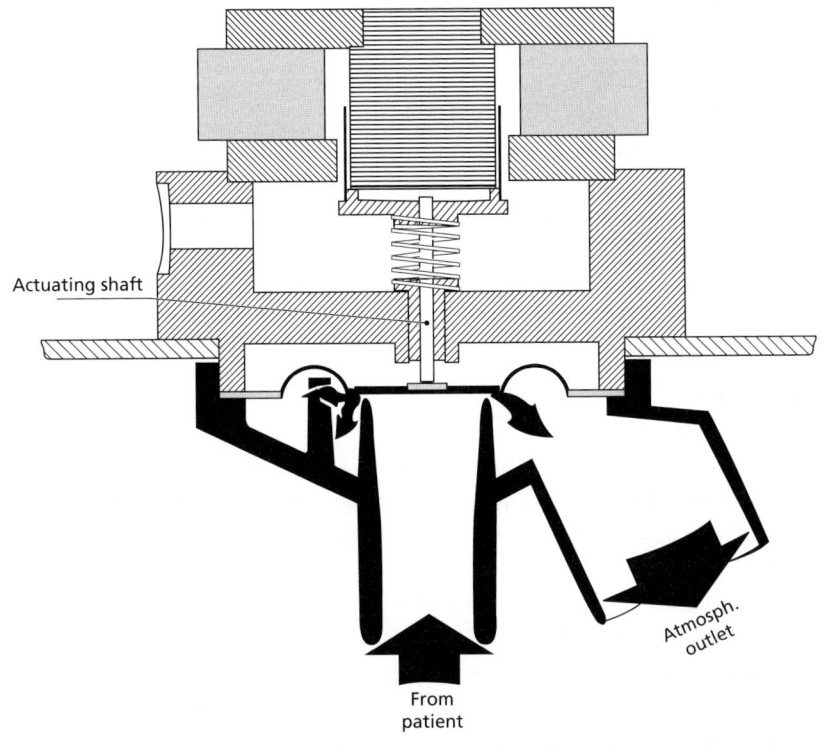

Fig. 26.12 Expiratory valve of Veolar ICU ventilator.

Draeger Evita 4

This is a ventilator with similar principles of operation and capacity as the one described above, although it is more compact. It has a section containing pneumatic controls, an electronic compartment, and a detachable display screen with controls. Gases (usually oxygen and air) enter the machine from high pressure sources, the exact pressures of which are measured by sensors in the ventilator, whence the gases are fed directly into two proportional flow valves as described above, which control the relative flows of the two gases, after which the gases are blended directly into the patient circuit. A central microprocessor uses the information on the upstream individual gas pressures to adjust the function of the proportional flow valves. The appropriate gas flow is delivered to the breathing system, thence to the patient. On expiration, the gas is returned to the ventilator via a diaphragmatic expiratory valve like the one shown above, and thence to a hot wire anemometric flow measurement device, and so to atmosphere. PEEP or CPAP is applied to the system by using

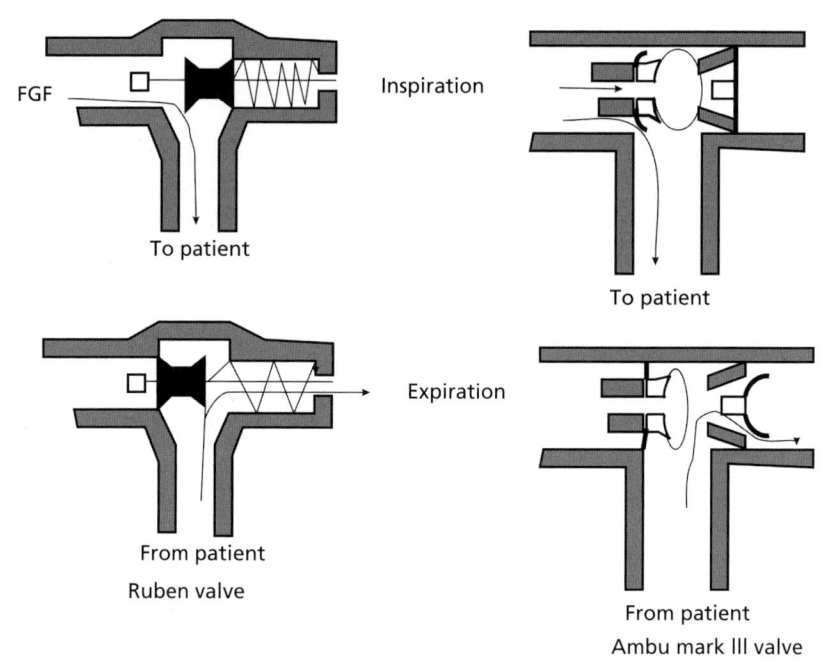

Fig. 26.13 Two examples of patient valves in manual resuscitators.

the oxygen supply to apply a controlled pressure to the downstream side of the expiratory valve.

Manual resuscitators

On occasions, a very much simpler device is needed to allow ventilation by hand. This consists basically of a self inflating reservoir bag with a supply of fresh gas, and a non-rebreathing valve. The design of this valve is crucial to the successful function of the resuscitator and two examples of their design are shown in the diagram in Figure 26.13.

Portable ventilators

There are now numerous compact ventilators on the market, which are useful for transporting patients. They are usually powered by a cylinder of oxygen, which is also the source of the gas to be delivered to the patient. Their mechanisms of action are similar to some ventilators already described. All have limitations of performance, but are adequate for short term use [Attebo et al. 1993, Heinrichs et al. 1989, Wong et al. 2000, Fludger et al. 2008].

Other methods of gas exchange
High frequency ventilation
[Chang *et al.* 1984, Crawford, 1986]

High frequency ventilation (HFV) is a term used to describe artificial ventilation at frequencies ranging from 1 to 20 Hz, depending on the mode of application. It uses tidal volumes, which are often smaller than the patient's deadspace volume. First used with some enthusiasm in the 1980s, it was initially thought to result in lower mean airway pressures than conventional ventilation. However, this subsequently proved not to be the case [Sherry *et al.* 1987], although the method has applications where conventional ventilation is inconvenient (e.g. chest surgery), or where each lung has different ventilatory requirements. Under these conditions, HFV is applied to one lung, while IPPV is applied to the other [Raphael *et al.* 1995].

High Frequency Positive Pressure Ventilation: High Frequency Positive Pressure Ventilation (HFPPV) is administered by a positive pressure device, with either a timed solenoid valve, or a mechanically driven rotating valve. An ordinary endotracheal tube is used and the ventilation frequency is usually 1–2 Hz.

High Frequency Jet Ventilation: High Frequency Jet Ventilation (HFJV) is widely used where bronchoscopic work is being carried out. A high velocity jet of gas, usually oxygen, attached to the inside of a rigid bronchoscope or endotracheal tube, results in air or gas entrainment, which provides an adequate inspired volume. The jet is sometimes manually controlled, or sometimes the frequency is controlled by a solenoid valve, when the frequency may be 2–10 Hz. Precise positioning of the gas injector is important if wide variation in performance of the device is to be avoided.

High Frequency Oscillation: High Frequency Oscillation (HFO) [Kolton 1984] uses a reciprocating mechanism, usually a piston pump, to produce an oscillatory gas flow. Unlike the other methods, expiration is not passive, but a well defined negative pressure change. A steady bias FGF is also required to maintain appropriate oxygen and carbon dioxide transfer gradients. The frequency of HFO is 3–50 Hz.

High Frequency Chest Wall Oscillation: High Frequency Chest Wall Oscillation (HFCWO) means that rapid reciprocating chest wall compression allows gas flow to occur to and from the lungs without tracheal intubation. This is usually achieved by applying a gas tight cuirass to the chest wall and applying a reciprocating pressure waveform to its inner surface. The *Hayek oscillator* is the modern development of this method [Petros *et al.* 1995].

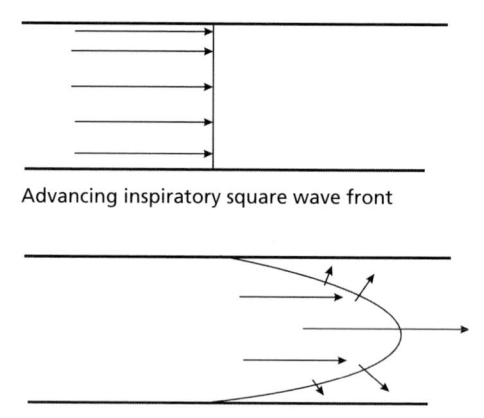

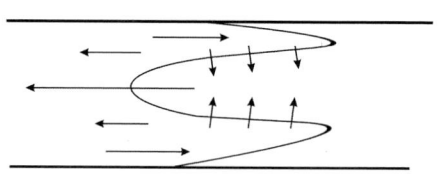

Fig. 26.14 Axisymmetric velocity profiles and axial gas mixing contribute to convective gas transport in High Frequency Ventilation (HFV).

Advancing inspiratory square wave front

Further advance of inspiratory front with axial mixing

Reversal of front on expiration with axial mixing

Mechanisms of gas transport in HFV

With high frequencies and small tidal volumes, gas transport by normal convective mechanisms plays some part [Scerer et al. 1989], but this cannot be the only mechanism. This is particularly true where different lung units have different time constants and where gas acceleration and inertance become significant variables. The presence of axisymmetric velocity profiles, shown in Figure 26.14, also contributes to convective gas transport in HFV; around bifurcations, inspiratory velocity profiles are peaked and expiratory profiles are flattened [Crawford 1986, Haselton et al. 1982], allowing oxygen transport to occur towards the periphery of the lung and carbon dioxide transport to occur from the periphery to the central bronchial tree. Another mechanism of transport in the convective range is *Pendelluft*, where gas is transported between lung units with different time constants, out of phase with the ventilatory cycle, shown in Figure 26.15.

Secondary mechanisms of transport [Smith 1990] may include:

- resonance – the frequency of ventilation approaches the natural resonant frequency of the structures of the lungs and chest wall, leading to volume amplification [Bourgain et al. 1994, Stott et al. 1995];

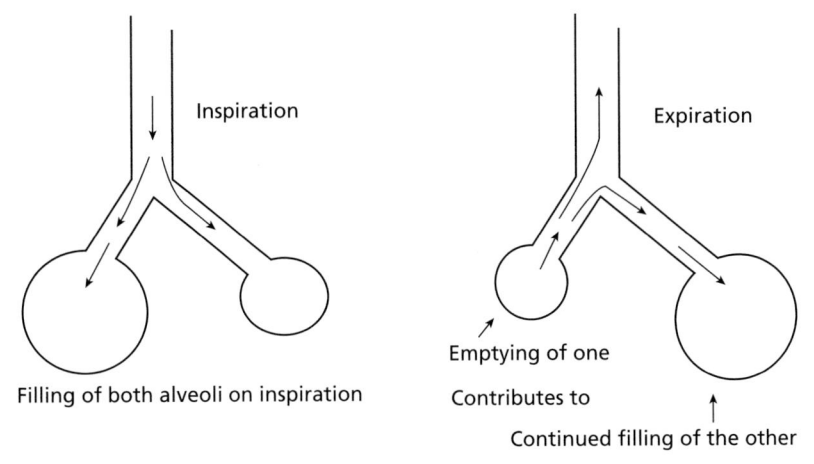

Fig. 26.15 *Pendelluft* as a mechanism of gas transport in HFV.

- *Taylor dispersion* – convection and radial dispersion;
- turbulence in the upper airways, which abolishes radial gas concentration gradients;
- laminar flow in the smaller, peripheral airways, where radial gas concentration gradients are preserved, which allows radial mixing to occur along the airway long axis; it can be calculated [Dorrington, 1989a] and confirmed that overall gas transfer is proportional to *frequency* $\times (V_T)^2$;
- molecular diffusion at alveolar level.

Extracorporeal Membrane Oxygenation (ECMO)

[Mudaliar 1991, Gattinoni *et al.* 1988]

There are occasions when the lung needs to be rested at FRC and oxygenation of the blood stream is achieved by a membrane oxygenator, much like those used in cardiac surgery. It is also possible to facilitate carbon dioxide removal by an extracorporeal membrane and to oxygenate with reduced IPPV via the lungs. Clearly, in either case, vascular access is required to facilitate either veno-arterial or veno-venous bypass.

The membrane can be either the *parallel plate* type or the *hollow fibre* type. Oxygen transport occurs according to the *advancing front* theory, in which the dimensionless constant known as the *Sherwood number*, a ratio of partial pressure changes between the membrane and the bloodstream, is an important concept. Because CO_2 exists in three forms in the blood, its transport through a membrane is more complex [Dorrington 1989b].

References

Attebo, L., Bengtsson, M. and Johnson, A. (1993) Comparison of portable emergency ventilators using lung model, *Br. J. Anaesth.*, **70**, pp. 372–77.

Barnes, P. K. (1982) Principles of lung ventilators and humidification. Chapter 48 in *Scientific Foundations of Anaesthesia*, William Heinemann, London, pp. 533–44.

Bourgain, J. L., Billard, V. and Desruennes, E. (1994) Resonant volume amplification during high frequency jet ventilation, *Br. J. Anaesth.*, **72**(suppl. 1), p. 10.

Campbell, R. M., Sheikh, A. and Crosse, M. M. (1996) A study of the incorrect use of ventilator disconnection alarms, *Anaesthesia*, **51**, pp. 369–70.

Chakrabarti, M. K. *et al.* (1990) A new infant oscillatory ventilator, *Br. J. Anaesth.*, **64**, pp. 374–75.

Chang, H. and Harf, A. (1984) High frequency ventilation–a review, *Respir. Physiol.*, **57**, pp. 135–52.

Classic paper (1995) The Manley ventilator, *Br. J. Anaesth.*, **50**, pp. 64–71.

Crawford, M. R. (1986) High frequency ventilation, *Anaesth. Int. Care*, **14**, pp. 281–92.

Davey, A. J. (2005) Automatic Ventilators. Chapter 11 in Davey, A. J. and Diba, A. (eds) *Ward's Anaesthetic Equipment*, 5th edn, Elsevier, pp. 241–71.

Dorrington, K. L. (1989a) Mechanical ventilation of the lungs. Chapter 4 in *Anaesthetic and Extracorporeal Gas Transfer*, Oxford Science Publications, Clarendon Press, Oxford, pp. 172–201.

Dorrington, K. L. (1989b) Oxygen transfer in extracorporeal lungs; Carbon dioxide transfer in extracorporeal lungs. Chapters 5 and 6 in *Anaesthetic and Extracorporeal Gas Transfer*, Oxford Science Publications, Clarendon Press, Oxford, pp. 203–67.

Fludger, S. and Klein, A. (2008) Portable ventilators. Continuing Education in Anaesthesia, *Critical Care and Pain*, **8**, pp. 199–203.

Gattinoni, L. *et al.* (1988) Extracorporeal support in acute respiratory failure, *Int. Care World.* **5**, pp. 42–47.

Haselton, F. R. and Scherer P. W. (1982) Flow visualization of steady streaming in oscillatory flow through a bifurcating tube, *J. Fluid Mech.*, **123**, pp. 315–33.

Heinrichs, W., Mertzluft, F. and Dick, W. (1989) Accuracy of delivered versus preset minute ventilation of portable emergency ventilators, *Crit. Care Med.*, **17**, pp. 682–85.

Kirplani, J., Santos-Lyn, R. and Roberts, R. (1988) Some infant ventilators do not limit peak inspiratory pressure reliably during active expiration, *Crit. Care Med.*, **16**, pp. 880–83.

Kolton, M. (1984) A review of high frequency oscillation, *Canad. Anaes. Soc. J.*, **31**, pp. 416–29.

Lewis, R. P. *et al.* (1991) The use of the Penlon Nuffield 200 in a monoplace hyperbaric oxygen chamber, *Br. J. Anaesth.*, **46**, pp. 767–70.

Lin, E. S. and Hanning, C. D. (1989) Artificial ventilation of the lungs. Chapter 19 in Nimmo, W. S. and Smith, G. (eds) *Anaesthesia*, Vol. 1, Blackwell Scientific Publications, Oxford, pp. 343–68.

Loh, L. and Venn, J. P. (1987) Classifying mechanical ventilators, *Br. J. Hosp. Med.*, **38**, pp. 466–70.

McCluskey, A. and Gwinnutt, L. (1995) Evaluation of the Pneupac Ventipac portable ventilator: comparison of performance in a mechanical lung and anaesthetised patients, *Br. J. Anaesth.*, **75**, pp. 645–50.

Mudaliar, M. Y. *et al.* (1991) Extracorporeal gas exchange for respiratory failure, *Hosp. Update*, **17**, pp. 410–415.

Mushin, W. W., Rendell-Baker, L., Thompson, P. W. and Mapleson, W. W. (1980) *Automatic Ventilation of the Lungs*, Blackwell Scientific Publications, Oxford, pp. 33–251.

Petros, A. J. *et al.* (1995) The Hayek oscillator. Nomograms for tidal volume and minute volume using high frequency oscillation, *Br. J. Anaesth.*, **50**, pp. 601–606.

Pryn, S. J. and Crosse, M. M. (1989) Ventilator disconnection alarm failures, *Anaesthesia*, **44**, pp. 978–81.

Raphael, J. H. and Atkinson, I. (1995) A combined high frequency oscillator and intermittent positive pressure ventilator, *Br. J. Anaesth.*, **50**, pp. 611–613.

Roth, S. *et al.* (1986) Excessive airway pressure due to malfunctioning anaesthesia ventilator, *Anesthesiology*, **65**, pp. 532–34.

Sara, C. A. and Wark, H. J. (1986) Disconnection: an appraisal, *Anesth. Int. Care*, **14**, pp. 448–52.

Scherer, P. W. *et al.* (1989) Convective mixing mechanism in high frequency intermittent jet ventilation, *Acta Anaesthesiol. Scand.*, **33**(suppl. 90), pp. 65–69.

Schreiber, P. (1985) *Safety Guidelines for Anaesthesia Systems*, North American Draeger, Telford, PA, pp. 31 and 51.

Sherry, K. M. *et al.* (1987) Comparison of haemodynamic effects of intermittent positive pressure ventilation with high frequency jet ventilation, *Anaesthesia*, **42**, pp. 1276–83.

Smallwood, R. W. (1986) Ventilators – reported classifications and their usefulness, *Anaesth. Intens. Care*, **14**, pp. 251–57.

Smith, B. E. (1990) High frequency ventilation: past, present and future? *Br. J. Anaesth.*, **65**, pp. 130–38.

Somner, R. M. *et al.* (1988) Hypoventilation caused by ventilator valve rupture, *Anesth. Analg.*, **67**, pp. 999–1001.

Spittal, M. J., Hunter, S. J. and Jones, L. (1991) The Pneupac hyperbaric variant HB: a ventilator suitable for use in a one man hyperbaric chamber, *Br. J. Anaesth.*, **67**, pp. 488–91.

Stott, J. R. (1995) Vibration. Chapter 14 in Ernsting, J. and King, P. (eds) *Aviation Medicine*, 2nd edn, Butterworth Heinemann, pp. 194–95.

Synnott, A., Wren, W. S. and Davenport, J. (1987) Peak intratracheal pressure during controlled ventilation in infants and children, *Anaesthesia*, **42**, pp. 719–62.

Tipping, T. R. and Sykes, M. K. (1991) Tracheal tube resistance and alveolar pressure during mechanical ventilation in the neonate, *Br. J. Anaesth.*, **46**, pp. 565–69.

Wong, L. S. S. and McGuire, N. M. (2000) Laboratory assessment of the Bird T-Bird ventilator performance using a lung model, *Br. J. Anaesth.*, **84**, pp. 811–817.

Chapter 27

Intravenous pumps and syringe drivers

Many infusions are given by gravity assisted, drip sets that give a flowrate dependent on the height of the reservoir above the patient, the length of the tubing, the bore of the IV cannula, the density and viscosity of the fluid being delivered, and the patient's venous pressure. However there is an increasing tendency to use programmable volumetric intravenous pumps and syringe drivers to deliver intravenous anaesthesia, fluids, patient controlled analgesia, epidural infusions and other drugs. Not only are they programmable, but they can also be adjusted to give desired flowrates or volumes. Some infusion devices are powered only by gravity, but the flowrate is controlled by a photoelectric drip rate detector in conjunction with a microprocessor controlled drip occlusion device. Other infusion devices use a stepper motor to control the rate of infusion. A stepper motor is designed so that the rotation is by a fixed amount per supplied electrical pulse, independent of the mechanical load it is carrying. The pulses are controlled by a microprocessor in the pump and the rate of infusion is dependent on the stepper motor's output.

Syringe drivers are designed to use a range of syringe sizes and some require special delivery tubing. The flow is a continuous, pulsatile flow and accuracy is 2–5%. Some syringe drivers are driven by clockwork motors, others by a battery powered motor that is intermittently on and off, depending on required flowrate. The driving mechanism is usually by a screw threaded rod connected to the syringe plunger. Other syringe drivers use a stepper motor connected to the screw threaded rod. Care should be taken not to position the syringe driver above the patient's venous cannula or the syringe may siphon a drug additional to that programmed on the driver, by virtue of the weight of the column of fluid in the tubing above the patient. Care should also be taken to avoid any bubbles in the syringe reaching the patient. Modern syringe drivers are usually sufficiently accurate over the desired range of infusion [Stokes *et al.* 1990]. However, there may be a delay before the drug is delivered to the patient as the parts attached to the syringe take up slack [O'Kelly *et al.* 1992]. Where the bolus to be delivered is small, say 0.5 ml as in a PCA pump (patient controlled analgesia), the accuracy of the delivered bolus becomes questionable [Jackson *et al.* 1991].

INTRAVENOUS PUMPS AND SYRINGE DRIVERS | 365

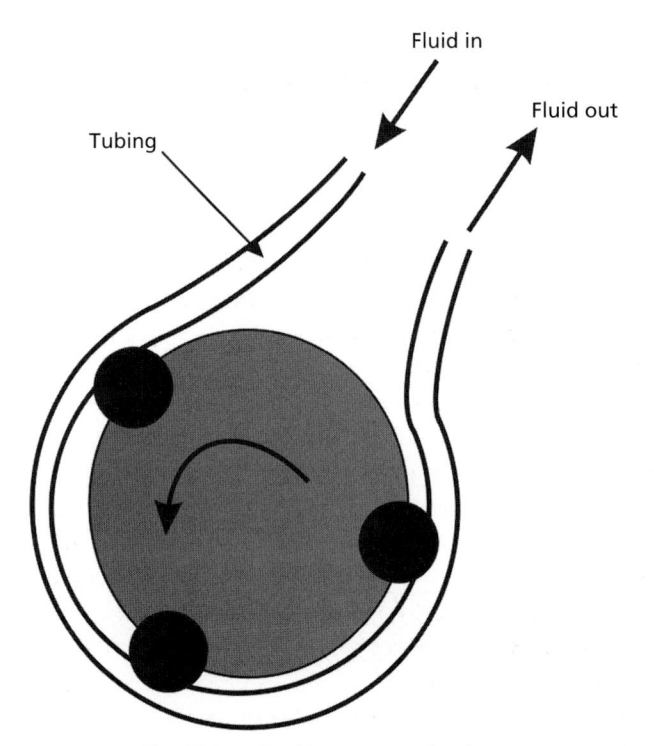

Fig. 27.1 Peristaltic pump mechanism.

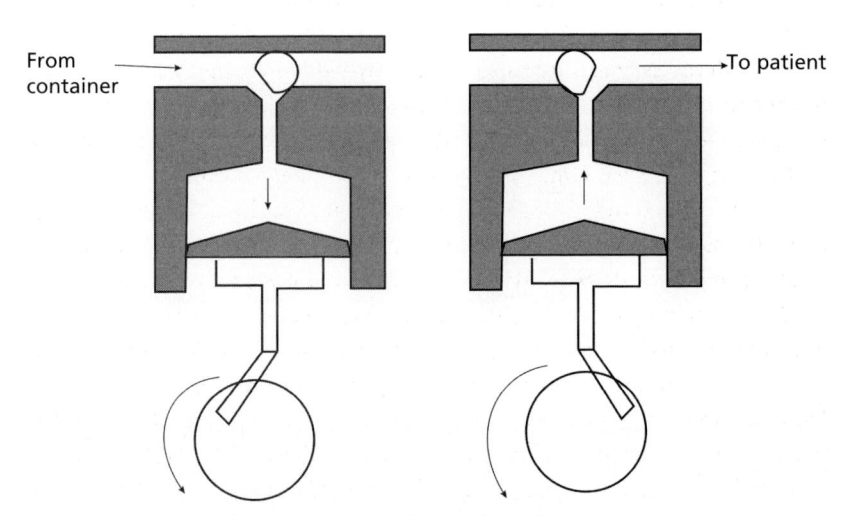

Fig. 27.2 Syringe volumetric infusion pump.

366 | INTRAVENOUS PUMPS AND SYRINGE DRIVERS

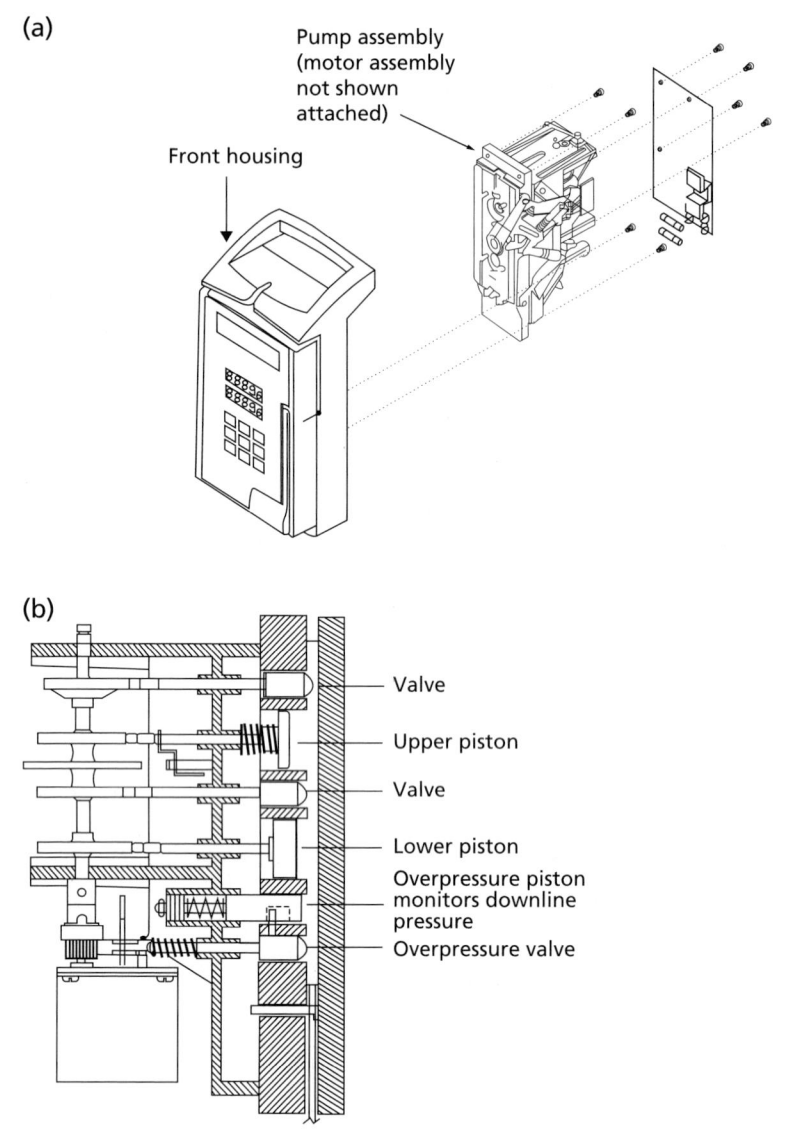

Fig. 27.3 Two sections to show structure of Sims Graseby volumetric pump.

Volumetric pumps enable constant volumetric delivery despite variation in resistance to flow and use either a peristaltic pump, a reservoir or syringe type cassettes to drive the flow and have an accuracy of 5–10%. The peristaltic pump principle is shown in Figure 27.1. The syringe cassette mechanism is shown in Figure 27.2. Both mechanisms are controlled by a stepper motor, controlled by

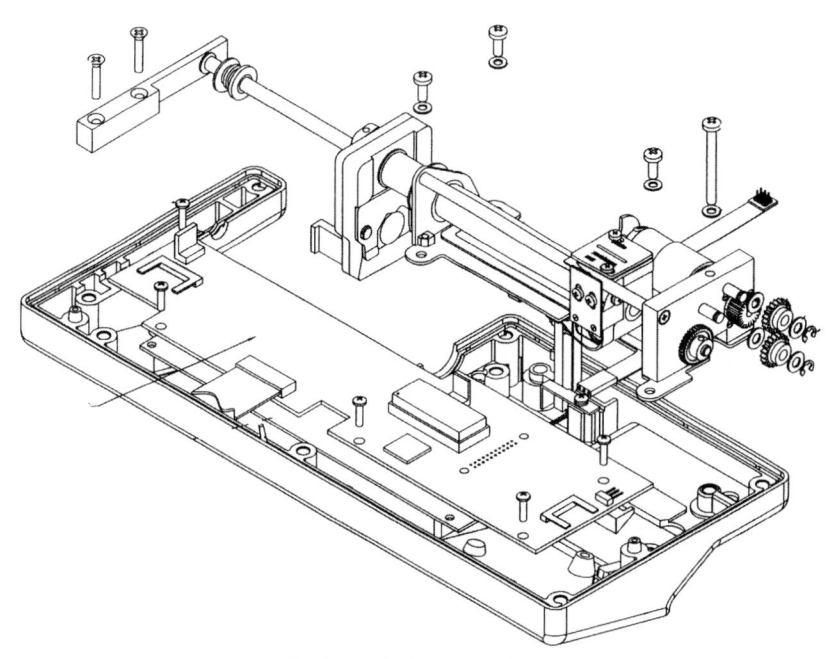

Fig. 27.4 Alaris syringe driver.

a microprocessor. Safety problems with these devices include: a rapid rate of infusion through not setting the device properly, infusion of air, power failure, disturbance of the infusion by a secondary infusion going through the same IV cannula, and software corruption [Hutchison *et al.* 1985].

Figure 27.3 shows an exploded diagram of the Sims Graseby volumetric pump. Figure 27.4 shows a similar diagram of the Alaris syringe driver, showing how different the construction between the two pumps is.

There now exist target controlled infusion pumps with advanced software that allow patient body parameters and desired drug plasma concentration to be entered, subsequent calculated plasma concentration to be displayed and manual adjustment to the pump to be made. The devices can accommodate different sized syringes and different drugs may be programmed for infusion, such as remifentanil and propofol; furthermore, in the case of propofol infusion, the pumps may also be programmed to deliver according to different pharmacokinetic models (Schnider, Marsh). This provides a great deal of choice to the anaesthetist, but also some potential confusion in the face of such choice, with an uncertainty for the feel for what the drug concentration might be at the effector site [Absalom *et al.* 2009]. Furthermore, considerable care should be exercised using such a device in a paralysed, ventilated patient, lest the pump should fail and render the patient aware.

A novel form of ambulatory PCA pump uses a miniature electrolytic cell to produce gas, which compresses the drug reservoir [O'Keefe *et al.* 1994].

References

Absalom, A. R., Mani, V., de Smet, T. and Struys, M. (2009) Pharmacokinetic models for propofol – defining and illuminating the devil in the detail, *Br. J. Anaesth*, **103**, pp. 26–37.

Hutchison, A., Yeoman, P. M. and Byne, A. J. (1985) Evaluation of the IVAC 560 volumetric pump. Clinical and laboratory study, *Anaesthesia*, **40**, pp. 996–99.

Jackson, I. J. B., Semple, P. and Stevens, J. D. (1991) Evaluation of the Graseby patient controlled analgesia pump, *Anaesthesia*, **46**, pp. 478–81.

O'Keefe, D. *et al.* (1994) Patient controlled analgesia using a miniature electrochemically driven infusion pump, *Br. J. Anaesth.*, **73**, pp. 843–46.

O'Kelly, S. W. and Edwards, J. C. (1992) A comparison of the performance of two types of infusion device, *Anaesthesia*, **47**, pp. 1070–72.

Stokes, D. N. *et al.* (1990) The Ohmeda 9000 syringe pump. The first of a new generation of syringe drivers, *Anaesthesia*, **45**, pp. 1062–66.

Further reading

Moyle, J. T. B. and Davey, A. (2005) *Ward's Anaesthetic Equipment*, 5th edn, WB Saunders Co. Ltd, pp. 409–426.

Chapter 28

Environmental safety

> This chapter contains: fire and explosion; atmospheric pollution and anaesthetic gases; scavenging systems.

Fire and explosion

For these to occur, there is a need for combustible material, oxygen and a source of ignition. The risk of these being present results from the use of high oxygen partial pressures and the use of inflammable anaesthetic agents or other inflammable materials.

If the pressure of any gas is increased, heat is liberated. If the gas is oxygen and this comes into contact with something flammable like oil or grease in a confined space, the heat liberated may cause an explosion. Hence oil or grease should be kept well away from pressurised oxygen sources. These include not only oxygen, but pressurised air and pressurised nitrous oxide, which can dissociate into nitrogen and oxygen.

Although modern anaesthetic volatile agents are non-flammable, ether and cyclopropane are flammable and may still be used in some parts of the world. Ethyl chloride, used to test sensory perception in local anaesthetic blocks and methyl alcohol for cleaning skin, are also flammable.

Ignition sources include sparks from static electricity, or faulty electrical apparatus from the diathermy machine or from mains plugs sparking when disconnected. To prevent static electricity causing ignition, not only should efforts be made to minimise the generation of static electricity, but also to discharge any static slowly to earth. There should therefore be an upper and a lower limit to the electrical resistance between the antistatic floor and earth, of between 5 MΩ and 20 kΩ respectively. All equipment capable of generating static electricity should make electrical contact with the floor through a medium made of antistatic (conducting) rubber. Staff footwear should also have antistatic rubber soles and the tubing of breathing systems should also be made of antistatic material.

Classification of anaesthetic proof equipment is based on the ignition energy required to ignite the most flammable mixture of ether and air. 'AP' standard equipment can be used between 5 and 25 cm from such an inflammable anaesthetic gas mixture escaping from a breathing system; furthermore its temperature should not exceed 200°C. *'APG'* standard is a more stringent one for anaesthetic proof equipment; it is based on the ignition energy required to ignite the most flammable mixture of ether and oxygen, which should be less than 1 µJ. The equipment temperature should be not more than 90°C and the equipment can be used within 5 cm of an escaping flammable gas mixture.

Atmospheric pollution and anaesthetic gases

There is a heightened awareness of the many ways in which human activity pollutes the atmosphere, and steps we should take to minimise, and perhaps even reverse the process. There are two compelling reasons to be concerned. The first is the gradual increase in atmospheric temperature due to the trapping of infrared radiation within the atmosphere by so-called 'greenhouse' gases, such as carbon dioxide and methane. The second cause for concern is the loss of the layer of ozone (O_3) at the top of the troposphere, which protects the Earth's surface from the damaging effects of solar ultraviolet radiation.

In its early evolution the original atmosphere of helium and hydrogen was lost from the Earth's gravitational field and replaced with the contents of outgassing volcanoes, such as CO_2 and N_2, and only very small amounts of O_2 were available to support life. Photosynthetic activity in plant life resulted in a build up of O_2 as a by-product from CO_2 consumption. An appreciable greenhouse effect developed due to the release of methane from decaying organic plant matter, even though CO_2 levels themselves were falling at the time. Until the start of the industrial revolution, atmospheric CO_2 levels remained in the 280–300 ppmv, and have risen to 350 ppmv since then, due to the burning of fossil fuels. In addition, methane levels continue to rise, as do the levels of other greenhouse gases, which include nitrous oxide and chlorofluorocarbons (CFCs), which include some anaesthetic volatile agents, and which have half lives of hundreds of years.

Light in the visible and infrared wavebands of inbound solar radiation penetrates the atmosphere, heating the Earth's surface. A proportion of this is reflected back and escapes the atmosphere, but infrared radiation is prevented from escaping by water vapour and CO_2 at the top of the troposphere, whose molecules absorb infrared radiation. This heats up the Earth's surface by as much as 25°C. While methane is present in low concentration, it is 25 times as effective in producing a greenhouse effect. This and other chlorofluorocarbons absorb infrared light in the 7–13 µm waveband, which is different from the wavebands of CO_2 and water absorption. This effectively adds extra panes of glass to the greenhouse, increasing atmospheric heating.

Oxygen successfully absorbs ultraviolet radiation and provides protection on the Earth's surface against excessive ultraviolet radiation, but ozone does this much more effectively. Ozone has a maximum concentration of 10 ppmv at an altitude of 100 000 feet, and is formed by short wave ultraviolet irradiation of oxygen. Below 40 000 feet the ultraviolet radiation is sufficiently weak not to catalyse this reaction, the concentration of ozone being 0.03 ppmv at sea level. Ozone itself is toxic to the human respiratory tract, causing bronchial irritation, a fall in vital capacity, FEV and diffusion capacity, and 10 ppmv can cause fatal pulmonary oedema.

Ozone levels vary throughout the year, depending on the local levels of ultraviolet radiation, but can be removed by the action of free radicals of chloride and nitric oxide. Chlorine is highly reactive, but does not normally penetrate the troposphere. CFCs do, and remain stable and pass freely through the troposphere to photodissociate and produce chloride radicals, each of which can destroy 10 000 molecules of ozone. Nitrous oxide will also pass through the troposphere to produce nitric oxide radicals with a similar effect on ozone. Ozone levels in the atmosphere fell from 300 units to 100 between 1960 and 1994, and continue to fall.

As anaesthetists we are therefore encouraged to use chloride free volatiles, and to avoid nitrous oxide, even though medical use of such CFCs is small compared with those from older refrigerator circuits and other industrial sources.

With this background occupational exposure limits were devised under COSHH (Control of Substances Hazardous to Health) regulations in 1988. Anaesthetic gases were reviewed by the Health and Safety Executive in the 1990s, to determine if there was any link between occupational exposure to nitrous oxide and spontaneous abortions. Although no conclusive evidence was drawn, it is thought that an underlying mechanism of cell damage could be methionine synthetase inhibition, leading to impaired synthesis of vitamin B_{12}, folate and DNA, which could lead to subacute combined degeneration of the spinal cord. The stated 8 hour weighted average occupational exposure limit for nitrous oxide is 100 ppmv. Equivalent ppmv exposure limits under COSHH for Halothane are 10, and for enflurane and isoflurane are 50. Manufacture of Halothane, Enflurane and Isoflurane will stop after 2030.

Scavenging systems

Environmental pollution from anaesthetic waste gases is a potential hazard to staff in the operating suite, as well as a longer term issue with respect to the wider environment. Local pollution can cause staff to have chronic side effects of the volatile agents. As indicated above, chlorinated fluorocarbons and nitrous oxide are both *greenhouse gases* and they interfere with the formation of *ozone*. Such gases from medical sources only account for 1% of gases from all sources. Scavenging systems can only help local pollution.

Scavenging systems are designed to transport waste anaesthetic gases from the breathing system to the atmosphere, in order to avoid pollution in the enclosed environment of the operating room. They should not affect the patient's oxygenation or ventilation adversely. Systems can be either *active* or *passive* and should contain a *collecting system, a transfer system, a receiving system* and *a disposal system*.

Active system

This is shown in Figures 28.1 and 28.2. It has a collecting system, consisting of a shroud, either around the expiratory valve of the breathing system or around the APL valve of the ventilator. This is connected to the transfer tubing by a 30 mm conical connector. At the other end of the transfer tubing is the receiving system, which consists of an open ended reservoir, shown in Figure 28.2, in which received expired gas passes upwards through a non-return valve assembly, visible through a clear tube, which is also allowed to entrain air

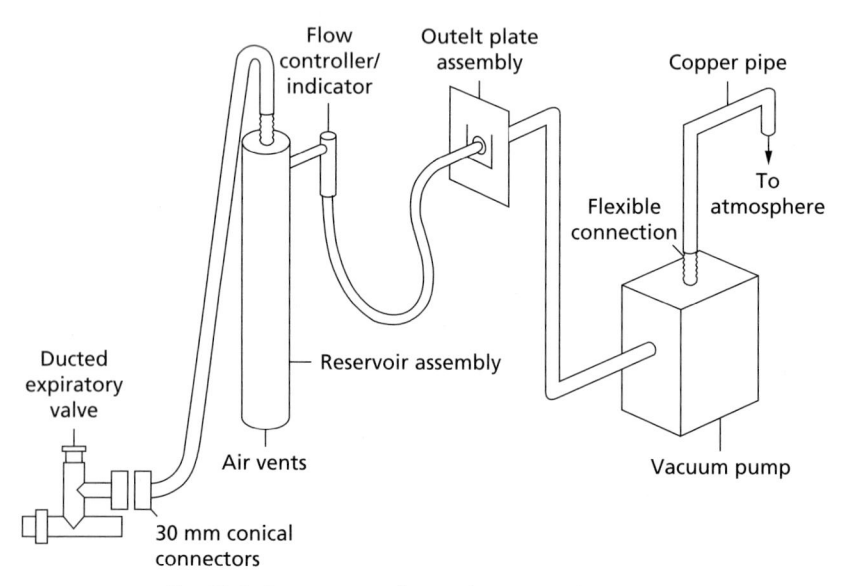

Fig. 28.1 Components of an active scavenging system.

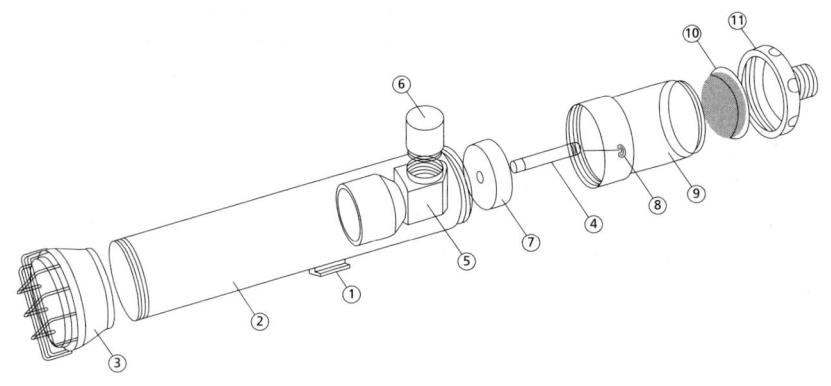

Fig. 28.2 The receiver of an active scavenging system.
Legend: 2 – tube; 3 – vented tube base; 4 – float rod; 5 & 6 – gas inlet; 7– float; 9 – flow tube; 10 – filter; 11 – cap.

from below. This entrainment allows uniform gas flow to the disposal system, which consists of a vacuum pump that delivers the gas to the atmosphere.

Passive system

This is shown in Figure 28.3 and consists of a collecting and transfer system similar to the active scavenging system. The receiving system consists of a reservoir bag associated with two spring loaded valves: a positive pressure

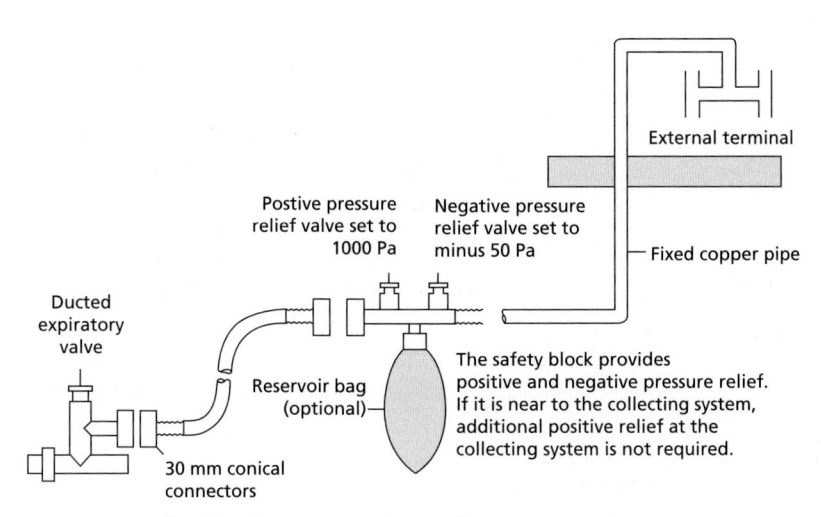

Fig. 28.3 Components of a passive scavenging system.

relief valve set to 1000 Pa, and a negative pressure relief valve set to −50 Pa. The disposal system is a wide bore copper pipe leading to the atmosphere.

Some systems need to be adapted to suit circumstances that might otherwise make scavenging gases difficult, i.e. when using a Jackson–Rees modification of an Ayre's T piece [Dhara *et al.* 2000].

References

Dhara, S. S., Pua, H. L. (2000) A non-occluding bag and closed scavenging system for the Jackson–Rees modified T piece breathing system, *Anaesthesia*, **55**, pp. 450–54.

Further reading

Al-Shaikh, B. and Stacey, S. (2002) Chapter 3 in *Essentials of Anaesthetic Equipment*, 2nd edn, Churchill Livingstone.

Davey, A. J. and Diba, A. (2005) Atmospheric pollution. Chapter 20 in *Ward's Anaesthetic Equipment*, 5th edn, pp. 395–406.

MacDonald, A. G. (1994) A short history of fires and explosions caused by anaesthetic agents, *Anaesthesia*, **72**, pp. 710–722.

MacDonald, A. G. (1994) A brief historical review of non-anaesthetic causes of fires and explosions in the operating room, *Anaesthesia*, **73**, pp. 847–856.

Mushin, W. W. and Jones, P. L. (1987) Initiation of explosions. Chapter 22 in *MacIntosh, Mushin and Epstein: Physics for the Anaesthetist*, 4th edn, pp. 543–585.

Chapter 29

Imaging and radiation

> This chapter contains: radioactive decay; production of X-rays; imaging (X-rays, computed axial tomography, nuclear medicine, positron emission tomography; biological effects of radiation, radiation protection; magnetic resonance imaging, hazards; lasers.
> The chapter links with: Chapter 3.

Introduction

This chapter explains in simple terms the background physics of imaging using standard X-rays, computed axial tomography (CT), nuclear medicine (including positron emission tomography-PET), and magnetic resonance imaging (MRI). It covers the basics of ionising radiation, and also discusses lasers, which are a form of non-ionising radiation (imaging using ultrasound is covered in Chapter 10).

X-rays, CT, aspects of nuclear medicine, and lasers are covered briefly. MRI is examined in more detail as this is a newer modality that is often difficult to comprehend, and in any case often involves the presence of the anaesthetist.

Radioactive decay

Some isotopes are naturally occurring but many of the radioactive nuclides used in medicine are produced artificially by a nuclear reactor or *cyclotron*. Each of these will provide isotopes that are useful for different purposes. Unstable radioactive nuclides achieve stability by radioactive decay, during which they can lose energy. This occurs in a number of ways. For example, atoms can lose energy by ejection of an *alpha* particle (an extremely tightly bound basic atomic structure of 2 protons and 2 neutrons, which is equivalent to a helium nucleus). This occurs if they have too many nucleons (protons or neutrons) and results in the atomic number being reduced by two and the atomic mass by 4. Other ways that unstable radionuclides decay include: emission of an electron (β^-) from the nucleus if the atoms have an excess of neutrons, or by, either emitting a positron (β^+) or capturing an electron if they are neutron deficient. Normally isotopes produced by a reactor will be neutron rich and

Table 29.1 Mechanisms of decay

Nuclear instability due to:	Mode of decay
excess neutrons	beta minus (β^-)
excess protons (deficiency in neutrons)	positron (β^+) or electron capture
excess neutrons and protons	alpha (α)

Fig. 29.1 Radioactive decay showing gamma radiation. The parent in this example could be molybdenum-99 (^{99}Mo). The daughter in the excited state would be technetium-99m or ^{99m}Tc. The ground state would be technetium-99 or ^{99}Tc.

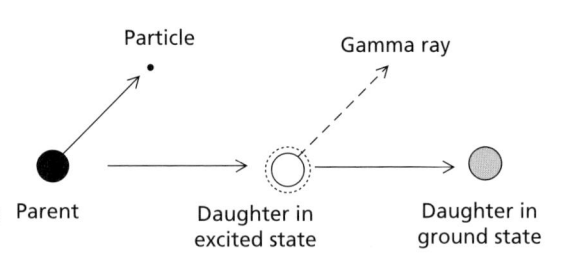

decay by emitting an electron and the cyclotron will tend to produce isotopes that are proton rich and the decay will then be by emitting a positron. This is illustrated in Table 29.1.

The new nuclide formed by the decay process (the daughter nuclide) may be left in an excited nuclear state and can release this excess energy by emission of gamma (γ) radiation as shown in Figure 29.1. This example is where the electron (β^-) has been emitted. The situation is more complex when a positron has been emitted. The emitted positron travels a short distance depending on the isotope (typically less than 1 mm) during which time it loses energy and interacts with an electron. This encounter annihilates both the electron and positron and the result is a pair of 511 keV gamma photons travelling in opposite directions. This pair of gamma rays are used in *positron emission tomography* (PET), discussed later in the chapter.

Gamma radiation is electromagnetic radiation like X-rays. They have similar energy and similar properties. Gamma rays come from the atomic nucleus and X-rays from changes in energy of electrons. Diagnostic radiology involves the use of X-rays produced in an X-ray tube, whilst nuclear medicine imaging detects gamma radiation emitted by a radioisotope that has been injected into the body. The amount of radioactivity is measured in terms of decay rate. A becquerel (Bq) is one disintegration per second.

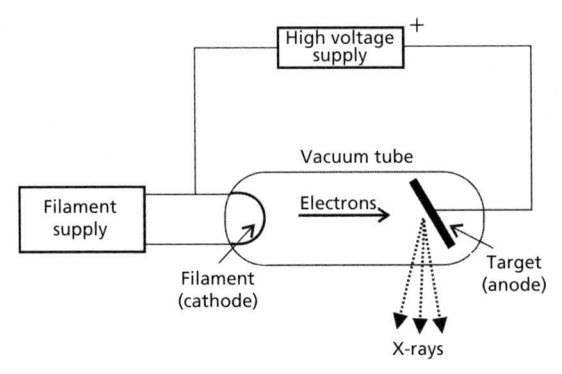

Fig. 29.2 Diagram of an X-ray tube and the power supplies.

Production of X-rays

X-rays are produced when fast moving electrons are suddenly stopped by impact on a target (often metal). The kinetic energy of the electrons is converted into X-rays (1%) and heat (99%). An X-ray tube (see Figure 29.2) consists of a negative electrode (cathode) which has a tungsten filament heated to greater than 2200°C, and a positive electrode (anode) that incorporates a metal target, usually made of tungsten. Electrons leave the cathode, and are accelerated towards the anode under the influence of a large potential difference of 50 000 to 140 000 volts, which is applied across the X-ray tube. These electrons bombard the target anode with a velocity of around half the speed of light.

X-rays are generated in two ways, emerging as discrete characteristic radiation and a continuous spectrum (the bremsstrahlung). Characteristic radiation is illustrated in Figure 29.3(a)–(c). When an accelerated electron from the filament collides with an electron in the K-shell (for example) of a tungsten atom in the target, an electron is ejected from this atom, leaving a *vacancy* in the K shell. The vacancy is most likely to be filled by an electron falling from the L-shell accompanied by the emission of a single X-ray photon. In this example the energy of the characteristic photon is equivalent to the energy difference between the L and K shells in the tungsten target. A characteristic photon can only be produced if the energy of the bombarding electron is greater than the energy required to eject a K-shell electron, and it will be characteristic of the particular target material. The mechanism for producing the continuous spectrum of X-ray energies is illustrated in Figure 29.3(d). A fast bombarding electron penetrates the K-shell and approaches close to the nucleus where it is deflected. It loses some or all of its energy in the interaction

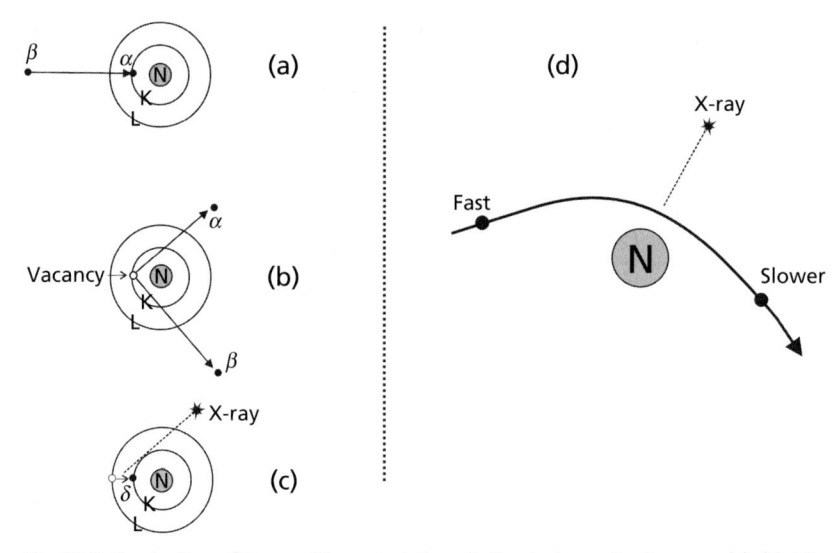

Fig. 29.3 Production of X-rays. Characteristic radiation is shown in diagrams (a)–(c). N is the nucleus of the atom. In (a), a bombarding electron (β) from the filament collides with an electron in the K shell (α). In (b), the α electron is ejected from the atom leaving a vacancy. In (c), the vacancy is filled by an electron δ falling from the L shell, which causes the emission of a single X-ray photon. In (d), N is the nucleus. The fast bombarding electron is deflected and slows down, losing some or all of its energy. This lost energy is carried away as a single photon of X-rays, as shown.

and the *lost* energy is carried away as a single X-ray photon. The process is repeated with various amounts of energy being lost in further interactions with other target nuclei. The net effect of all the electron interactions is a continuous spectrum of X-ray photon energies.

Imaging

Imaging using X-rays

When photons in an X-ray beam fall on to the patient, some pass through unchanged and some are absorbed. Overall the X-ray beam is attenuated exponentially as it travels through the body. When the beam emerges from the patient, it carries a pattern of intensity resulting from the thickness and composition of the organs in its path. The X-rays can be captured on a large flat phosphor screen, which converts the invisible X-ray image into a visible image of light. This image is recorded and then, in computed radiography (CR) and digital radiography (DR), is viewed on a high quality workstation.

Some X-ray equipment allows the display of a 'live' image on a monitor and fluoroscopy is an example of this type procedure. In this case the detector is either an image intensifier or a so called 'flat-plate' digital detector. Some computed tomography (CT) procedures (see below) also use 'real-time' image display, e.g. cardiac CT angiography.

Computed axial tomography imaging

Conventional X-ray imaging produces a two-dimensional image of a three-dimensional object, which can be limiting for diagnosis. In computer axial tomography (CT, or sometimes CAT), a more detailed image of a thin slice of tissue can be obtained by rotating an X-ray tube around the patient. A narrow beam of X-ray is used and the intensity of the transmitted beam is recorded at each location around the body using a bank of highly sensitive detectors. The resultant signals are fed into a computer, which reconstructs a slice of the human body within seconds. The image is formed by a mathematical process. This is most commonly *filtered back projection* but increasingly iterative reconstruction methods are being used. As the demand for combined imaging modalities and cone beam CT increases, there is a need for better processing algorithms, i.e. accurate and more efficient. In general the doses received by the patient undergoing CT examination are significantly higher than for conventional X-rays.

Nuclear medicine imaging

Whereas CT and X-rays give anatomical information about the patient, nuclear medicine imaging gives dynamic information about organ function. The imaging can be two-dimensional, which is called scintigraphy, or three-dimensional, which is called Single Photon Emission Computed Tomography (SPECT). A more complex form of nuclear medicine is PET, which is discussed later. In scintigraphy and SPECT, a radiopharmaceutical, normally produced by a nuclear reactor, is administered to the patient, usually by intravenous injection. This follows biological pathways and is designed to concentrate on the organ of interest. The role of the radionuclide is to signal the location of the radiopharmaceutical by the emission of gamma rays, as discussed earlier. As given in the example, Figure 29.1, ^{99m}Tc is the most commonly used radionuclide. It emits a gamma ray with an energy of 140 keV and has a half-life of 6 hours. Following injection of a radionuclide, the patient acts as a source of radiation, the emerging gamma rays are detected by a *gamma camera*, and an image of the organ of interest is displayed. Depending on the isotope used, a patient can remain radioactive for a long period, ranging from hours to several days. The gamma camera consists of a number of sodium iodide crystals (or more recently, solid state devices) in a chamber. The crystals will produce light pulses when a gamma photo strikes them, and this pulse is then detected by electronic light detectors by the computer. For the 2D image, this will sit over the region of interest near the patient, and for the 3D image, the camera will normally rotate around the patient, and computer reconstruction will create the image.

Positron emission tomography

PET is a more complex form of nuclear medicine already described. It is a 3D image process that again involves functional processes in the body, but it involves different isotopes that have much shorter half lives, and are normally produced by a cyclotron. The PET process involves the pairs of 511 keV gamma rays already described. The radionuclides used in PET scanning are typically isotopes with short half lives such as carbon 11 (half life 20 minutes), nitrogen-13 (10 minutes), oxygen-15 (2 minutes) and fluorine-18 (110 minutes). These radionuclides are incorporated either into compounds normally used by the body such as glucose, water or ammonia. The most commonly used nuclide is fluorine-18 in the form of fluorodeoxyglucose (FDG). The masses of the isotopes are small and so are their half lives. Therefore the cyclotron must be within a suitable travelling distance of the PET facility. The gamma camera in the case of PET is slightly different in that it normally consists of a complete ring of detectors around the patient. The PET process relies on the organ of interest in the body emitting the dual gamma rays in opposite directions, and the principle of coincidence detection will localise the object of interest. In simplistic terms, a detector will detect the gamma photon and then, at this same time, all the other detectors in the ring will be scanned to determine if they have detected a photon as well. The dual detection will be assumed to have come from the same location, but the gamma photons will have travelled in opposite directions. The place of emission will be along the line joining these two detectors. (This is shown in Figure 29.4.) PET imaging is proving very useful in neurology and oncology applications and, in particular, in the assessment of cerebral and cardiac function. It is very useful in tumour staging, in the early detection of cancer, and in the discrimination of benign and active tumours.

Biological effects of radiation and radiation protection

Radiation can cause damage to the human body because the energy it carries causes ionisation (usually of water). Ions are produced that react with other

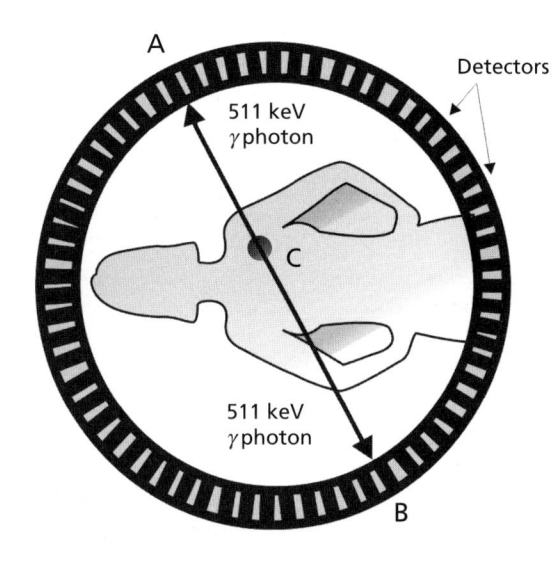

Fig. 29.4 PET imaging. The diagram shows the emission of the gamma rays at C, travelling in opposite directions. They will be detected by the gamma detectors at A and B. The diagram shows a complete ring of multiple detectors all around the patient.

water molecules in the body leading to free radicals and hydrogen peroxide. These produce further reactions with molecules such as DNA, which can lead to cellular and biological damage either immediately or after some time.

Radiation protection

The three basic rules of radiation protection are shielding, distance and time (length of exposure). One should either move away from the source of the radiation, or put an absorber between the source and the person. If, for example, a 185 kBq source from a radiotherapy machine fell on the floor about 1 m away from a person, then this person would receive a lethal dose of radiation in about 2 minutes. However, if the source were placed 4 m away, then the lethal dose would take around 32 minutes. About 30 mm of lead would be required to give the same protection as moving 4 m away. At low energies, lead screening is very effective; for example for a protective screen in a diagnostic X-ray room, 2.5 mm of lead is typically required.

Radiation units

The absorbed dose is measured in terms of the energy absorbed per unit mass of tissue. Energy is measured in joules (J) and mass in kg and the unit of dose is the *gray* (Gy) where 1 Gy = 1 J kg^{-1}. Radiation energy is conveniently measured in electron volts and 1 J = 6.2×10^{18} eV. A dose of 1 Gy means that 6.2×10^{18} eV of energy has been absorbed in 1 kg of tissue.

Table 29.2 Typical figures for X-rays and γ waves doses for different conditions

Condition	Effective dose
Typical whole body dose due to background radiation in 1 year (UK average)	2.2 mSv
Statutory dose limit to general population in 1 year	1 mSv
Statutory dose limit to staff who work with radiation	20 mSv
Dose exposure that will cause nausea, sickness and diarrhoea in some people	0.5 Sv (500 mSv)
Dose exposure that will kill many people in the few months following exposure	5 Sv (5000 mSv)

The unit of equivalent dose, i.e. the dose that gives the same risk of damage to tissue whatever the type of radiation, has a special unit which is the sievert (Sv). An equivalent dose of 1 Sv is equal to 1 J kg^{-1} multiplied by a constant that depends on the type of radiation. For example, for X-rays and γ waves this constant is 1 while for α particles it is 20. In most hospital situations the constant is 1. Typical doses from X-rays and γ waves are shown in Table 29.2.

Measurement methods

Various measurement devices are used for personal radiation dosimetry including the following: film badges (films exposed within screens inside a holder), thermoluminscent dosimeters, laser stimulated phosphor badges and electric dosimeters. A personal dosimetry badge is worn by the user for a set time during which the badge is affected by the amount of radiation falling on to it. It is then processed and provides a retrospective measure of the radiation dose received by that person. Electronic dosimeters give a real-time instantaneous read out and are useful when a measurement of a dose to an individual is required immediately. Although electronic dosimeters in the past were commonly based on ionisation chambers, solid state devices are now widely used.

Magnetic resonance imaging

Magnetic resonance imaging (MRI) is an imaging technique that uses the behaviour of atomic nuclei in strong magnetic fields to produce high quality images of the human body. A patient is placed in a static magnetic field and radio-wave energy is applied. When the radio-wave transmission is turned off, the patient re-emits radio-waves, due to magnetic resonance that occurs in the atomic nuclei of body tissues. The signals from the body are detected and used to construct an image. The largest magnetic resonance signal is produced by the nuclei of hydrogen atoms (free or attached to other molecules, mainly water or fat) due to the high biological abundance of this element in the body. The nucleus of a hydrogen atom consists of a single unpaired proton and it therefore possesses the property of 'spin', essential for the magnetic resonance effect. MRI is able to measure the hydrogen content of individual volume elements, or *voxels* (e.g. a few cubic millimetres) in a slice of the patient and represent it as a shade of grey in the corresponding pixel on the image screen. This section provides a simplified introduction to the subject.

Unpaired protons *spin* continuously like a spinning top around an axis, and this causes them to behave like *bar magnets*. Normally all these individual magnets point (i.e. the direction of north–south pole axis) in a random fashion and combine to give a zero total magnetic effect. However, if a patient lies in a coil (as shown in Figure 29.5) and DC is passed through it (making it into an electromagnet and producing a strong magnetic field of 0.15–3 tesla) the magnetic field causes the *bar magnets* to turn and point along the direction of the field in one of two stable directions: either in the direction of the applied magnetic (spin-up) or in the opposite direction (spin-down). For various reasons, there are slightly more magnets pointing spin-up than spin-down since this represents the lower energy state. MRI depends on detecting this difference although the excess number of spin-up over spin-down protons is very small. For example in a cubic millimetre of water, there are about 7×10^{19} protons of which around 2×10^{14} will be excess spin-up. The resulting magnetic field

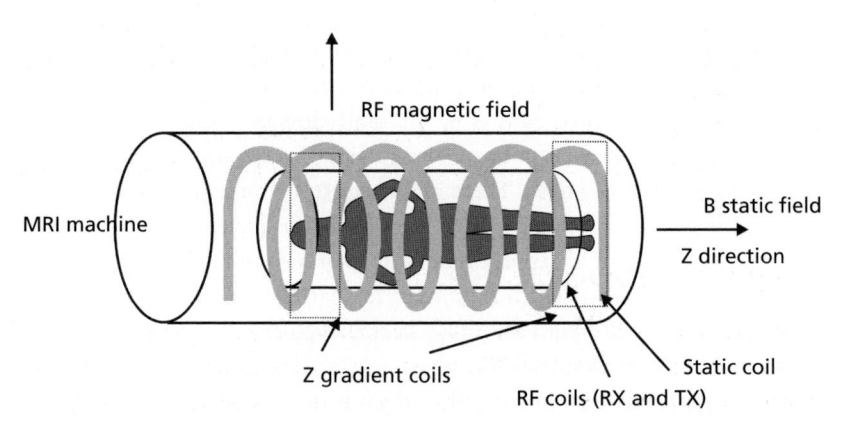

Fig. 29.5 Diagram of the coils in an MRI machine. RF = radiofrequency, RX = receiver coils, TX = transmitter coils.

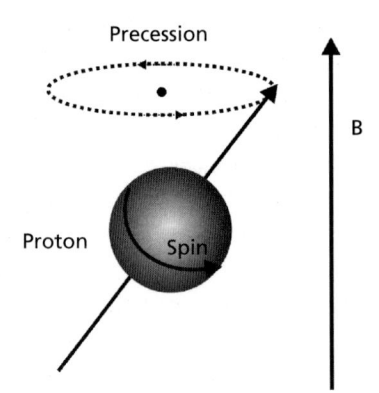

Fig. 29.6 The precession of a spinning proton. The magnetic field direction is shown as B.

is not directly detectable because the direction is the same as the applied field, which would swamp this new field. As well as lining up the protons in a regular manner, the magnetic field causes the spinning protons to *precess*. This effect can be likened to a spinning top as it gradually slows down – it begins to tilt and rotate around another axis (sometimes referred to as wobble). This is illustrated in Figure 29.6. This shows the proton spinning around its own axis, and the separate precession axis.

This precession frequency, the *Larmor* frequency is proportional to both the strength of the magnetic field and a certain property of the nuclei called the *gyromagnetic* ratio. As an example, in a magnetic field of 1 tesla, the proton (hydrogen nucleus) has a Larmor frequency of 42.6 MHz, which is in the radiofrequency range. When a magnetic field is applied, there is a net magnetisation in the direction of the magnetic field, but no magnetisation in any other direction, since all the protons are precessing independently in a random manner (out of phase). There are two potential signals but neither are yet useful: the *excess* spinning effect in the direction of the magnetic field cannot be measured as it is in the same direction and thus swamped by it, and the magnetic effect in other directions is zero because the protons precessions are out of phase. There still needs to be a method of measuring the MR signal (and obtaining an image) and this is carried out by supplying radiofrequency energy via appropriate coils, as shown in the Figure 29.5. This energy, the frequency of which is *tuned* to the Larmor frequency (resonance), is applied by a radiofrequency pulse lasting typically 1 ms and causes all the protons to precess in step (in phase). As a result, their combined spins produce a measurable magnetic pulse which is in a different direction to the applied static magnetic field. Viewed perpendicular to the applied field, this net magnetisation appears to oscillate due to the rotations of the protons. The use of a radiofrequency pulse to flip the direction of the net magnetisation, enables the magnetic resonance signal to be measured. When the radiofrequency pulse is temporarily switched off (the *off* part of the pulse), the circularly precessing magnetic field will be still present for up to 3 seconds as many of the precessions will be still in phase. This radiofrequency energy induces a voltage in the tuned receiver coils (Figure 29.5) and the effect is analogous to a radio transmitter and receiver, both tuned to a specific station. The received signal controls the pixel brightness (that makes up the image), and is dependent on the number of protons per cubic millimetre (the voxel), the gyromagnetic ratio (the magnetic properties) of the nucleus and the strength of the static field.

Only mobile protons give out signals and those immobilised in bone do not. Air in sinuses, for example, having no hydrogen produces no signal and always appears black on an image. Tissues do not vary greatly in the proton density but fat has a higher proton density than other soft tissues and grey matter has greater proton density than white matter.

The MR signal is greatest immediately after the 1 ms radiofrequency pulse has *stopped* and it decays away due to two different processes: *spin-lattice* and *spin-spin* relaxation. Both are affected in different ways by properties of the human anatomy, and knowledge of the relaxation properties of different tissues allows images to be obtained with enhanced contrast. Water and CSF have long relaxation times (also urine, amniotic fluid, and other solutions of salts). In contrast, bone, tendons, and teeth have very short relaxations times. Different solutions in the body also have different relaxation properties [Farr 1998]. With an understanding of the properties of different materials in the body, the contrast of each pixel can be improved to produce sharper, crisper images. Other machine related factors can also influence the image quality and contrast, such as radiofrequency pulse repetition rate, echo time and pulse sequences.

Spatial encoding

An MR scan is made up of a series of parallel slices (Figure 29.7) imaged in turn. To obtain the image, it is necessary to identify the signals coming from individual pixels within each slice (and therefore voxels). Three different perturbations are applied to the magnetic field, one for each of the three orthogonal directions, and this allows spatial information to be localised within the body in three dimensions. Each direction is considered separately in the description below. The following method is used to select the signal specific to an axial slice through the body.

Since the precessing frequency is proportional to the strength of the magnetic field, applying a magnetic field gradient in the direction along the

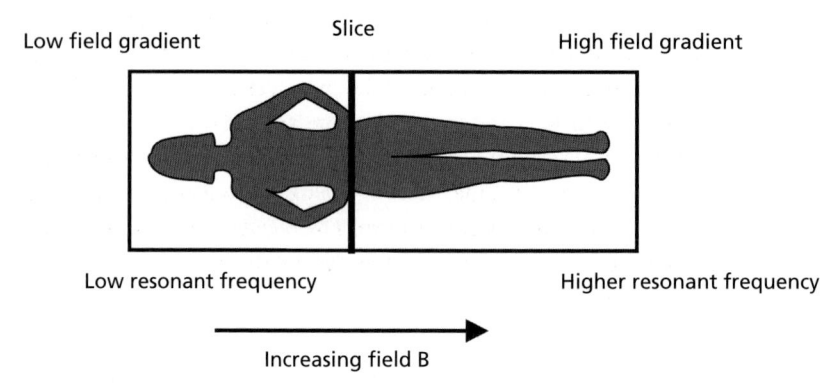

Fig. 29.7 Diagram showing slice selection. There is increasing field strength from left to right, and this causes increasing resonant frequency from left to right of the diagram.

patient, allows axial position to be uniquely related to signal frequency. This is done by supplying DC through a pair of gradient coils which are in addition, but close proximity to, the main static coil (Figures 29.5 and 29.6). The field is made to vary from the head to the foot of the patient with a constant gradient. For example if the field is diminished at the head of the patient and augmented at the toe, accordingly protons at the head will precess more slowly than those in the middle and those at the toe will precess faster. There is therefore a corresponding gradient in the resonant frequency of the protons in the Z direction (see Figures 29.7 and 29.8). The selected slice will have a

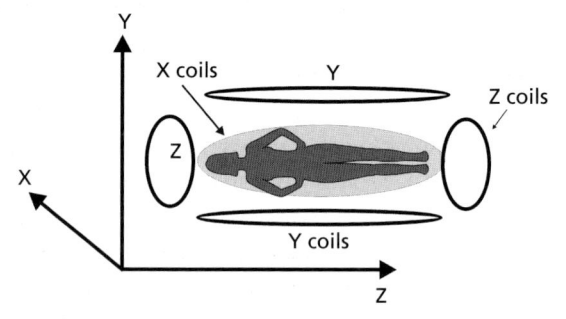

Fig. 29.8 The various gradient coils, X, Y and Z. The Z coils cause a gradient change along the patient. The Y coils are over and under the patient, and cause a gradient change in the Y direction. The X coils (one each side of the patient) cause a gradient change in the X direction.

unique narrow frequency range and the radiofrequency transmitter is tuned to generate a pulse that contains this small range of frequencies. Only protons in a certain thin slice of the patient will be excited by this transmission, and will respond by producing an MR signal as described previously.

In a similar same manner, the other directions (i.e. X and Y) can be selected. The Y coil, shown in Figure 29.8, identifies lines within the slice and the X coil determines actual pixels with the line. Knowing the location of each pixel, the image can gradually be created by putting together all the pixels.

Hazards

MRI does not involve ionising radiation, and up to the present time there has been an absence of serious adverse health effects due to the magnetic field. In 2008 the Health Protection Agency (HPA) recommended a relaxation of the safety guidelines for patient exposure to static magnetic fields. The revised limit for patients and volunteers undergoing MRI procedures permits static magnetic field strengths of up to 4 T in routine operating mode.

Various national and international bodies have proposed exposure limits for workers in static and time varying electric, magnetic and electromagnetic. For example in 2002 the National Radiological Protection Board in UK recommended a limit on magnetic flux density of 0.2 T continuous exposure throughout the working day while permitting short exposures to the head at 2 T and limbs at 5 T. Most recently, the EU Physical Agents (Electromagnetic Fields) Directive 2004/40/EC has been designed to protect the health and safety of workers exposed to electromagnetic fields. The Directive, approved by the European Commission (EC) in 2004, is due to be implemented into national law in 2012.

The switching of the gradient magnetic fields can create induced *eddy* currents (induced currents produced by changing magnetic fields as in transformers), which might be induced in any biological conductor, e.g. nerve fibres, which can result in involuntary muscular contraction, breathing difficulties and possible VF. Particular care is required for patients with heart disease and MRI may not be possible on a patient with an implantable pacemaker. When very strong fields are present, staff and patients may experience flashes of light on the retina and sensations of taste. The magnetic gradient fields should not build up too quickly to avoid shock-like symptoms due to severe nerve stimulation: no faster than 20 T s^{-1} and in practice, 1 to 5 T s^{-1} is typical. As a precaution, MRI should not be carried out during the first trimester of pregnancy.

Microwave heating may occur with radiofrequency fields, especially at the higher frequency associated with strong static fields. It is usually compensated by vasodilatation. However both the cornea and testis have a lower blood supply and may be at risk. The specific absorption ratio is the radiofrequency energy deposited per unit mass of tissue and should not exceed 2.0 W kg^{-1} for the whole body. HPA (2008) recommend a limit on the rise in core body temperature of 0.5°C and an upper temperature limit of 39°C for the trunk and the head. The radiofrequency may also affect the workings of pacemakers, computers and vacuum related devices such as gamma cameras in the immediate vicinity. MR compatible monitoring equipment should be used in the MRI room and patient leads must be redirected so that induced currents are minimised.

Mechanical attraction of ferromagnetic objects varies as the square of the magnetic field and the inverse cube of the distance. The fringe field, which can extend for a few metres, may convert scissors, scalpels and ferromagnetic gas cylinders into projectiles. Therefore non-magnetic materials should always be used. Joint and dental prostheses are firmly fixed and should present no problem. Patients should remove hair grips, ear-rings and mascara containing iron oxide. The fringe fields can also affect watches, credit cards and computers and disks. Free access of the public must be limited to areas where the field strength is less than 0.5 mT.

Special features of MRI

The special features of MRI include:

- ionising radiation is not involved in producing images;
- images can be obtained simultaneously in a number of places at any angle;
- soft tissue contrast is high;
- bone and air do not produce artefacts;
- it is non-invasive.

Lasers

As explained in Chapter 3 (see also Figure 3.7), electrons can be considered to be distributed in a number of concentric shells around the nucleus that have different energies. If electrons move from one level or shell to another, energy is consumed or liberated in steps or quanta.

If external energy, above a certain threshold, is applied to suitable gases, liquids or solids, and therefore to the atoms of these materials themselves, the atoms can leave the *ground state energy* level (resting level) and achieve an *excited* level. The level of excitation depends on the amount of applied energy being supplied by, for example, heat, light or electricity. In the excited state, the electrons move to higher energy orbits, further away from the nucleus. Once an electron moves to a higher energy orbit, it eventually wants to return to the ground state. When this happens, it releases a photon – a particle of light.

The laser is a device that controls the way that energized atoms release protons. *Laser* is an acronym: *L*ight *A*mplification by *S*timulated *E*mission of *R*adiation. In any type of laser, the lasing medium, which gives the laser its wavelength or colour, is *pumped* to get the atoms into an excited state. Typically, intense flashes of light or electrical discharges pump the lasing medium and create a large collection of excited state atoms (atoms with higher than normal number of electrons in excited states). It is necessary to have a large collection of atoms in the excited state, which requires a large energy input. There is then a *population inversion*, which is where the normal ground level electrons become dominant in the higher shells. The electrons have absorbed energy to get to the higher levels and all the atoms will have electrons at the same higher levels. These electrons subsequently are encouraged to relax to a lower energy state and release energy in the form of photons of light of a specific wavelength or colour. For example, two identical atoms with electrons in the same higher level, pumped from the same energy source, will release photons with identical wavelengths. Laser light has the following properties:

- The light is monochromatic: it contains one specific wavelength;
- The light released is coherent: all the wavelengths are in phase;
- The light is highly directional.

To achieve these properties a process called stimulated emission has to occur. A photon emitted by an atom has a certain wavelength, energy and phase. If this photon then interacts with another atom and electron in the same excited state, stimulated emission can occur. The first photon can stimulate or induce atomic emission, such that the subsequent photon from the second atom has the same frequency and direction as the incoming photon.

The cavity containing the lasing material also contains a pair of mirrors. Photons are reflected by the mirrors and so travel back and forth through the lasing medium, constantly stimulating other electrons to make the downward energy jump. In this way a cascade effect occurs. Since one of the mirrors is not completely silvered, the laser light is able to pass through once an appropriate density of light is achieved. To give an example, Figure 29.9 illustrates how a ruby laser works. An energy source (pump), such as a high intensity flash from a mercury vapour discharge lamp, emits radiation into the ruby rod. The light excites atoms in the ruby and some of these atoms emit photons of energy as they fall back to the lower shell. Some of these photons travel in a direction parallel to the crystal axis, and are reflected by the mirrors. As they continue

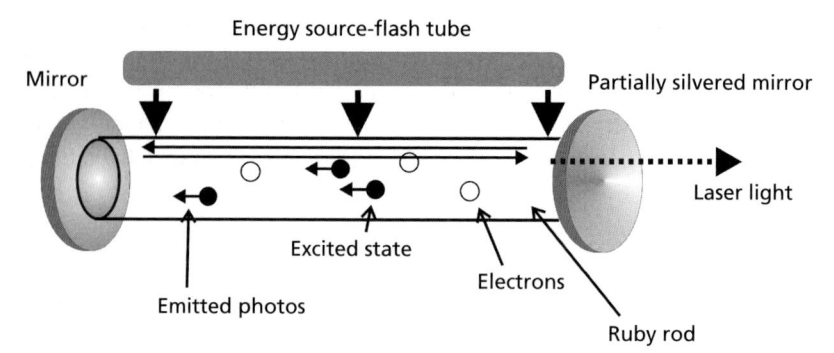

Fig. 29.9 Simplified diagram of a ruby laser. See text for details.

to travel back and forth through the crystal, they stimulate emissions in other atoms and a cascade effect occurs.

There are many types of laser and the lasing medium can be solid, liquid, gas or semiconductor. Examples are:

- solid state lasers, e.g. ruby with a 694.3 nm wavelength;
- gas lasers e.g. CO_2 (far infrared) 10 600 nm;
- dye lasers; complex organic dyes in liquid solution; they are tuneable between 570 and 630 nm;
- semiconductor lasers; low power, current through PN (semiconductor) junction.

The laser can be classified into four classes:

- class 1: these lasers cannot emit laser radiation at known hazard levels;
- class 2: these are low-power visible lasers, at a radiant power not above 1 mW;
- class 3: these are intermediate and moderate power lasers, and are hazardous only for intrabeam viewing;
- class 4: these are high power lasers (> 500 mW continuous), which are hazardous to view and are a skin hazard.

Biomedical Applications

All biomedical applications of laser energy depend on tissue absorption. However, the particular response depends on where the absorption occurs [Cornelius 1983]. There are three ranges of radiant energy (infrared, visible and ultraviolet) that produce different tissue responses. Infrared induces molecular vibration, leading to heat effects. Visible light produces photochemical effects. Ultraviolet radiation can produce molecular bond dissociation, as well as skin burns. CO_2 lasers are normally used for surgical applications, such as in gynaecology, and argon-ion gas lasers can be used for ophthalmology. Solid state lasers such as neodymium (Nd): yttrium-aluminium garnet (YAG) have dermatology uses.

References

Cornelius, W. A. (1983) Applications of lasers in medicine, *Australas. Phys. Eng. Sci. Med.*, **6**, 106–114.

Farr, R. F. and Allisy-Roberts, P. J. (1998) Magnetic resonance imaging. In *Physics for Medical Imaging*, Harcourt Publishers Ltd, Edinburgh, pp. 215–251.

Health Protection Agency (HPA) (2008) *Protection of Patients and Volunteers Undergoing MRI Procedures.*

Further reading

Gibson, K. F. and Kernohan, W. G. (1993) Lasers in medicine-a review, *Journal of Medical Engineering and Technology*, **17**(2), 51–57.

Farr, R. F. and Allisy-Roberts, P. J. (1998) *Physics for Medical Imaging*, Harcourt Publishers Ltd, Edinburgh.

Parker, R. P., Smith, P. H. S. and Taylor, D. M. (1978) *Basic Science of Nuclear Medicine*, Churchill Livingstone.

Chapter 30

Cleaning and sterilisation of equipment

Although a proportion of anaesthetic and surgical equipment is disposable nowadays, there is still a significant amount of cleaning and sterilisation required, and with the emergence of new organisms, the methods used have come under closer scrutiny. It is worth noting, in passing, that hand cleanliness of staff coming into contact with patients has also come under close scrutiny in recent years. Cleaning of equipment involves the physical removal of as much of the infectious agent as possible, usually using water and a detergent, and is an important precursor to disinfection or sterilisation. It can be done manually where automated devices are unavailable. Ultrasonic washers are sometimes used, as are irrigation pumps for flushing out the lumina of tubes.

There is a difference in definition between disinfection and sterilisation. Disinfection is merely the killing of non-spore producing micro-organisms. It kills most bacteria except mycobacteria and spores, and it kills some fungi and some viruses. A higher level of disinfection ensures the destruction of mycobacteria, and most fungi and viruses. Sterilisation is required to kill all micro-organisms, including spores, fungi and viruses. Prions are, however, resistant to most sterilisation procedures. To disinfect or sterilise the modes of heat, chemicals or radiation are used.

Moist heat is much more efficacious at coagulating bacterial protein than dry heat, which requires higher temperatures for longer periods to guarantee effect. Moist heat achieves this by increasing the permeability of the organism's cellular structure to the heat. A hot water washer or low temperature steam applied to instruments for fifteen minutes kills bacteria, but not spores. Higher temperatures are achievable by pressurising the steriliser (Boyle's law). The modern autoclave uses steam at 134°C and 2 bar, when 3½ minutes is sufficient to kill all organisms, providing the steam can reach the instruments; however, to dry the equipment, the steam is removed and replaced with sterile air, the total cycle time being 10 minutes. Rubber and plastic materials degenerate after some time with this regime, and a combination of 121°C for fifteen minutes, or 115°C for thirty minutes can be used instead. If the equipment being sterilised is damaged by superheated steam, a low temperature of 73°C at 290 mmHg gauge pressure for two hours can be used to kill bacteria, and the addition of formaldehyde to the steam deals with spores.

Chemical sterilisation is useful for instruments such as endoscopes, which may not tolerate steam. Chemicals, which kill by protein coagulation, only act on the surface of materials, and some of them react with metals; they may impregnate rubber and plastic, which may be a source of tissue irritation on subsequent use of the equipment. Formaldehyde may be used for endoscopes and catheters, but is hard to get rid of after the sterilisation process. Ethylene oxide is an agent that is bactericidal, and toxic to humans. It is effective against all organisms and penetrates well, but requires at least eight hours for effectiveness. It is explosive at a concentration above 3% in air, and requires 10% CO_2 and a humidity of between 30 and 50% for safety. Although slow, it is an efficacious method of sterilising delicate equipment, such as ventilators, other respiratory equipment and prostheses. To get rid of the ethylene oxide after the sterilisation process, six vacuum cycles are recommended.

A number of liquids are used for sterilisation, for topical use and otherwise. 1–5% phenol is used for cleaning surfaces, and should not be used on objects that contact the patient's skin; it does not kill spores. 0.5–2% iodine in alcohol irritates skin, although 10% povidone-iodine is less so. 70–80% ethyl alcohol is more effective than absolute alcohol. 0.5% chlorhexidine in ethyl alcohol, sometimes with cetrimide added, is effective for skin sterilisation. 2% glutaraldehyde solution in sodium bicarbonate is used for endoscopes and kills bacteria in fifteen minutes and spores in three hours.

Gamma irradiation is used to sterilise equipment because it does not leave the equipment radioactive. A dose of 2.5 Mrad is bactericidal. Clearly careful control of radiation dosing is required. It is normally used to sterilise surgical gloves, surgical instruments and orthopaedic implants.

Further reading

Davey, A. J. and Diba, A. (2005) Cleaning, disinfection and sterilization. Chapter 23 in *Ward's Anaesthetic Equipment*, 5th edn, Elsevier, pp. 435–45.

Davies, N. J. H. and Cashman, J. N. (2006) Anaesthetic equipment. Chapter 2.1 in *Lee's Synopsis of Anaesthesia*, pp. 96–98.

Chapter 31

Medical training using simulators

This chapter contains: simulation centres and environments; model driven simulators, mathematical modelling; scenario based training; surgical skills laboratories; hybrid training; virtual reality training.
The chapter links with: Chapters 1 and 11.

Introduction

Training and education using simulation has been used extensively in many high risk industries including aviation, nuclear power, military and rail. Repeated exposure to simulated crises and events has meant that, for example, airline crews are well prepared to face a rare disaster when it happens in real life. The use of simulation and simulators in medicine, to train and educate healthcare professionals has gained increasing attention in recent years and many simulation centres have now been set up in the UK. The Bristol Medical Simulation Centre, which opened in 1997, was the first training centre of its kind in the UK. There are now over 70 similar centres in the UK and many more with manikins in simpler settings, and hundreds of centres throughout the world [Department of Health 2010]. These offer a similar concept to that which the high risk industries use, where training for medical emergencies using sophisticated manikins are used in realistic medical settings, and task trainers are used to teach, for example, practical surgical skills. Many potential accidents in medicine are due to human error and communication problems [(Kohn *et al.* 1999, Department of Health 2009)]. Simulators can help train teams to function optimally using *human factors* style teaching. Simulation could also be a practical solution to several current educational issues. These include the challenges faced by educational institutions in securing clinical placements, the decrease in social acceptance of trainees learning on patients, the drive to maximise patient safety, and the dramatic decrease in training time being available to junior doctors due to the reduction in hours through the European Working Time Directive.

Simulation centre environments

The simulations centres consist of a number of different designated rooms. Simulated operations and team training can be carried out in the operating room. This room is made as close as possible to the modern operating room. It contains real equipment such as ventilators, defibrillators, patient monitors, trolleys and drip stands. A control room is next to the operating room, with a one way viewing window. This is where the manikin is controlled and where the simulation training is viewed and video recorded. Tutors in the control room communicate with key personnel in the clinical rooms using radio headsets. Audio-visual facilities are located in the control room, and any of the simulation rooms can be recorded. The operating room can be configured to be other clinical rooms, such as ITU.

A de-briefing conference room is essential to the simulation process. This can be connected to the simulation room by a one-way viewing window or viewed using the audio visual facilities. There is normally a multi-bedded ward included. This is provided with monitors, ventilators and similar equipment to that provided in the real situation.

Both the operating room and the multi-bedded wards have manikin-based simulators included in them and this is what makes this type of centre unique. These simulators can be at many levels of complexity with the *model driven* (driven by mathematical models) being the most complex. They are capable of functioning on their own. Non-model driven systems are still very useful, but are less complex and are operated more by the tutor or operator.

Model driven patient simulators

The model-driven high fidelity human patient simulators (METI Florida, US) (an example of a model-driven paediatric simulator is shown in Figure 31.1) are a vast improvement on desk top computer simulators (for example *Gas Man* [Philips 1986]). The human is represented by a correct sized whole body manikin (either adult or child), and human physiology and pharmacology are modelled realistically by computers that drive it. Furthermore, many of the ways in which the medical team interact with a patient can be simulated. The manikin can be cannulated, monitored (ECG, blood pressure, oximetry), intubated and ventilated, and have *intravenous* (in reality, saline) drugs delivered to it.

The most advanced patient simulators have real gas exchange. Anaesthetic gases are recognised by the gas analyser in the manikin associated electronics cabinet, and intravenous drugs are recognised by a bar-code reader, or input via the computer console. Each drug syringe has a bar code on it that relates

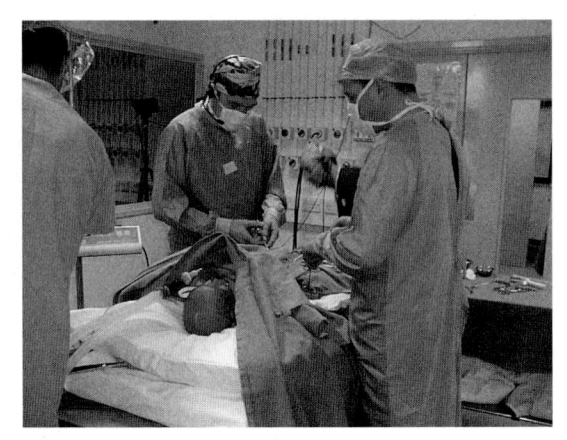

Fig. 31.1 An example of a paediatric human patient simulator in a simulated medical operating theatre.

to the strength and type of drug. The syringe actually contains saline. When the syringe is *injected* into the patient, the bar code reader will relay to the computer what drug and concentration is being supplied. A flowmeter will note the volume of saline supplied to the manikin, so that the computer can provide the correct dose response.

Mathematical modelling

Mathematical models simulate the patient's physiology and pharmacology and provide responses in real time to the therapeutic interventions to the patient. The models rely on software to predict the manikin's responses and have great flexibility for changing the model parameters. The human patient simulator currently contains thirteen different models [*van Meurs et al.* 1997, 1998], but they can be grouped into three groups, the cardiovascular (the heart systems), the respiratory (the lung systems), and distribution and pharmacology (drug systems). The complete models are shown in Table 31.1.

The models receive either inputs from the outside world, e.g. gas flow measurements from the lungs, or take outputs from another model. Many of the models, where appropriate, interact with each other. Most of the model parameters can be changed by user control from the control room terminal, and this can override the physical inputs.

Most of the models provide real outputs, e.g. breath sounds, ECG, various pulses around the body, pulse oximetry signals, blood pressure, thermo-dilution cardiac output, and temperature.

Table 31.1 Human patient simulator mathematical models

Cardiovascular	Respiratory	Pharmacology and reflex control
Uncontrolled cardiovascular system (blood volumes, pressures, flows, wedge)	Lung and chest wall mechanics (gas volumes, pressures, flows)	Distribution of anaesthetic gases
Myocardial oxygen balance	Respiratory rhythm (including respiratory muscle pressure generator)	Pharmacokinetics of intravenous drugs
Cardiac rhythm (including contractility and ECG generators)	Pulmonary gas exchange	Pharmacodynamics and control of circulation
Thermodilution cardiac output	Distribution of respiratory gases	Pharmacodynamics and control of breathing
		Pharmacodynamics of neuromuscular blockade and reversal

Patient types

As the simulator is model driven, the vast number of parameters that control a typical patient can be changed at the start of a simulation session to create many different patient types. The simulator comes supplied with many people types. Standard Man (70 kg, 25 yr old male, standard fitness, non-smoker, standard height, standard physiology), Standard Junior, Stanette (female Stan), and truck driver (very obese, heavy smoker, unfit, short male) are some examples. The basic physiological models can be changed to create different types and many interesting people can be created.

Scenario scripts

The simulator can be allowed to run freely on its own for tutorials on, for example, physiology and pharmacology. The models and other parameters of the simulation can be controlled in real time by the user interface. However, the simulator is also a script-controlled system and computer programs (similar to a high level languages, like C, Pascal) can be used to control and modify physiological parameters and give commands to the simulator to react or respond in a certain manner. The same sequence can be used time and time again for *fair* testing of trainees.

For example, a simple critical incident scenario such as anaphylaxis (idiosyncratic, abnormal, and potentially fatal reaction to a normally innocent

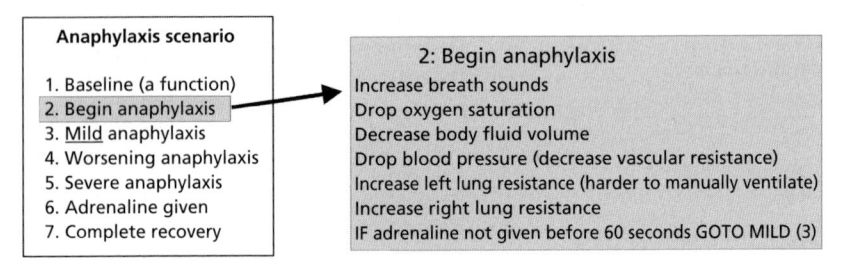

Anaphylaxis scenario	2: Begin anaphylaxis
1. Baseline (a function) 2. Begin anaphylaxis 3. Mild anaphylaxis 4. Worsening anaphylaxis 5. Severe anaphylaxis 6. Adrenaline given 7. Complete recovery	Increase breath sounds Drop oxygen saturation Decrease body fluid volume Drop blood pressure (decrease vascular resistance) Increase left lung resistance (harder to manually ventilate) Increase right lung resistance IF adrenaline not given before 60 seconds GOTO MILD (3)

Fig. 31.2 An example of a simplified anaphylaxis script.

material, drug or foodstuff) uses a script as shown in Figure 31.2. The scripts are written by the course leader and can range between complex and very simple. Each stage or 'state' of the script contains several procedures to produce the effect, and the *begin analphylaxis* state is expanded to show this. The scenario will start at the baseline and will either progress by the operator manually advancing it or can be sequenced automatically. If a Boolean *condition* is met, such as 'IF time >60 s', or 'IF drug given', then the algorithm will automatically advance to the next state.

There are numerous crisis events that can be programmed, for example malignant hyperthermia, asthma and major haemorrhage. These can be as complex as the imagination and skill of the course designer allows. For example a severe crisis can be made to develop. Oxygen supply can actually run out during a Caesarean section operation, complications can occur, there is a fire alarm, power failure, etc.

Scenarios

The scenario computer script must be accompanied by a *play* or scenario to give some background so that the medical *team* and controllers know what is happening. For example, to go with the anaphylaxis script, there can be the following scenario in a GP course:

The delegates are told that an 18 yr old man has been rushed into the emergency room having accidentally eaten a sweet containing peanuts. The man has a known allergy to peanuts and carries an *Epipen* (a source of readily available injectable adrenaline). The delegates now attend the manikin, which has oral itching and swelling, which rapidly progresses to severe stridor (upper airway obstruction). The patient cannot be intubated and requires a surgical airway. The delegates have then to deal with this situation!

At most simulation centres, the human patient simulators are situated in various realistic medical environments, such as fully equipped operating theatres, multi-bedded wards, and side rooms. Further facilities normally include audio-visual equipment to allow relay of live simulations to a conference room, to video recorders, or for video conferencing. Simulated episodes may be replayed and analysed by the trainees with the tutor.

The educational opportunities are many with human patient simulators. Trainees can be taught at the bedside by the tutor who can run the simulator from a handset. Equally, the simulator can be run from a hidden control room so that the trainees are unaware how the tutor is programming the simulator. This allows a range of educational packages to be developed. These range from simple human physiology and pharmacology tutorials to the practice of emergency protocols that involve the whole operating team, e.g. the response to a fire in the theatre complex. In all situations the advantage of simulation is to allow trainees to investigate human physiology and responses or practise the management of medical crises without danger to patients.

Surgical skills labs

Another area where simulation can be useful is in surgery. The surgical skills laboratory can contain many different devices to teach and provide objective evaluations of the trainee surgeons' technical abilities. These simulation devices range from the simple plastic device that teaches medical students how to stitch wounds, to complex sophisticated devices that combine mechanical model with computer stations. Both can be used to teach basic skills and surgical tasks through repetitive proctored challenges, and will enable detection and analysis of surgical errors and near miss incidents without risk to patients. These simulators include an endoscopic sinus surgery simulator for procedural training, and the Minimally Invasive Surgical Trainer-Virtual Reality (MIST-VR) system for basic surgical skills training. Measurements include time-to-completion of tasks, number of errors, economy of motion, and the tracking of the movement of the trainees' limbs.

Hybrid training

Hybrid training is where surgical trainers are combined in either manikins or strapped to actors, but in a realistic environment, so that team training using human factors and crisis events can be practised during 'real surgery' [Kneebone *et al.* 2005].

Virtual reality training

Simulation task and team training can be provided in a complete virtual world (such as Second Life, Kamel Boulos *et al.* 2007) that contains virtual patients, and virtual medical devices within a very realistic three-dimensional clinical environment. Advanced computer models mean that the virtual patients behave very realistically, although there is no actual physical hands on experience.

References

Department of Health (2009) *150 Years of the Annual Report of the Chief Medical Officer*, Department of Health, London.

Department of Health (2010) *NHS Simulation Provision and Use Study Summary Report*, Department of Health, London.

Kneebone, R. L., Kidd, J., Nestel, D., Barnet, A., Lo, B., King, R., Yang, G. Z. and Brown, R. (2005) Blurring the boundaries: scenario-based simulation in a clinical setting, Medical Education, **39**(6), pp. 580–87.

Kohn L. T., Corrigan J. and Donaldson M. S. (1999) *To Err is Human: Building a Safer Health System*, Institute of Medicine (U.S.), Committee on Quality of Health in America.

Kamel Boulos, M. N., Hetherington, L. and Wheeler, S. (2007) Second Life: an overview of the potential of 3-D virtual worlds in medical and health education, *Health Information & Libraries Journal*, **24**(4), 233–45.

Philip, J. H. (1986) Gas Man® - An example of goal oriented computer-assisted teaching which results in learning, *Int. J. Clin. Mon.*, **3**, 165–73.

van Meurs, W. L., Good, M. L. and Lampotang, S. (1997) Functional anatomy of full-scale patient simulators, *J. Clin. Monit.*, **13**, 317–24.

van Meurs, W. L., Nikkelen, E. and Good, M. L. (1998) Pharmackinetic-pharmacodynamic model for educational simulations, *IEEE Transactions on Biomedical Engineering*, **45**(5), 582–90.

Further reading

Henson, L. C. and Lee, A. C. (eds) (1998) *Simulators in Anesthesilogy Education*, Plenum Press, New York and London.

Riley, R. H. (2008) *Manual of Simulation in Healthcare*, Oxford University Press, Oxford.

Index

Note: Page numbers in *italics* refer to Figures and Tables.

10–20 system, EEG electrode placement 249, *250*

AAI pacing mode 273
absolute blood flow measurement 157–8
absolute humidity 126
absolute zero 119–20, 123
absorbed dose, SI unit *48*
acceleration plate, mass spectrometry 229
active electrode, diathermy *281*
active scavenger systems 372–3
adiabatic compression 120
advancing front theory 361
AEP/2 monitor 267
aepEX system 267
Airtraq optical laryngoscope *318*
airway laser surgery, endotracheal tubes *312*
airway management devices (AMD) 301–2
  artificial airways 302
    cuffed oropharyngeal airways (COPA) 308
  endotracheal tubes 308–13
  facemasks 303–4
  I-gel *307*
  laryngeal mask airways (LMA) 304–7
airway monitoring 164–5
airway resistance
  effects on flow and tidal volume
    constant flow generators *342*
    constant pressure generators *340*
  ventilation 338, *340*
Alaris syringe driver *367*
alarm devices 162–3
  in diathermy equipment 285
  disconnection alarms 165
  oxygen supply 164
  in ventilators 344
A-line monitors 267
alpha activity, EEG 248
alpha particles 375
alternating current (AC) 67–8
  frequency, physiological importance 95–6
altitude, pressure changes 109–10
aluminium, electrode potential *84*
Ambu mark III valve *358*
amount of substance, SI unit *48*
amperes (A) *48*, 60–1

amplifiers
  biological signal processing 81
  operational 76–7
  patient isolated differential amplifier 79–81
  single ended 77–9, *78*
anaerobic threshold 236–7
anaeroid barometer 201–2
anaesthetic depth assessment *see* depth of anaesthesia monitoring
anaesthetic machine
  pressure regulators 293–6, *295*
  safety checks 296–9
  safety devices 296, *297*
anaesthetic proof equipment, classification 369–70
analogues 188–9
analogue-to-digital converter (ADC) 81–3
analysis of variance (ANOVA) 39–40
anaphylaxis scenario script *398*
anatomical deadspace measurement, Fowler's method 210, *211*
ancillary equipment, safety checks 299
ångstrom (Å) *48*
AOO pacing mode 273
AP standard 369–70
APG standard 370
argon-ion lasers 391
arithmetic mean 26
arterial blood pressure monitoring 166–7
  see also blood pressure measurement
arterial stenosis, Doppler frequency shift spectrum *155*–7
arterial waveform *178*, 189
artificial airways 302
ascending bellows type ventilators 344, 346
astronauts, decompression sickness 132
atmosphere, evolution 370
atmospheric pollution 370–1
atmospheric pressure 108
  variation with altitude 109–10
atomic mass number 55
atomic number 55
atomic structure 55–6
atrial triggered pacing 272
attenuation 75
attribute data 22
audio amplification 78

auditory evoked responses (AER) 253, *254*
  acquisition 225–6
  averaging *226*
  depth of anaesthesia monitoring 255, 267
auscultation, blood pressure measurement 171–2
auscultatory gap, Korotkoff sounds *171–2*
autoclaves 392–3
averages 26
aviation, decompression sickness 132
awareness indication 261
  see also depth of anaesthesia monitoring
axisymmetric velocity profiles *360*
Ayre's T piece *323*, 329

B type equipment, electrical safety classification 93
  symbol *94*
baffles, vaporisers *140*
bag-in-bottle ventilators 345–6
  Datex-Ohmeda ADU ventilator 346–8
Bain breathing system 327–9, *328*
  safety checks 298
band-pass filters 73–4
Baralyme, CO_2 absorbers 333
base excess 241–2
bases of logarithms 14, 16
batteries 61
Bear Cub ventilator 345
becquerels (Bq) 376
Beer–Lambert law 216–17
bell shaped curve *see* Normal (Gaussian) distribution
Bell spirometer *203–4*
bellows and cone design, vaporisers *141*
'bends' 132
Bernoulli's equation 106–7
beta activity, EEG 248
beta particles 375–6
BF type equipment, electrical safety classification 93
  symbol *94*
bicarbonate, standard 242
bicoherence 264
bilirubin, and pulse oximetry 220
bimetallic strips, vaporisers *140–1*
binary data 22
biological signal processing *81–2*
biphasic effect, MPF 252, *253*
biphasic waveforms, defibrillators 275, *277–8*
bipolar electrodes, diathermy 283–4
bipolar leads, pacemakers 271
bispectral index (BIS) monitors 167, 263–4, *265*
black-box concept *50*
blood, non-Newtonian nature 102
blood–gas partition coefficient 133

blood flow
  absolute measurement 157–8
  electrical analogue 71–2
blood flow waveform, frequency components *52–3*
blood gas analysis 239
  derived variables 241–2
  intravascular 243
  pCO_2 electrode 241
  pH electrode 239–*40*
  pO_2 electrode 239
  sources of error 242
  temperature effects 242
  transcutaneous 242–3
blood pressure measurement 170, 189
  alternative cuff sites 176
  causes of error 172
  cuff morbidity 176
  CVP 190
  direct 177–8
    components of monitoring system 178–82
    resonance and damping 182–8
  non-invasive 170–1
    auscultation 171–2
    cuff size 175
    Doppler ultrasound 175
    oscillometry 173–5, *174*
    oscillotonometry 172–3
    palpation 171
    plethysmography *176–7*
blood pressure monitoring 166–7
blood pressure transducers, calibration 53–4
blood velocity, Doppler velocimetry *154*
Bodok seals 293
body plethysmograph *209*–10
Bohr's equation 212
boiling 138
bougies 320
Bourdon gauge *202*
Boyle's bottle 141, *142*
Boyle's law 111, 118
bradycardia, pacemakers 270–4
breathing systems 322
  circle system 330–5
  Humphrey ADE system 329
  Mapleson A systems 323–6
  Mapleson B and C systems 326
  Mapleson classification *323*
  Mapleson D, E and F systems (T pieces) 326–9
  safety checks 298
  Venturi systems 330
bremsstrahlung 377
bridge circuits 65–6
Bristol Medical Simulation Centre 394
bronchoscopic surgery, high frequency jet ventilation (HFJV) 359

bubbles, enhancement of ultrasound
    visualisation 153
buffer base 241
Bunsen solubility coefficient 133
burst suppression, EEG 248

calculus
    differential equations 13–14
    differentiation 8–9
    integration 10–12
calibration, transducers 53–5
candela (cd) *48*
capacitance, stray 91–2
capacitative transducers 87
capacitive based gauges, direct blood pressure
    measurement *181–2*
capacitors *68*, 69
capillary pressures 190
capnographs 166
carbon dioxide
    atmospheric levels 370
    gas cylinders *292*
carbon dioxide absorber, circle breathing
    systems 333–4
carbon dioxide lasers 390, 391
carbon dioxide monitoring 165
carbon monoxide, diffusion capacity
    of lung 113
carboxyhaemoglobin (HbCO), absorption
    spectrum *217*, 220
cardiac output measurement
    echocardiography 195–6
    pulmonary artery catheters 192–5
    pulse contour analysis 197–8
    thermodilution curve *12–13*
    transthoracic electrical impedance 198–9
    ultrasound 107
cardiac rate and rhythm, monitoring
    165–6
cardiac vectors *245–6*
categorical data 22–3
    frequency tables 44–5
catheters
    intubation aids 320
    sterilisation 393
centiles 27
central range of data 27
central venous cannulation, needle
    positioning verification 152–3
central venous pressure 190
cerebral function monitor (CFM) 261–2
CF type equipment, electrical safety
    classification 93–4
    symbol *94*
charge, electric 60
Charles' law 118, *120*
chemical sterilisation 393
chemical transducers 87

Chi squared (X^2) distribution 45
children
    circle breathing systems 335
    endotracheal tube length 311
    endotracheal tube size 310
    FGF requirements for T pieces *330*
    ventilators 345, 351–2
    see also neonates
chlorofluorocarbons (CFCs)
    greenhouse effect 370
    ozone depletion 371
circle breathing systems 330
    arrangement of components *331*
    arrangements *332*
    CO_2 absorber 333–4
    fresh gas flows 334
    overflow valve placement 332
    paediatric use 335
    placement of FGF inlet 331–2
    reservoir bag placement 332–3
    safety checks 298
circuit breakers 97–8
circulation, pressure distribution *189*
circulation monitoring 165–7
Clarke polarographic electrode 234–6, 239
class 1 equipment, electrical safety
    classification 90–*1*
class 2 equipment, electrical safety
    classification 91
    symbol *94*
class 3 equipment, electrical safety
    classification 91
cleaning of equipment 392
CO_2 *see* carbon dioxide
coagulation diathermy 282, 283
Coanda effect *107*
coaxial Mapleson A (Lack) breathing
    system *325*
coaxial Mapleson D (Bain) breathing
    system 327–8
coherent averaging 253–4, *255*
coils, current-carrying 62–3
colour coding, gas cylinders *292*
colour flow Doppler imaging 155
combined gas law 118, 120
Combitube 301, 308
common mode rejection ratio (CMRR) 80
complex numbers 70
compliance 338–9
    effects on flow and tidal volume
        constant flow generators *342*
        constant pressure generators *340*
Compound A 334
compressed air 290–1
    gas cylinders *292*
compressed spectral array (CSA) monitor
    262–3
compression of gases 120

computed axial tomography (CT, CAT) 379
conduction of heat 125
confidence intervals 34
constants, integration 10–11
contingency tables 44–5
continuous (analogue) data 23
convection 125
Cook catheter 320–1
copper, electrode potential *84*
corrected flow time (FTc), echocardiography 195
correlation 42–*3*
correlation coefficient 42
cosecant (cosec) 5
cosine (cos) 5
cotangent (cotan) 5
coulombs (C) *48*, 60, 61
cowls, vaporisers *140*
cricothyrotomy tubes 312
critical damping 184, 188
critical flow rate (critical velocity) 102–3
critical incidents
    classification *161*
    development *163*
    impact of monitoring 160–2
critical pressure 122
critical region, hypothesis testing 34
critical temperature 122, 134
critically damped waveform, defibrillators *275*
crystallization, latent heat of 123
cuff morbidity, blood pressure measurement 176
cuff pressures, endotracheal tubes 311
cuff sites, blood pressure measurement 176
cuff size, blood pressure measurement 172, 175
cuffed endotracheal tubes 310–11
cuffed oropharyngeal airways (COPAs) 301, 308
current, electric 61
current density 280–1
    safety issues 99
curves
    area under 10–12
    maxima and minima *9*
cut-off frequency, filters 75
cutting diathermy 281, *282*, 283
cyanosis, detection 160–1, 216
cycling modes, ventilators *340*, 343
    comparison of effects *344*
cyclopropane, flammability 369
cyclotron 375, 376, 380

Dalton's law of partial pressures 111
damped natural frequency 185–6
damped resonant systems, electrical analogue 188–9

damping, direct blood pressure measurement 183–8
damping ratio 184, 188
Danmeter monitors 267
data
    categorical 22–3
    numerical 23
data displays 23–5
data variability 26–7
Datex-Engstrom device 228
Datex-Ohmeda ADU ventilator 346–8, *347*
Datex-Ohmeda anaesthetic machine, integrated vaporiser 142–6, *145*
DDD pacing mode 273
de-activation state, EEG 248
decibels 15
decompression sickness 132
defibrillation damage protected equipment, symbol *94*
defibrillator waveforms 274–5
defibrillators 274
    biphasic waveforms *277*–8
    monophasic circuits *276*–7
degrees of freedom 34
delay time, respiratory gas analysers 222, *223*
delta activity, EEG 248
demand led pacing 272
density 56
    relationship to flow characteristics 103–4
density spectral array (DSA) 263
depth of anaesthesia, effect on EEG 249
depth of anaesthesia monitoring 167, 261
    auditory evoked response monitors 267
    bispectral index monitors 263–4
    cerebral function monitor 261–2
    compressed spectral array monitor 262–3
    ECG-based 267–8
    emerging devices 268
    entropy value of EEG 264–6
    Narcotrend monitor 266
    patient state analyser 267
descending bellows type ventilators 344, 346
desflurane, saturated vapour pressure (SVP) *139*
desflurane vaporiser 138
desiccation diathermy 283
desynchronised state, EEG 248
dew point 127
diagnostic tests 38, *39*
diamagnetic molecules 233
diastolic blood pressure
    Korotkoff sounds 172
    oscillometry *174*–5
diastolic decay *178*
diathermy 280
    blended modes 283
    coagulation 282

current density 99, 280–1
cutting 281
desiccation 283
electrodes
  bipolar 283–4
  monopolar 283
  fulguration 282
  safety 283, 284
    stray capacitance 284–5
  as source of interference 285–6
  waveforms *282*
dichotomous data 22
differential amplifiers 79–81
differential equations 13–14
differentiation 8–9
  digital 246
differentiators 76
diffusion 112–13
diffusion capacity of lung 113
diffusion hypoxia 113
digital differentiation 246
digital signal processing *82–3*
dimensional analysis 48–9
dimensions 47
Dinamap oscillometer 173–4
direct blood pressure measurement 177–8
  monitoring system components
    hydraulic coupling 178–9
    transducers 179–82
  resonance and damping 182–8
direct current (DC) 66–7
  electrolysis 99–100
disconnection alarms 165
discrete data 23
disinfection 392
diving, decompression sickness 132
domestic mains electricity *89–90*
DOO pacing mode 273
Doppler effect *153*
Doppler ultrasound
  absolute blood flow measurement 158
  blood pressure measurement 175
  cardiac output measurement 195–6
  colour flow imaging 155
  frequency shift spectrum *156*
Doppler velocimetry 154
  applications 156
  duplex *154*
  frequency shift spectrum 155–7
dose equivalent, SI unit *48*
dosimetry badges 382
double lumen ETTs *313*
doubly insulated equipment 91
Draeger anti-hypoxic mechanism *297*
Draeger Evita 4 ventilator 357–8
Draeger Vapor vaporiser 141–2, *144*
draw-over vaporisers 146
drift 50–1

duplex Doppler 155
dye lasers 390
dynamic compliance 338–9
dynamic viscosity (η) 48–9, 101–2

e (natural exponential) 16–17
earth connection 89
earth faults *92*, 94–5
earth free electrical supplies 98
Earth Leakage Current 92
earthing of electrical equipment 90
ECG *see* electrocardiogram
echocardiography
  cardiac output measurement 195–6
  cardiac structure and function 196
eddy currents, risk in MRI 387
EEG *see* electroencephalogram
Einthoven's triangle *245*
electric charge 60
  SI unit *48*
electric current 60–1
  alternating and direct 66–8
  induction of magnetic fields 62–3
  Ohm's law 63
  SI unit *48*
electric field exposure, safety limits 387
electric potential, SI unit *48*
electrical circuits *63–4*
  bridge circuits *65–6*
  capacitors and inductors 68–9
  voltage dividers *64*
  water analogy 61–2
electrical currents, physiological effects *95*, 280
electrical double layer 85
electrical power 58–9
electrical safety
  circuit breakers *97–8*
  class 1 equipment 90–*1*
  class 2 equipment 91
  class 3 equipment 91
  current density effect 99
  DC and electrolysis 99–100
  earth faults 94–5
  earth free supplies 98
  heating effects 98–9
  leakage currents *90*, 91–3
  microshock 95–7
  physiological effect of electric currents *95*
  static electricity 100
  types B, BF, and CF equipment 93–4
electrical safety symbols *94*
electricity 60
  domestic supply 67, 68, *89–90*
  transmission heat loss 89
electrocardiogram (ECG) 244–6
  artefacts and interference 248
  cardiac vectors *245–6*

electrocardiogram (*Continued*)
  depth of anaesthesia monitoring 267–8
  digital differentiation 246–7
  indices 247
  monitoring 165–6
  signal amplification 79–81
electrode arrangements
  ECG monitoring 245
  EEG 249, *250*
electrode potentials *84*
electrodes 84–5
  for diathermy 280–*1*, 283–4
  in electromyography 257
  impedances 85
  use on humans 85–6
electroencephalogram (EEG) 248–9
  analysis 249–50
    frequency domain analysis 250–2
    time domain analysis 250, *251*
  data collection 249
  depth of anaesthesia monitoring
    bispectral index monitor 263–4
    cerebral function monitor 261–*2*
    compressed spectral array monitor 262–*3*
    entropy values 264–6
    Narcotrend monitor 266
    patient state analyser 267
  influencing factors 248–9
  see also evoked responses
electrolysis 99–100
electromotive force (emf) 61
electromyogram
  electrodes 257
  nerve stimulator 257
  response measurement 257
    electrical responses 258–*9*
    mechanical responses 258
  stimulations 256–7
electron emission, radioactive decay 375–6
electron shells 55, *56*
electronic dosimeters 382
electrons 55
emergency oxygen bypass control, safety checks 297–8
enclosed afferent reservoir (EAR) breathing system 326
enclosure Leakage Current *see* Touch Current
endoscopes, sterilisation 393
endotracheal tubes (ETTs) 308, *309*
  in airway laser surgery *312*
  airway resistance 310
  avoidance of kinking 311
  confirmation of placement 165, 312
  cuff pressures 311
  cuffs 310–11
  double lumen *313*

  length 311
  paediatric *310*
energy 57–8, *59*
  SI unit *48*
enflurane, occupational exposure limit 371
Entonox
  effect of low temperatures 134
  gas cylinders *292*
  lamination (Poynting effect) 289–90
Entonox supply 289–*90*
entropy value, EEG 264–6
epochs, EEG 250–*1*
error types, hypothesis testing 34, *35*
ether, flammability 369
ethyl alcohol, skin sterilisation 393
ethyl chloride, flammability 369
ethylene oxide, sterilisation of equipment 393
evidence based medicine 46
evoked responses 252–5
  acquisition 255–6
  auditory evoked response monitors 267
  electromyogram 256–9
exercise testing 236–7
expected frequencies 45–6
expiration, ventilators 343–4
explosion risk 369–70
  ethylene oxide 393
exponential functions 18–21
exponential processes 13
extracorporeal membrane oxygenation (ECMO) 361
extrapolation 4

F ratio 40
F test 37
facemasks 303–4
false positives/negatives, hypothesis testing 34, *35*
Fast Fourier Transform (FFT) 82
Fathom monitor 267–8
feedback, operational amplifiers 77
fibre optic laryngoscopes 316, *318*, 320
fibre optics *319*
  applications 318–19
  reflection and refraction of light *319*–20
Fick principle 193–4
Fick's law of diffusion 112–13
filling ratio 291
film badges 382
filtered back projection 379
filters
  simple band-pass 73–4
  simple high pass 76
  simple low pass 74–6
Finapres *176*–7
fire risk 369–70
first differential 8
fixed rate pacing 271

Fleisch pneumotachograph 205–7
flexible LMA 305
flow generators 339, 341–2
   electrical analogue *342*
   internal resistance 343
flow measurement 104–7
   Fleisch pneumotachograph 205–7
   hot wire anemometry 208
   peak expiratory flowmeter 212, *213*
   pitot tube *207*–8
   Rotameter *212–14*
   ultrasonic flowmeter *208*
   Wright's respirometer 204–5, *206*
flow time (FT), echocardiography 195, *196*
fluid columns, pressure measurement 108–9
fluid flow 101–4
fluid velocity profile, laminar flow *102*
fluorodeoxyglucose (FDG) 380
Fluotec 3 vaporisers 142
flux density 63
foetal haemoglobin (HbF), absorption spectrum 220
force 57
   SI unit *48*
forced expiratory volume in one second (FEV_1) 204, *205*
forced vital capacity (FVC) 204, *205*
forced warm air blankets 126
formaldehyde, sterilisation of equipment 393
four-dimensional ultrasound 158
Fourier analysis 7
Fourier series 53
Fowler's method, anatomical deadspace measurement 210, *211*
fractional saturation 220
frequency
   AC current, physiological importance 95–6
   SI unit *48*
   ultrasound 148–9
frequency distributions, shape 28–9
frequency domain analysis, EEG 250–2
frequency domain signals 51
frequency polygons 30
frequency response 51–3
   effect of damping *186*–7
frequency shift spectrum, Doppler *155–7, 156*
frequency spectra 51–2
frequency tables 44–*5*
fresh gas flow inlet, circle breathing systems 331–2
fresh gas flows (FGF)
   circle breathing systems 334
   Mapleson A breathing systems 324, 325, 326
   Mapleson D breathing systems *328*–9
   T pieces 326, 329, *330*

F-type isolated floating applied part 93, 94
fuel cells 236
fulguration 282
functional gas storage property, T pieces *327*
functional residual capacity 204
   measurement
     body plethysmograph *209*–10
     dilution and washout method 210–12, *211*
     helium dilution technique 212
functional saturation 220
fundamental frequency 51
fuses 90, 97

gamma camera 379–80
   PET *381*
gamma radiation 376
   sterilisation of equipment 393
   typical radiation doses *382*
   see also positron emission tomography (PET)
gas analysers *see* blood gas analysis; respiratory gas analysis
gas cylinders 291–3
   Bodok seals *293*
   colour coding *292*
   dimensions *292*
   gas capacities *292*
   ISO pin index system *292*–3
   pressure gauges 203
gas driven nebulisers 128
gas exchange, patient simulators 395
gas lasers 390
gas laws 117–20
gas pressure measurement 201–3
gas supplies
   Entonox 289–*90*
   medical compressed air 290–1
   nitrous oxide 289
   oxygen 287–9
   safety checks 297–8
gas transport mechanisms, high frequency ventilation (HFV) 360–1
gas volume and flow measurement
   Fleisch pneumotachograph 205–*7*
   hot wire anemometry 208
   pitot tube *207*–8
   spirometer 203–4
   ultrasonic flowmeter *208*
   Vitalograph 204, *205*
   Wright's respirometer 204–5, *206*
gases
   effect of compression 120
   effect of heating 120–1
   heat and moisture exchangers 126
   perfect 121
   pressure–volume curve *121*
   pressurised, fire and explosion risk 369
   resistance to flow 201

gases (*Continued*)
　solubility in gases 134
　solubility in liquids 131–3
Gay-Lussac's law 118, *120*
geometric mean 26
glutaraldehyde solution, sterilisation of equipment 393
goodness of fit 43
gradient coils, magnetic resonance imaging (MRI) *386*
Graham's law of diffusion 113
graphs
　parabolas *2–3*
　polynomials *3–4*
　rectangular hyperbolae *4–5*
　straight line *1–2*
gravity 56
gray (Gy) *48*, 381
greenhouse effect 370
Guedel oropharyngeal airway *302*
gyromagnetic ratio 384

haemoglobin, absorption spectra *217*
Hagen–Poiseuille equation 49, 103
half lives ($t_{1/2}$) 18–*20*
halothane
　occupational exposure limit 371
　solubility 133
halothane vaporisers, service intervals 142
harmonics 7, 51, *52*, 53
Hayek oscillator 359
heart block 270
heart rate variability (HRV), depth of anaesthesia monitoring 267–8
heat
　application to gases 120–1
　eponymous effects 124–5
　heat transfer 125
　relationship to work 117
　SI unit 123
heat and moisture exchangers (HME) 126, 128
heat exchangers, warming intravenous fluids 126
heat loss minimisation 125–6, 167
　humidification 128
heat of vaporisation 135
heating effect of electric currents 98–9
heating effect of MRI 387–8
helium dilution technique, FRC measurement 212
Henderson–Hasselbach equation 241
Henry's law 131
hertz (Hz) *48*
high frequency chest wall oscillation (HFCWO) 359
high frequency jet ventilation (HFJV) 359
high frequency oscillation (HFO) 359

high frequency positive pressure ventilation (HFPPV) 359
high frequency ventilation (HFV) 359
　gas transport mechanisms 360–1
high pass filters 76
high power ultrasound 158
histograms 24, *25*
hot wire anemometry 208
humidification 127–8
humidification systems 128
humidity 126–7
　measurement 127
Humphrey ADE breathing system 329
hybrid training 399
hydraulic coupling, direct blood pressure measurement 178–9
hydrogen nucleus, spin property 383–4
hygrometers 127
hyperbaric chambers, ventilators 352
hyperbaric pressure 110
hyperbolae *4–5*
hyperthermia, malignant 167
hypobaric pressure 109–10
hypothermia, blood gas analysis 242
hypothesis testing 34–*5*
　non-parametric methods
　　Mann–Whitney U test 37–*8*
　　Wilcoxon signed rank sum test 36–7
　parametric methods, Student's t-tests 35–6
hypoxia, anti-hypoxic safety devices 296, *297*
hysteresis 50, *51*, 58

I-gel airway 301, *307*
ignition sources 369
imaging
　CT 378–9
　MRI 382–5
　　hazards 387–8
　　spatial encoding 385–7
　　special features 388
　PET 380
　scintigraphy 379–80
　SPECT 379–80
　X-rays 378–9
impedance 69–70
　electrodes 85
implantable cardiac defibrillators (ICDs) 274
independent events, probability 29–30
inductance gauges, direct blood pressure measurement 180–*1*
inductive resistance (reactance) 69
inductive transducers 86–7
inductors 68–*9*
inertance 338
infrared gas analysers
　accuracy 228
　calibration 228

photocells 228
principle 226
sources of error 226–8
inhibit mode pacemakers 272
inputs 50
inspiration–expiration loop, hysteresis *51*
insulation
    double, class 2 equipment 91
integration 10–12
integrators 75–6
intensive care ventilators 352
    Draeger Evita 4 357–8
    Veolar ICU ventilator 352–3
        air–oxygen mixer 353, *354*
        electronic control system 356
        expiratory valve 356, *357*
        flow sensor 356
        functional diagram *353*
        gas reservoir 353–4
        servo-controlled flow valve 354–6, *355*
interference, diathermy as source 285–6
interference fringes 223–*4*
intermittent blower ventilators 349–50
    Penlon Nuffield 200 ventilator *350*–1
    Pneupac ventilator 351–2
internal resistance, ventilators 343
interpolation 4
inter-quartile range 27
intra-arterial blood pressure monitoring
    166–7
    see also direct blood pressure measurement
intravascular blood gas analysers 243
intravenous fluids, warming 126
intravenous pumps 364
    ambulatory PCA 368
    peristaltic pump mechanism *365*
    target controlled infusion pumps 367–8
    volumetric 366–7
intubating LMAs 301, 305, *306*
intubation aids
    bougies and catheters 320–1
    fibre optics 318–20
    laryngoscope 316–18
inverting amplifier circuit 77, *78*
iodine, sterilisation of equipment 393
ionisation 56
ISO pin index system 292–*3*
isobestic points 217
isoflurane
    occupational exposure limit 371
    saturated vapour pressure (SVP) *139*
isotonic fluids 115
isotopes 55
    sources 375

Jackson Rees' modification T piece *323*, 329
Jakob–Creutzfeldt disease 316
jaundice, pulse oximetry 220

Joule effect 124
joules (J) *48*, 57

Kelvin temperature scale 123–4
kilograms (kg) *48*, 56
kinematic viscosity (v) 102
kinetic energy 57–8, *59*
kinetic energy changes, ventilation 338
Korotkoff sounds *171*–2

Lack breathing system *325*
laminar flow 101
    fluid velocity profile *102*
    role in high frequency ventilation 361
lamination, Entonox 289–90
Larmor frequency 384
laryngeal mask airways (LMAs) 301,
        304–7, *305*
laryngoscope *317*
    disposable blades 318
    fibre optic 320
    ventilation jets 318
lasers
    basic principles 388–90
    biomedical applications 390–1
    classes 390
    types 390
latency, evoked responses 253
latent heats 123
    of vaporisation 135
latex sensitivity, airway management
        devices (AMD) 301
lead, electrode potential *84*
lead screening, radiation protection 381
leakage currents *90*, 91–3
least squares fit 43
length, SI unit *48*
LiDCO 198
light-emitting diodes (LEDs), pulse oximetry
        217–18
limits, integration 10–12
line of best fit *41*
linear regression 43
linear relationships 1
linear scales 23, *24*
linear systems 50
liquid crystal displays, temperature
        measurement 124
litres (L) 56
live lead 89
local anaesthetic nerve blocks, nerve
        stimulators 257
logarithmic scale 23
logarithms 14–16
    natural 17
log–log plots 16
long Bain breathing system 329

low pass filters 74–6
luminous intensity, SI unit *48*
lung filling, exponential nature 18, *19*

Macintosh laryngoscope 316, *317*
Magill breathing system 323–4
Magill laryngoscope 316
magnetic field exposure, safety limits 387
magnetic resonance imaging (MRI)
　basic principles 382–5
　breathing systems 329
　diagram of coils *383*
　gradient coils *386*
　hazards 387–8
　slice selection *386*
　spatial encoding 385–7
　special features 388
magnetic sector mass spectrometry 229
magnetism 62–3
magnets, in pacemaker control 274
mains interference *79*
　ECG 248
　reduction 85
malignant hyperthermia 167
manikins 394
　see also simulators
Manley MP3 ventilator 348–9
Mann–Whitney U test 37–*8*
manometers 108
　gas pressure measurement 203
manual resuscitators *358*
Mapleson A breathing systems 323–6
Mapleson B and C breathing systems *323*, 326
Mapleson D, E and F breathing systems *323*, 326–9
Mapleson's classifications
　semi-closed rebreathing systems *323*
　ventilators 339
mass 56
　SI unit *48*
mass spectrometry 228–9, *230*
　basic principles 229–30
　interpretation problems 230–1
　magnetic sector mass spectrometry 229
　quadrupole mass spectrometer 229–*30*
　sources of error 229
mathematical modelling, simulators 396
maxima of curves *9*
MC masks 303
McCoy laryngoscope 316, *317*
mean 26
　of a Normal (Gaussian) distribution 30
　sample and population 32–4
measurements, dimensions and units 47–9
median 26
median power frequency (MPF)
　EEG 251–*2*
　　biphasic effect *253*

medical compressed air 290–1
　gas cylinders *292*
mercury manometer 170
meta-analysis 41–2
methaemoglobin (Met Hb), absorption spectrum *217*, 220
methane, greenhouse effect 370
methyl alcohol, flammability 369
microshock 95–7
Miller modification breathing system 325–6
millimetres of mercury (mmHg) 48
minima of curves *9*
minute volume dividers 343, 348
　Manley MP3 ventilator 348–9
mode 26
model-driven simulators 395–6
　mathematical modelling 396–*7*
　patient types 397
　scenario scripts 397–8
　scenarios 398–9
modulation 81
moist heat, effectiveness in sterilisation 392
molecular leak, mass spectrometry 229
monitoring
　AAGBI recommendations 163
　airway and ventilation 164–5
　alarms 162–3
　blood pressure 166–7
　cardiac rate and rhythm 165–6
　data displays 162
　depth of anaesthesia 167
　impact on safety 160–2
　main requirements 160
　neuromuscular function 168
　oxygen supply 163–4
　oxygenation of the patient 164
　postanaesthesia care 168
　safety checks 299
　temperature 167
　tissue perfusion 166
　see also auditory evoked responses (AER); depth of anaesthesia monitoring; electrocardiogram (ECG); electroencephalogram (EEG); electromyogram
monitoring system components, direct blood pressure measurement
　hydraulic coupling 178–9
　transducers 179–82
monophasic defibrillator circuits *276–7*
monopolar electrodes, diathermy 283
motion, Newton's laws 57
MRI *see* magnetic resonance imaging
Murphy's eye, endotracheal tubes 308, *309*
myocardial contractility, assessment from arterial waveform *178*

Narcotrend monitor 266
nasopharyngeal airways 302, *303*
nasotracheal tubes 311
natural logarithms 17
nebulisers 128, *140*
needle positioning, ultrasound visualisation 152–3
needle valves 294
negative end expiratory pressure (NEEP) 344
negative skewness *29*
neonates
  endotracheal tubes 310
  facemasks 303
  pulse oximetry 220
  ventilators 345
nerve stimulation, MRI 387
nerve stimulators 257
neuromuscular function monitoring 168, 256–9
neutral lead 89, 90
neutrons 55
Newton valve 351
Newtonian fluids 102
newtons (N) *48*, 56, 57
Newton's laws of motion 57
nickel, electrode potential *84*
nitrogen, solubility in body tissues 132
nitrogen meter 233
nitrogen washout method, FRC measurement 210–12, *211*
nitrous oxide
  blood–gas partition coefficient 133
  critical temperature 122, 134–5
  diffusion hypoxia 113
  gas cylinders 291, *292*
  occupational exposure limit 371
  oil–gas partition coefficient 133
  ozone depletion 371
  solubility 132–3
nitrous oxide supply 289
nominal data 23
non-invasive blood pressure (NIBP) measurement 170–1
  auscultation 171–2
  cuff size 175
  Doppler ultrasound 175
  oscillometry 173–5, *174*
  oscillotonometry 172–3
  palpation 171
  plethysmography *176*–7
non-inverting amplifier circuit 77, *78*
non-linear regression 44
non-linear scales 23, *24*
non-linear systems 50
non-parametric hypothesis testing
  Mann–Whitney U test 37–*8*
  Wilcoxon signed rank sum test 36–7

Normal (Gaussian) distribution 30–2, *31*
Normal distribution tables 34
nuclear medicine imaging 379–80
nucleus (atomic) 55
null hypothesis 34
numerical data 23
Nyquist rate 82–*3*

occupational exposure limits 371
Ohmeda anti-hypoxic linkage mechanism *296*
Ohm's law 63
oil–gas partition coefficients 133–4
oncotic pressure 115, 190
one sample t-test 35
one way analysis of variance 39–40
operational amplifiers 76–7
  basic circuits *78*
optical fibres *see* fibre optics
ordinal data 23
orifice meter flow measurement device *104*
oropharyngeal airways 302
oscillometry 173–5, *174*
oscillotonometry 172–*3*
osmolality 115
osmolarity 114–15
osmometers 115
osmosis 113–15, *114*
osmotic pressure *114*
Ostwald solubility coefficient 133
outliers 27
outputs 50
overflow valve placement, circle breathing systems 332
overshoot, direct blood pressure measurement 184–5, *188*
Overton–Meyer hypothesis 133–4
oxygen
  critical temperature 122, 134
  gas cylinders *292*
  pressure reduction 293–4
  saturated vapour pressure (SVP) 138
oxygen analyser, safety checks 297
oxygen concentrator 288–9
oxygen failure warning devices 296
oxygen saturation, pulse oximetry 216–20
oxygen supply 287–9
  monitoring 163–4
oxygenation levels, effect on EEG 249
oxygenation of the patient, monitoring 164
ozone depletion 370

P values 34
P wave 244, *245*
pacemaker codes 272–3
pacemakers
  advanced types 272

pacemakers (*Continued*)
  for bradycardia 270–4
  components 270–1
  fixed rate pacing 271
  interference 273–4, 285–6
  physiological 273
  rate-adaptive 273
  risks from MRI 387
  safe mode 274
  for tachyarrhythmias 274
  unipolar and dipolar leads 271
paediatric laryngoscopes *317*
  disposable blades 318
paediatric ventilators 345, 351–2
paired t-test 36
palpation, blood pressure measurement 171
parabolas *2–3*
parallel connection, electrical circuits 64
parallel Lack breathing system 325
paramagnetic oxygen analyser 233–*4*, *235*
  alternating magnetic fields 234, *235*
parametric hypothesis testing 35–6
partial pressure 110–12
  effect on diffusion *112*
partition coefficient 133–4
pascals (Pa) *48*
passive electrode, diathermy *281*, 284
passive scavenging systems *373*–4
patient controlled analgesia
  accuracy 364
  ambulatory pumps 368
patient isolated differential amplifier 79–81, *80*
patient oxygenation, monitoring 164
patient state analyser (PSI) 267
patient types, model-driven simulators 397
patient valves, manual resuscitators *358*
pCO$_2$ electrode, blood gas analysis 241
peak expiratory flowmeter 212, *213*
peak velocity, echocardiography 195–6
Peltier effect 125
Peñaz technique, blood pressure measurement *176–7*
Pendelluft 360, *361*
Penlon Nuffield 200 ventilator *350*–1
percentiles 27
perfect gas 121
peripheral arterial waveforms 189
peripheral nerve stimulators 168
peristaltic pump mechanism *365*
PET *see* positron emission tomography
pH 16, 239
pH electrode, blood gas analysis 239–*40*
phenol, sterilisation of equipment 393
photoelectric transducers 87
physiological deadspace 210
physiological pacemakers 273
PiCCO 197

pie charts 25
piezoelectric crystals, ultrasound production *149*
piezoelectric gas analysis 232–3
piezoelectric transducers 87
pin index system, gas cylinders 292–*3*
piped medical vacuum 291
pitot tube *207*–8
plasma osmolarity 115
plasma proteins, oncotic pressure 115
platinum, electrode potential *84*
plethysmographic signal, pulse oximetry 219
plethysmography 165, 166, *176*–7
Pneupac ventilator 351–*2*
pO$_2$ electrode, blood gas analysis 239
Poiseuille's equation, dimensional analysis 48–9
polarisation impedance 85
polarography 234–6
Polio laryngoscope 316
polynomials *3–4*
population inversion, electrons 389
population mean 32–3
portable ventilators 358
positive end expiratory pressure (PEEP) 343–4
positive pressure ventilation 201
positive skewness 29
positron emission, radioactive decay 375–6
positron emission tomography (PET) 376, 380, *381*
postanaesthesia care 168
postsynaptic potentials 248
post-tetanic facilitation (PTF), electromyogram 256–7
potential energy 57–8, *59*
povidone iodine 393
power
  electrical 61
  SI unit *48*
  ventilators 342–3
powers 14, 58–9
Poynting effect, Entonox 289–90
precession, spinning protons *384*
pregnancy, MRI 387
preoperative exercise testing 236–7
pressure 57
  atmospheric 108
  gas laws 117–20
  hypobaric and hyperbaric 109–10
  partial 110–12
  SI unit *48*
pressure cycling, ventilators 343, *344*
pressure distribution, circulation *189*
pressure generators 339–41
  electrical analogue *341*
pressure gradients, fluid flow 102–3

pressure measurement 108–9
  gases 201–3
pressure regulators 293–6, *295*
pressure transducers 66
pressure–volume curve *121*
pressurised gases, fire and explosion risk 369
prions 392
probability 29–30
probability distributions 30
  see also Normal (Gaussian) distribution
propofol infusion 367
Proseal LMA 301, 305, *306*
protons 55
pseudocritical temperature, Entonox 289–90
pulmonary artery catheters 192–5
pulmonary artery occlusion pressure (PAOP) 192, 193
pulmonary capillary wedge pressure (PCWP) 192
pulsatility index (PI) 156
pulse contour analysis 197–8
pulse generators, pacemakers 270–1
pulse oximetry 161, 162, 164, 216
  absorption in different tissue components *218*
  AC and DC components 218, *219*
  calibration curve 218–*19*
  causes of error 219–20
  LEDs 217–18
  plethysmographic signal 219
  postanaesthesia care 168
  spectroscopy 216–17
pulsed echo ultrasound *151*–2

QRS wave 244, *245*
quadrupole mass spectrometer 229–*30*

R^2 value, regression 43
radians 6
radiation
  biological effects 380–1
  measurement methods 382
  typical doses *382*
radiation protection 381
radiation units 381–2
radiative heat loss 125
radioactive decay 375–6
radiofrequency pulse, MRI 384–5
radionuclides 55
Rae tube *309*, 311
Raman spectroscopy 231–*2*
  characteristic frequency shifts *232*
range of data 27
Raoult's law 115
rate-adaptive pacemakers 273
Rayleigh refractometer 224, *225*
Rayleigh scattering 231

reactance 69
receiver operating characteristics (ROC) curves 38–9, *40*
rectangular hyperbolae 4–*5*
rectilinear biphasic waveform *278*
reflection, fibre optics *319*–20
refraction, fibre optics *319*–20
refractive index (μ) 223
refractometry 223–4, 225
  portable refractometer 224–*5*
  Rayleigh refractometer 224, *225*
regional anaesthesia, needle positioning verification 152–3
Regnault hygrometer 127
regression 43–4
reinforced LMA *305*
relative humidity 126–7
relaxation properties, MRI 385
Rendell–Baker facemask 303
reservoir bag placement, circle breathing systems 332–3
residual current devices (RCDs) 97
residual variation 43
residual volume *204*
resistance, Ohm's law 63
resistance to flow 103
  see also airway resistance
resistance transducers 86
resistive index (RI) 156
resistive strain gauges, direct blood pressure measurement 179–80
resistor–inductance–capacitance (RLC) circuits *276*
resistors, series and parallel connection 64
resonance 73
  direct blood pressure measurement 183
  role in high frequency ventilation 360
resonant frequency 185
respiratory gas analysis 222–3
  fuel cells 236
  infrared spectroscopy 226–8
  mass spectrometry 228–9
  nitrogen meter 233
  paramagnetic oxygen analyser 233–*4*
  piezoelectric gas analysis 232–3
  polarography 234–6
  Raman spectroscopy 231–*2*
  refractometry method 223–*5*
  ultraviolet gas analysis 233
  zeroing and calibration 223
respiratory mechanics, ventilation 338–9
respiratory sinus arrhythmia based monitors 267–8
respiratory tract, humidification 128
respiratory work 58
respirometry 209
  Fleisch pneumotachograph 205–*7*
  hot wire anemometry 208

respirometry (*Continued*)
  pitot tube *207*–8
  spirometer 203–4
  ultrasonic flowmeter *208*
  Vitalograph 204, *205*
  Wright's respirometer 204–5, *206*
Response Entropy (RE) 266
response time, respiratory gas analysers *222, 223*
Reynold's number (*Re*) 103
root mean square (RMS) value, AC 67
Rotameter *105*–6, 212–*14*
  needle valve 294
R–R intervals 246–7
Ruben valve *358*
ruby laser 389–*90*

safety
  atmospheric pollution 370–1
  diathermy 283, 284
    stray capacitance 284–*5*
  fire and explosion risk 369–70
  impact of monitoring 160–2
  MRI 387–8
  occupational exposure limits 371
  oxygen failure warning devices 296
  scavenging systems 371–4
  ventilators 344
  see also electrical safety
safety checklist 296–9
sample mean 32–3
sampling flow rate, respiratory gas analysers 222
sampling of waveforms, Nyquist rate 82–3
saturated vapour pressure (SVP) 126–7, 134, 138
  variation with temperature *139*
scales, linear and non-linear 23, *24*
scatter diagrams *24*
  correlation *43*
scavenging systems 371–2
  active *372*–3
  passive *373*–4
  safety checks 299
scenario scripts
  anaphylaxis script *398*
  simulators 397–8
scenarios 398–9
scintigraphy 379–80
SCOTI (Sonomatic Confirmation of Tracheal Intubation) device 312
secant (sec) 5
Sechrist ventilator 345
Seebeck effect 124
semiconductor lasers 390
semi-log plots 16
semi-permeable membranes 113
  osmosis 113–15

sensitivity of tests 38, *39*
series connection, electrical circuits 63, 64
sevoflurane
  oil–gas partition coefficient 133
  reaction with soda lime 334
  saturated vapour pressure (SVP) *139*
shear stress (τ) 101
Sherwood number 361
SI (Système International) units 47–*8*
sievert (Sv) *48*, 382
signal processing 81–*2*
  digital processing 82–*3*
silver, electrode potential *84*
silver chloride (AgCl) electrodes 84–5
simple harmonic motion *7*, 13
Sims Graseby volumetric pump *366*
simulation centre environments 395
simulators 394
  hybrid training 399
  model-driven 395–6
  patient types 397
  scenario scripts 397–8
  scenarios 398–9
  surgical skills labs 399
  virtual reality training 399–400
sine (sin) 5
sine wave generators, use in calibration *54*
sine waves 6–*7*
single emission photon computed tomography (SPECT) 379–80
single ended amplifiers 77–9, *78*
single twitch stimulation, electromyogram 256
skewed distributions *29*
skin colour, and pulse oximetry 220
skin–electrode impedance 85–6
skin sterilisation 393
smokers, pulse oximetry 220
soda lime
  CO_2 absorbers 333
  reactions with volatile agents 333–4
solid state lasers 390, 391
solubility
  of gases in gases 134
  of molecules in liquids 131–3
  partition coefficient 133–4
solubility coefficients 133
somatosensory evoked responses 255
  see also evoked responses
spatial encoding, MRI 385–7
specific heat 122
specific heat capacity (SHC) 99
specificity of tests 38, *39*
SPECT *see* single emission photon computed tomography
spectral edge frequency (SEF), EEG 251–*2*
spectroscopy
  infrared 226–8
  light absorption 216–*17*

spin-lattice relaxation, MRI 385
spin property, protons 383–4
spinning disc nebulisers 128
spin-spin relaxation, MRI 385
spirometer 203–4
splitting ratio, vaporisers 139–40
square wave, frequency components 51–2
square wave changes, respiratory gas analysers 222–3
ST segment 247
standard atmosphere (atm) 48
standard bicarbonate 242
standard deviation 28, 33
   of a Normal distribution 30, 32
standard error of the mean (SEM) 33
star point, transformers 89
State Entropy (SE) 266
static electricity 60, 100
   prevention of ignition 369
stepper motors 364
sterilisation of equipment 392–3
sternal notch device, absolute blood flow measurement 158
stimulated emission, lasers 389
stone fragmentation, high power ultrasound 158
straight line graphs 1–2
strain gauges 65–6
   direct blood pressure measurement 179–80
   gas pressure measurement 202–3
stray capacitance 91–2
   diathermy 284–5
stress, SI unit 48
stroke volume, assessment from arterial waveform 178
Student's t-tests 34, 35–6
subacute combined degeneration of the spinal cord 371
sublimation 122
suction devices, piped medical vacuum 291
sum of squares, ANOVA 40
supraglottic airways 301
surgical skills labs 399
syringe drivers 364
   ambulatory PCA 368
   target controlled infusion pumps 367–8
   volumetric pumps 366–7
syringe volumetric infusion pump 365
systolic blood pressure, oscillometry 174–5

T pieces 326
   Ayre's T piece 329
   Bain breathing system 327–9, 328
   FGF requirements for children 330
   functional gas storage property 327
   Jackson Rees' modification 323
T wave 244, 245
tachyarrhythmias, pacemakers 274
tangent (tan) 5
target controlled infusion pumps 367–8
tau (τ) (time constant) 18–20
Taylor dispersion, role in high frequency ventilation 361
^{99m}Tc 379
Tec5 vaporiser 141, 143
temperature 123–4
   effect on blood gas analysis 242
   human body 124
   keeping patients warm 125–6, 128
   measurement 124
   SI unit 48
temperature compensation, vaporisers 140–1
temperature monitoring 167
temperature scales 123–4
tension equilibrium 133
Tensymeter 177
thalamic discharges 248
thermal energy 57
thermistors 124
thermocouples 124
thermodilution curve, cardiac output measurement 12–13
thermodilution method 194
thermodynamics, laws of 117
thermometry 124
theta activity, EEG 248
Thompson effect 125
threshold values, diagnostic tests 38
thumb movements, measurement of evoked muscle responses 258
tidal volume 204
time, SI unit 48
time cycling, ventilators 343, 344
time domain analysis, EEG 250, 251
time domain signals 51
tissue perfusion monitoring 166
TOF (train of four) ratio 256
Torricellian vacuum 108
total lung capacity 204
Touch Current (previously Enclosure Leakage Current) 92, 93
trachea device, blood flow measurement 158
tracheostomy tubes 312
training
   hybrid training 399
   simulation centre environments 395
   simulators 394
      model-driven 395–6
      patient types 397
      scenario scripts 397–8
      scenarios 398–9
      surgical skills labs 399
   virtual reality training 399–400
train-of-four (TOF) stimulation, electromyogram 256

transcutaneous blood gas analysers 242–3
transducers 86–7
  calibration 53–5
  direct blood pressure measurement 179–82
transfer functions 50
transformed data, geometric mean 26
transformers 63, 70–*1*
  substations 89
transit time, respiratory gas analysers 222
transoesophageal Doppler 195–6
transthoracic electrical impedance 198–9
trichloroethylene, reaction with soda lime 333–4
trigonometry 5–7
triple point 122
  water 123–4
truncated exponential biphasic waveform *277*
tubing, absorption of volatile agents 146
turbulence 101
  critical flow rate (critical velocity) 102–3
  effect of viscosity and density 103–4
  role in high frequency ventilation 361
two-dimensional B-scanning ultrasound *151*–2
two sample t-test 35–6
two way analysis of variance 39

ultrasonic flowmeter *208*
ultrasonic nebulisers 128
ultrasound
  applications 152
    absolute blood flow measurement 157–8
    Doppler 153–7
    visualisation of needle positioning 152–3
  cardiac output measurement 107
  enhancement of visualisation 152–3
  future directions 158
  generation 148, *149*
  high power 158
  period and wavelength *150*
  properties
    attenuation 150
    resolution 150
    wavelength, period and frequency 148–*50*
  two-dimensional B-scanning *151*–2
ultraviolet gas analysis 233
underdamped biphasic waveform *277*
unipolar leads, pacemakers 271
units 47–8
universal gas law 4

vacuum, wall suction 291
vacuum insulated evaporator (VIE) 287–*8*
valence shells 55–6

vaporisation 134–7
  heat loss 125
  latent heat of 123
vaporisers *136*, 139
  basis of vapour production 137–9
  Boyle's bottle 141, *142*
  contact maximization *140*
  Draeger Vapor 141–2, *144*
  draw-over 146
  integration into anaesthetic machine 142–6
  practical tips 142
  retrograde re-entry 141
  safety checks 298
  splitting ratio 139–40
  Tec5 141, *143*
  temperature compensation 140–*1*
variability of data 26–7
variable orifice, constant pressure flowmeters 212–14
variable pressure drop, fixed orifice flowmeter (Fleisch pneumotachograph) 205–*7*
variance 27–8
VDD pacing mode 273
velocity, dimensions 47
velocity profile, laminar flow 9, *10*
ventilation
  high frequency (HFV) 359
  respiratory mechanics 338–9
ventilation monitoring 164–5
ventilator circuits, airway pressure measurement 203
ventilator types
  bag-in-bottle 345–8
  intensive care ventilators 352
    Draeger Evita 4 357–8
    Veolar ICU ventilator 352–7
  intermittent blowers 349–52
  minute volume dividers 348–9
ventilators
  classification 339–42
  cycling modes *340*, 343
    comparison of effects *344*
  expiration 343–4
  internal resistance 343
  paediatric 345, 351–2
  portable 358
  power 342–3
  safety checks 298
  safety features 344
  testing 345
  work and energy 58
ventricular fibrillation (VF)
  as result of microshock 96
  see also defibrillators
ventricular pacing 272
Venturi breathing systems 330

venturi effect, applications 107
venturi flowmeter *106–7*
Venturi mask 303–4
Veolar ICU ventilator 352–3
   air–oxygen mixer 353, *354*
   electronic control system 356
   expiratory valve 356, *357*
   flow sensor 356
   functional diagram *353*
   gas reservoir 353–4
   servo-controlled flow valve 354–6, *355*
virtual reality training 399–400
viscosity, relationship to flow characteristics 103–4
visual evoked responses 254–5
   see also evoked responses
vital capacity *204*
Vitalograph 204, *205*
voltage 61
   alternating and direct current 66–8, *67*
   Ohm's law 63
   water analogy 61–2
voltage dividers *64*, 65
voltmeters 62
volts (V) *48*
volume cycling, ventilators 343, *344*
volume measurement, gases
   spirometer 203–4
   Vitalograph 204, *205*
volumetric pumps 366–7
von Reckinghausen's oscillotonometer 172–3
VOO pacing mode 273
voxels 383
VVI pacing mode 272

water
   critical temperature 122, 135
   saturated vapour pressure (SVP) 138, *139*
   triple point 123–4
water analogy, voltage 61–2
watts (W) *48*, 58
waveform sampling, Nyquist rate 82–*3*
waveforms, frequency components 51–3
wavelength, ultrasound 148–9
webers 62
Wee detector 312
weight, difference from mass 56
Welch test 37
wet and dry bulb hygrometer 127
Wheatstone bridge 65
wicks, vaporisers *140*
Wilcoxon signed rank sum test 36–7
Windkessel model 197
within-subject differences 36
work 57–8, *59*
   relationship to heat 117
   SI unit *48*
Wright's respirometer 204–5, *206*

X-ray tube *377*
X-rays 376
   imaging 378–9
   production 377–*8*
   typical radiation doses *382*

zeolites 288–9
zinc
   electrode potential *84*